Methods in Molecular Biology™

Series Editor
John M. Walker
School of Life Sciences
University of Hertfordshire
Hatfield, Hertfordshire, AL10 9AB, UK

For other titles published in this series, go to
www.springer.com/series/7651

Plant Reverse Genetics

Methods and Protocols

Edited by

Andy Pereira

Virginia Bioinformatics Institute, Virginia Tech, Blacksburg, VA, USA

Humana Press

Editor
Andy Pereira, Ph.D.
Virginia Bioinformatics Institute
Virginia Tech
Blacksburg, VA
USA
pereiraa@vbi.vt.edu

ISSN 1064-3745 e-ISSN 1940-6029
ISBN 978-1-4939-6127-6 ISBN 978-1-60761-682-5 (eBook)
DOI 10.1007/978-1-60761-682-5
Springer New York Dordrecht Heidelberg London

Printed on acid-free paper

Humana Press is part of Springer Science+Business Media (www.springer.com)

Preface

Plant biology is at the crossroads, integrating the data from genomics into knowledge and understanding of important biological processes. With the generation of genome sequence data from model and other plants, databases are filled with sequence information of genes with no known biological function. While bioinformatics tools can help analyze genome sequences and predict gene structures, experimental approaches to discover gene functions need to be widely implemented. This book deals with plant functional genomics using reverse genetics methods, namely, from gene sequence to plant gene functions. The methods developed and described by leading researchers in the field are high-throughput and genome-wide in the models Arabidopsis and rice as well as other plants to provide comparative functional genomics information.

This book describes methods for the analysis of high-throughput genome sequence data, the identification of noncoding RNA from sequence information, the comprehensive analysis of gene expression by microarrays, and Metabolomic analysis, all of which are supported by scripts to aid their computational use. A series of mutational approaches to ascribe gene function are described using insertion sequences such as T-DNA and transposons as well as methods for the silencing and overexpression of genes. The cataloging of developmental mutant phenotypes as well as analysis of functions using specific phenome screens described can be adapted to any lab conditions. The integration of the diverse comparative functional genomics information in a database, such as Gramene, provides the capabilities for an understanding of how plant genes work together in a systems biology view.

Blacksburg, VA *Andy Pereira*

Contents

Contributors

MADANA M. R. AMBAVARAM • *Virginia Bioinformatics Institute, Virginia Tech, Blacksburg, VA, USA*

GYNHEUNG AN • *Department of Plant Molecular Systems Biotechnology and Crop Biotech Institute, Kyung Hee University, Yongin 446-701, Republic of Korea*

JENNY M. ARMENTA • *Virginia Bioinformatics Institute, Virginia Tech, Blacksburg, VA, USA*

MICHAEL A. AYLIFFE • *CSIRO Plant Industry, Canberra, ACT, Australia*

RAMESH S. BHAT • *University of Agricultural Sciences, Dharwad, Karnataka, India*

SHU CHEN • *Taiwan Agricultural Research Institute, Wufeng, Taichung, Taiwan*

XIAOFEI CHENG • *Plant Biology Division, The Samuel Roberts Noble Foundation, Ardmore, OK, USA*

CHYR-GUAN CHERN • *Taiwan Agricultural Research Institute, Wufeng, Taichung, Taiwan*

DIEGO F. CORTES • *Virginia Bioinformatics Institute, Virginia Tech, Blacksburg, VA, USA*

S. P. DINESH-KUMAR • *UC Davis Genome Center, 1319 Genome and Biomedical Sciences Facility, 451 Health Sciences Drive, Davis, CA 95616, USA*

MING-JEN FAN • *Department of Biotechnology, Asia University, Wufeng, Taichung, Taiwan*

STEFAN DE FOLTER • *Laboratorio Nacional de Genómica para la Biodiversidad (Langebio), Centro de Investigación y de Estudios Avanzados del Instituto Politécnico Nacional (CINVESTAV-IPN), Irapuato, Mexico*

TAYLOR P. FRAZIER • *Department of Biology, East Carolina University, Greenville, NC, USA*

AMAL HARB • *Virginia Bioinformatics Institute, Virginia Tech, Blacksburg, VA, USA*

ANDREW HAYWARD • *Department of Molecular, Cellular, and Developmental Biology, Yale University, New Haven, CT, USA*

MIEKO HIGUCHI • *RIKEN Plant Science Center, Yokohama Kanagawa, Japan*

YUE-IE C. HSING • *Institute of Plant and Microbial Biology, Academia Sinica, Taipei, Taiwan*

SHENG-CHUNG HUANG • *Taiwan Agricultural Research Institute, Wufeng, Taichung, Taiwan*

TAKANARI ICHIKAWA • *RIKEN Plant Science Center, YokohamaKanagawa, Japan; Gene Research Center, Tsukuba University, Tsukuba, Japan*

PANKAJ JAISWAL • *Department of Botany and Plant Pathology, Oregon State University, Corvallis, OR, USA*

JONG-SEONG JEON • *Graduate School of Biotechnology & Plant Metabolism Research Center, Kyung Hee University, Yongin, Korea*

SUNG-RYUL KIM • *National Research Laboratory of Plant Functional Genomics, Division of Molecular and Life Sciences, POSTECH Biotech Center, Pohang University of Science and Technology, Pohang, Korea*
YOUICHI KONDOU • *RIKEN Plant Science Center, Yokohama, Kanagawa, Japan*
ARJUN KRISHNAN • *Virginia Bioinformatics Institute, Virginia Tech, Blacksburg, VA, USA*
MING-HSING LAI • *Taiwan Agricultural Research Institute, Wufeng, Taichung, Taiwan*
SHRINIVASRAO P. MANE • *Virginia Bioinformatics Institute, Virginia Tech, Blacksburg, VA, USA*
NAYELLI MARSCH-MARTÍNEZ • *Laboratorio Nacional de Genómica para la Biodiversidad (Langebio), Centro de Investigación y de Estudios Avanzados del Instituto Politécnico Nacional (CINVESTAV-IPN), Irapuato, México*
MINAMI MATSUI • *RIKEN Plant Science Center, Yokohama, Kanagawa, Japan*
ANNEMARIE H. MEIJER • *Clusius Laboratory, Institute of Biology (IBL), Leiden University, Leiden, The Netherlands*
PEDRO MENDES • *Virginia Bioinformatics Institute, Virginia Tech, Blacksburg, VA, USA; School of Computer Science, University of Manchester, Manchester, UK; Department of Cancer Biology, Wake Forest University School of Medicine, Winston-Salem, NC, USA*
THERO MODISE • *Virginia Bioinformatics Institute, Virginia Tech, Blacksburg, VA, USA*
SAROJ K. MOHAPATRA • *Virginia Bioinformatics Institute, Virginia Tech, Blacksburg, VA, USA*
KIRANKUMAR S. MYSORE • *Plant Biology Division, The Samuel Roberts Noble Foundation, Ardmore, OK, USA*
PIETER B. F. OUWERKERK • *Sylvius Laboratory, Institute of Biology (IBL), Leiden University, Leiden, The Netherlands*
MEENU PADMANABHAN • *Department of Molecular, Cellular, and Developmental Biology, Yale University, New Haven, CT, USA*
ANDY PEREIRA • *Virginia Bioinformatics Institute, Virginia Tech, Blacksburg, VA, USA*
REVONDA M. POKRZYWA • *Virginia Bioinformatics Institute, Virginia Tech, Blacksburg, VA, USA*
ANTHONY J. PRYOR • *CSIRO Plant Industry, Canberra, ACT, Australia*
PASCAL RATET • *Institut des Sciences du Vegetal, CNRS, Gif sur Yvette Cedex, France*
CLEMENCIA M. ROJAS • *Plant Biology Division, The Samuel Roberts Noble Foundation, Ardmore, OK, USA*
MUTHAPPA SENTHIL-KUMAR • *Plant Biology Division, The Samuel Roberts Noble Foundation, Ardmore, OK, USA*
VLADIMIR SHULAEV • *Department of Horticulture, Virginia Bioinformatics Institute, Virginia Tech, BlacksburgVA, USA; Department of Cancer Biology, Wake Forest University School of Medicine, Winston-Salem, NC, USA*
JOEL L. SHUMAN • *Virginia Bioinformatics Institute, Virginia Tech, Blacksburg, VA, USA*
BRUNO W. SOBRAL • *Virginia Bioinformatics Institute, Virginia Tech, Blacksburg, VA, USA*

MILLION TADEGE • *Plant Biology Division, The Samuel Roberts Noble Foundation, Ardmore, OK, USA*
ARUNEE TRISIRIROJ • *Institute of Plant and Microbial Biology, Academia Sinica, Taipei, Taiwan*
NARAYANA M. UPADHYAYA • *CSIRO Plant Industry, Canberra, ACT, Australia*
ZARIR VAGHCHHIPAWALA • *Plant Biology Division, The Samuel Roberts Noble Foundation, Ardmore, OK, USA*
FU-JIN WEI • *Institute of Plant and Microbial Biology, Academia Sinica, Taipei, Taiwan*
JIANGQI WEN • *Plant Biology Division, The Samuel Roberts Noble Foundation, Ardmore, OK, USA*
CHENG-CHIEH WU • *Institute of Plant and Microbial Biology, Academia Sinica, Taipei, Taiwan*
SU-MAY YU • *Institute of Molecular Biology, Academia Sinica, Taipei, Taiwan*
BAOHONG ZHANG • *Department of Biology, East Carolina University, Greenville, NC, USA*
QIAN-HAO ZHU • *CSIRO Plant Industry, Canberra, ACT, Australia*

Chapter 1

Analysis of High-Throughput Sequencing Data

Shrinivasrao P. Mane, Thero Modise, and Bruno W. Sobral

Abstract

Next-generation sequencing has revolutionized biology by exponentially increasing sequencing output while dramatically lowering costs. High-throughput sequence data with shorter reads has opened up new applications such as whole genome resequencing, indel and SNP detection, transcriptome sequencing, etc. Several tools are available for the analysis of high-throughput sequencing data. In this chapter, we describe the use of an ultrafast alignment program, Bowtie, to align short-read sequence (SRS) data against the Arabidopsis reference genome. The alignment files generated from Bowtie will be used to identify SNPs and indels using Maq.

Key words: Next-generation sequencing, Short-read sequences, Alignment programs, Bowtie, Maq

1. Introduction

Next-generation sequencers from Roche/454, Illumina, Applied Biosystems and Helicos have revolutionized biological research by greatly increasing sequencing output while dramatically lowering costs. Roche/454 produces ~400 bp sequence reads suitable for de novo sequencing and medium throughput applications, while Illumina and ABI produce short-read sequences (SRSs) typically ranging from 35 to 80 bp in length suitable for resequencing and high-throughput applications. SRS technologies provide endless opportunities for genomics, comparative genome biology, medical diagnostics, etc. Some of the examples include genome resequencing to detect SNPs and mutations within populations (SNP-seq), sequencing of closely related species, methylome profiling, DNA-protein interactions (ChIP-seq), transcriptome sequencing (RNA-seq), mRNA expression profiling (DGE), and small RNA identification and profiling.

Since SRS technology produces enormous amounts of very short reads, assembly tools developed for Sanger sequencing data

Andy Pereira (ed.), *Plant Reverse Genetics: Methods and Protocols*, Methods in Molecular Biology, vol. 678,
DOI 10.1007/978-1-60761-682-5_1,

cannot be directly applied to assemble SRS data because the algorithms rely on longer reads and different sequencing error characteristics. Although several assemblers have been developed to assemble smaller genomes, they are not well suited to handle large eukaryotic genomes. Recently, several tools for efficiently mapping/aligning the SRSs to reference genomes of any arbitrary length have been developed. Table 1 provides a list of tools currently available for mapping. These tools can be used for resequencing, identification of SNPs and indels, identification of small RNA, mRNA transcripts, and alternate splicing.

In this chapter, we focus on analyzing resequencing data using Bowtie and Maq. Bowtie is an ultrafast, memory-efficient short-read aligner. It aligns SRSs to the human genome at a rate of over 25 million 35-bp reads per hour. It works best with short reads although it can support reads up to 1,024 bp in length. Currently, Bowtie does not support colorspace data (from ABI SOLiD), but this will be added in future releases. Bowtie provides alignment parameters similar to Maq and SOAP but can run at much faster speeds than both. Although Maq is much slower than Bowtie at mapping reads to a reference sequence, it has more sequence analytical tools. For example, Maq can produce consensus sequences from alignments and also has tools for SNP discovery.

2. Materials

This section contains a list of prerequisite hardware and software for mapping the reads to the reference genome. In addition to requirements, the formats of the input and output files are described. As mentioned previously, we use Bowtie and Maq. These software are open source and free to use under the GNU public license.

2.1. Downloading the Software

Bowtie can be downloaded from http://bowtie-bio.sourceforge.net/. Maq can be downloaded from http://maq.sourceforge.net/. Source code and binary releases are available for Windows, Linux/Unix, and Mac platforms.

2.2. Installing Bowtie

The software was tested on a 2.66 GHz Two Dual-Core Intel Xeon Mac Pro with 4 GB RAM and 8 core AMD Opteron Linux machine with 64 GB RAM. The software system requires the following:

(a) A regular desktop computer should be sufficient for bacterial genomes. For eukaryotic genomes, at least 2 GB of RAM is needed.

(b) Available disk space should be more than approximately five times the size of input files.

Table 1
List of next-generation sequence alignment software

Package	Description	Reference
ABySS	ABySS is a de novo sequence assembler as well as mapper designed for very short reads	(1)
BFAST	Blat-like Fast Accurate Search Tool	(2)
BLASTN	BLAST's nucleotide alignment program compares reads against a database. Slow and inaccurate for short reads	(3)
BLAT	BLAST-Like Alignment Tool. Can handle one mismatch in initial alignment step	(4)
BWA	BWA is a fast light-weighted tool that aligns short sequences. Supports colorspace reads	(5)
Bowtie	Ultrafast, memory-efficient short-read aligner	(6)
ELAND	Efficient Large-Scale Alignment of Nucleotide Databases	
Exonerate	Pairwise alignment of DNA/protein against a reference	(7)
GMAP	GMAP (Genomic Mapping and Alignment Program) for mRNA and EST Sequences	(8)
GenomeMapper	A short-read mapping tool designed for accurate read alignments	–
MAQ	Mapping and Assembly with Qualities. Supports colorspace reads	–
MOM	MOM or maximum oligonucleotide mapping is a query matching tool that captures a maximal length match within the short read	(9)
MOSAIK	Quickly aligns reads using a hashing scheme. Has an assembly step. Suited for 454 reads	(10)
MUMmer	Rapid whole genome alignment of finished or draft sequences	(11)
MrFAST and MrsFAST	Map short reads to reference genome assemblies. Robust to indels and MrsFAST has a bisulphite mode	–
Novoalign	Gapped alignment of single end and paired end Illumina reads	–
QPalma	Alignment tool targeted to align spliced reads produced by Illumina/454	(12)
RMAP	Assembles Illumina reads to a FASTA reference genome	–
SHRiMP	Assembles to a reference sequence. Supports colorspace reads	–
SLIDER	Uses the "probability" files instead of Illumina sequence files as an input for alignment to a reference sequence	(13)
SLIM Search	Ultrafast blocked alignment	–
SOAP	SOAP (Short Oligonucleotide Alignment Program) is a program for gapped and ungapped alignment of short oligonucleotides onto reference sequences	(14)
SOCS	Short Oligonucleotides in Color Space. Efficient mapping of ABI SOLiD sequence data to a reference genome.	(15)

(continued)

Table 1 (continued)

Package	Description	Reference
SSAHA2	Sequence Search and Alignment by Hashing Algorithm. Quickly, find near exact matches in DNA or protein databases using a hash table	(16)
SWIFT	A software collection for fast index-based sequence comparisons	(17)
SXOligoSearch	SXOligoSearch is a commercial platform. Aligns Illumina reads against a range of Refseq RNA or NCBI genome builds for a number of organisms. Web-based.	–
SeqMap	Maps large amount of oligonucleotide to the genome. Supports 5 or more bp mismatches/indels	–
Vmatch	A versatile software tool for efficiently solving large scale sequence matching tasks	(18)
ZOOM	Zillions Of Oligos Mapped. Maps 15–240 bp long reads to reference genome	–
gnumap	The Genomic Next-generation Universal MAPper. A fast mapping program also tries to align reads from nonunique repeats using statistics	–

– Unpublished

(c) The GCC compiler is needed if installing programs from source code.

Binary files can be copied to an appropriate executable directory. To install from the source, unzip the downloaded installation file. Change to the source directory and run:

```
$ make
```

Once it compiles without errors, copy the bowtie* executable files to the bin directory. You may need admin privileges to do this (see Notes 1 and 2).

2.3. Installing Maq

1. Download the Maq program from maq.sourceforge.net. An example of a Linux command to use is (see Note 3):

   ```
   $ wget http://internap.dl.sourceforge.net/sourceforge/maq/maq-X.XX.X.tar.bz2
   ```

   where X.XX.X denotes version number.

2. Unpack the downloaded file using the command as shown below:

   ```
   $ tar -xjvf maq-X.XX.X.tar.bz2
   ```

   There should be a new folder named maq-X.XX.X in the current working directory.

3. Change directory into maq-X.XX.X:

   ```
   $ cd maq-X.XX.X
   ```

4. Type at the shell prompt:

```
$ gedit INSTALL
```

Read the installation instructions.

5. Type at the shell prompt:

```
$ ./configure
$ make
$ sudo make install
```

Depending on the GCC compiler and some required library files, the installation should proceed without any errors (see Note 4).

3. Methods

The dataset described below, from *Arabidopsis thaliana* 1,001 genomes project, was used for this demonstration. The sequencing project was done by Max Planck Institute for Developmental Biology, using the Illumina Genome Analyzer platform. The library used in the sequencing project was Tsu-1. The following files were downloaded from ftp://ftp.arabidopsis.org/home/tair/Sequences/whole_chromosomes: chr1.fas, chr2.fas, chr3.fas, chr4.fas, chr5.fas, chrC.fas, chrM.fas. The sequencing run chosen, SRR013335, was performed in May 2008. A file containing read sequences with quality scores, SRR013335.fastq, was downloaded from this NCBI ftp site: ftp://ftp.ncbi.nlm.nih.gov/sra/static/SRX000/SRX000704/.

The steps outlined below show how to use the Bowtie and Maq programs to assemble a consensus sequence based on a reference genome. Since Bowtie is faster at alignments than Maq, we will use Bowtie for alignments and then use Maq to assemble the consensus sequence. Maq will also be used to predict SNPs using the same dataset.

1. First, we are going to create a new folder in our home directory called thailana_workspace.

```
$ cd ~
$ mkdir thailana_workspace
```

2. Change the directory into thailana_workspace folder.

```
$ cd thailana_workspace
```

3. Create the following folders: genome, reads, index, maq, assemblies.

```
$ mkdir genome reads index maq assemblies
```

4. Download the *Arabidopsis thaliana* chromosomes with the following command and save them to the genome directory:

```
$ wget ftp://ftp.arabidopsis.org/home/tair/
Sequences/whole_chromosomes/*.fas -P genome/
```

5. Change the directory to genome folder.

```
$ cd genome
```

6. We are going to build an indexed file for Bowtie for Arabidopsis chromosomes by running the bowtie-build utility command; the resulting index file will be named Thaliana. The bowtie-build accepts as inputs the chromosomes fasta files separated by a comma followed by the output name for the index.

```
$ bowtie-build chr1.fas,chr2.fas,chr3.fas,
chr4.fas,chr5.fas,chrC.fas,chrM.fas ./index/
Thaliana
```

   The building of the index will take a few minutes to run depending on the system. The process will output six files in the index directory.

7. The next step involves downloading and unpacking the read file from the NCBI read archive.

```
$ cd ../
$ wget ftp://ftp.ncbi.nlm.nih.gov/sra/static/
SRX000/SRX000704/SRR013335.fastq -P reads/
```

8. Run the Bowtie alignment program and specify a number of processors for faster alignments by using the option –p (see Note 5). Since Arabidopsis has five chromosomes, use the option – refout to split the alignments per chromosome. Since we also included the chloroplast and mitochondria, there will be seven output files of type map. Also it might be useful to print a list of reads that were not aligned to any of the chromosomes by adding the –un option and the name of the file.

```
$ bowtie -t index/Thailana reads/SRR013334.fastq
SRR013335.map -p 2 -un unmappedReads.txt
```

   The program will produce a similar output to the one shown below:

   Time loading forward index: 00:00:01
   Time loading mirror index: 00:00:01
   Seeded quality full-index search: 00:32:18

   Reported 23322363 alignments to seven output stream(s)

   Time searching: 00:32:20
   Overall time: 00:32:21

   In the current directory are the following new files:

   ref00000.map : reads aligned to chromosome 1
   ref00001.map : reads aligned to chromosome 2
   ref00002.map : reads aligned to chromosome 3
   ref00003.map : reads aligned to chromosome 4
   ref00004.map : reads aligned to chromosome 5
   ref00005.map : reads aligned to chloroplast
   ref00006.map : reads aligned to mitochondria

unmappedReads.txt : reads that did not align to any of the above

9. Since the program Maq has many postalignment analytical tools, we can use it to further process our data for SNPs and create consensus sequences from the .map files. Thus, we need to convert the *.map files to a format that is usable in Maq. We also need to first convert the reference chromosome fasta files to binary fasta format (bfa) that is usable in Maq. The command for this task is Maq fasta2bfa . This command accepts two inputs: reference sequence in fasta format and the output file name just as shown below:

```
$ maq fasta2bfa genome/chr1.fas genome/chr1.
bfa
$ maq fasta2bfa genome/chr2.fas genome/chr2.
bfa
$ maq fasta2bfa genome/chr3.fas genome/chr3.
bfa
$ maq fasta2bfa genome/chr4.fas genome/chr4.
bfa
$ maq fasta2bfa genome/chr5.fas genome/chr5.
bfa
```

10. Now, change the map files to a format usable in Maq using the bowtie-maqconvert command. This command accepts three inputs *in this order*: the map file, the output file name, and the corresponding reference sequences file in bfa format.

```
$ bowtie-maqconvert ref00000.map maq/chr1.map
genome/chr1.bfa
$ bowtie-maqconvert ref00001.map maq/chr2.map
genome/chr2.bfa
$ bowtie-maqconvert ref00002.map maq/chr3.map
genome/chr3.bfa
$ bowtie-maqconvert ref00003.map maq/chr4.map
genome/chr4.bfa
$ bowtie-maqconvert ref00004.map maq/chr5.map
genome/chr5.bfa
```

11. Assemble the alignments into consensus sequences and save the assemblies in the folder assemblies.

```
$ maq assemble assemblies/chr1.cns genome/
chr1.bfa maq/chr1.map
```

The program will output a series of statistics to the screen, similar to these shown below:

[cal_het] harmonic sum: 1.000000
[cal_het] het penalty: 26.99 vs. 26.99
[cal_het] 3 differences out of 20 bases: 29.64 vs. 29.64
[cal_het] 1 differences out of 20 bases: 47.20 vs. 47.20

```
[assemble_core] Processing Chr1 (30427671 bp)...
S0 reference length: 30427671
S0 number of gaps in the reference: 164359
S0 number of uncalled bases: 20111445 (0.66)
...
```

Run the following commands for other chromosomes:

```
$ maq assemble assemblies/chr2.cns genome/chr2.bfa maq/chr2.map
$ maq assemble assemblies/chr3.cns genome/chr3.bfa maq/chr3.map
...
```

Here, the program Maq outputs a file with type cns or consensus. The contents of the file cannot be read directly and must be further processed to extract information such as SNPs, alignments, and consensus sequence.

12. In this step, we will extract the consensus sequence from the chr1.cns file. In Maq, there is no direct way of converting chr1.cns to fasta format. The file can only be converted to fastq format. The command for conversion to fastq format is cns2fq. This command accepts one input: the consensus file. The output from the program must be redirected to a file using the > operator.

```
$ maq cns2fq assemblies/chr1.cns > assemblies/chr1.fastq
```

The file chr1.fastq should be about 59 Mb in size. Now, open this file in a text editor to view its contents.

```
$ gedit assemblies/chr1.fastq
```

The FASTQ standard format is divided into four lines as shown in Table 2. The first line contains the chromosome name or reference sequence name. The second line contains the sequence while the third line contains a "+" symbol signifying the end of the sequence and beginning of the quality scores.

Table 2
Fastq file format

FASTQ standard format	chr1.fastq file line #	Contents
Line 1	1	@Chr1
Line 2	2–507,129	ncctaaaccccaaaccccaaaccctaaacctctgaatcctt-nnnnnnnnnnnnnnnn...
Line 3	507,130	+
Line 4	507,131–1,014,258	!+/&93-6(.6??,?????=??????;??:??????????!!!!!!!!!!!!!!!!!! ...

There can be multiple sequences in a fastq file. The second column in the table shows the equivalent lines in our file. As one scrolls down the file, the consensus sequence has regions of unknown sequences denoted by n's. The lowercase nucleotides represent either a region of low coverage or the presence of repeats, while uppercase nucleotides are regions where the sequences of nucleotides have a high probability of being correct. About midway through the file, there is a "+" that denotes the end of the consensus sequence and the beginning of the quality scores on the next line. The quality scores are in ascii format and are decoded by programs for processing (see Note 6).

13. The command cns2snp extracts all the SNP information encoded in the consensus file. This is done in a similar way to cns2fq where we used the ">" symbol.

```
$ maq cns2snp assemblies/chr1.cns > assemblies/
chr1.snp
```

The chr1.snp can be viewed by opening it in a text editor.

```
$ gedit assemblies/chr1.snp
```

The format of the file is a tab-delimited file with the following columns shown in Table 3.

In our file chr1.snp, the first SNP occurs at position 11 with a base change from T to C. Although the mapping quality score

Table 3
Anatomy of the SNP result file

Column #	Description
1	Chromosome number
2	Position on chromosome
3	Nucleotide on reference genome
4	Nucleotide on our consensus sequence
5	Consensus sequence base Quality score similar to Phred
6	Read depth
7	Mean number of hits of reads in this position
8	Maximum mapping quality of reads on this position
9	Minimum quality of 3 bp flanking this position on either side
10	The second best possible nucleotide at this position
11	Log-likelihood ratio between the second and third best possible nucleotide
12	The third best possible nucleotide at this position

in column 8 is higher than the minimum recommended 40, the read depth at this position is very low at 1. This means that we do not know for certain if the predicted SNP is correct since only one read covered this position. In addition, the SNP on chromosome position 46,912 is probably true due to a higher read depth of 13 and a higher base quality score of 66.

4. Notes

1. The Maq installation needs zlib library otherwise Maq will not compile. To correct this problem, download the zlib files for your system using any of the following commands that suits your system:

   ```
   $ sudo yum install zlib zlib-devel # Redhat
   like system,
   $ sudo apt install zlib zlib-devel # Ubuntu
   like system
   ```

2. Or use your package/software manager to install zlib library. After installation of the zlib library, Maq should install seamlessly. The commands for Bowtie and Maq cannot be interchanged since the order of the commands is fixed. For example, these two commands do not mean the same thing. The second one will generate an error:

   ```
   $ bowtie-maqconvert ref00000.map maq/chr1.map
   genome/chr1.bfa
   $ bowtie-maqconvert maq/chr1.map ref00000.map
   genome/chr1.bfa
   ```

   An error will be generated that maq/chr1.map does not exist since the first input to the command accepts the mapped file ref00000.map.

3. Wget is a computer program that retrieves content from web servers.
4. Installing programs in default executable directories in Unix-like system requires admin privileges.
5. If your operating system is 64-bit, compile 64-bit version of bowtie since it is 50% faster than the 32-bit version. Compiled binaries for the 64-bit version are available from the website. If you are building from sources, you may need to pass the -m64 option to g++ to compile the 64-bit version; you can do this by using the argument BITS = 64 to the make command; e.g., make BITS = 64 bowtie.
6. In addition, there are other tools online that can convert fastq format to fasta format such as the fastx tool at this website: http://hannonlab.cshl.edu/fastx_toolkit. However,

for the fastx toolkit to work, the option –o must be added when using the Bowtie-maqconvert utility. The fastx toolkit does not yet accept the fastq format based on the new map format but the old one that restricted read bases to 63.

Acknowledgments

Funding gratefully acknowledged from Virginia Bioinformatics Institute, Virginia Tech, to BWS.

References

1. Simpson, J. T., Wong, K., Jackman, S. D., Schein, J. E., Jones, S. J., and Birol, I. (2009) ABySS: A parallel assembler for short read sequence data. *Genome Res* **19,** 1117–23.
2. Homer, N., Nelson, S. F., and Merriman, B. (2008) *(Unpublished).*
3. Altschul, S. F., Gish, W., Miller, W., Myers, E. W., and Lipman, D. J. (1990) Basic local alignment search tool. *J Mol Biol* **215,** 403–10.
4. Kent, W. J. (2002) BLAT--the BLAST-like alignment tool. *Genome Res* **12,** 656–64.
5. Li, H., and Durbin, R. (2009) Fast and Accurate Short Read Alignment with Burrows-Wheeler Transform. *Bioinformatics* **25**(14), 1754–60.
6. Langmead, B., Trapnell, C., Pop, M., and Salzberg, S. L. (2009) Ultrafast and memory-efficient alignment of short DNA sequences to the human genome. *Genome Biol* **10,** R25.
7. Slater, G. S., and Birney, E. (2005) Automated generation of heuristics for biological sequence comparison. *BMC Bioinformatics* **6,** 31.
8. Wu, T. D., and Watanabe, C. K. (2005) GMAP: a genomic mapping and alignment program for mRNA and EST sequences. *Bioinformatics* **21,** 1859–75.
9. Eaves, H. L., and Gao, Y. (2009) MOM: maximum oligonucleotide mapping. *Bioinformatics* **25,** 969–70.
10. Hillier, L. W., Marth, G. T., Quinlan, A. R., Dooling, D., Fewell, G., Barnett, D., Fox, P., *et al.* (2008) Whole-genome sequencing and variant discovery in C. elegans. *Nat Methods* **5,** 183–8.
11. Kurtz, S., Phillippy, A., Delcher, A. L., Smoot, M., Shumway, M., Antonescu, C., and Salzberg, S. L. (2004) Versatile and open software for comparing large genomes. *Genome Biol* **5,** R12.
12. De Bona, F., Ossowski, S., Schneeberger, K., and Ratsch, G. (2008) Optimal spliced alignments of short sequence reads. *Bioinformatics* **24,** i174–80.
13. Malhis, N., Butterfield, Y. S. N., Ester, M., and Jones, S. J. M. (2009) Slider--maximum use of probability information for alignment of short sequence reads and SNP detection. *Bioinformatics* **25,** 6–13.
14. Li, R., Yu, C., Li, Y., Lam, T. W., Yiu, S. M., Kristiansen, K., and Wang, J. (2009) SOAP2: an improved ultrafast tool for short read alignment. *Bioinformatics.* 25(15):1966–7
15. Ondov, B. D., Varadarajan, A., Passalacqua, K. D., and Bergman, N. H. (2008) Efficient mapping of Applied Biosystems SOLiD sequence data to a reference genome for functional genomic applications. *Bioinformatics* **24,** 2776–77.
16. Ning, Z., Cox, A. J., and Mullikin, J. C. (2001) SSAHA: a fast search method for large DNA databases. *Genome Res* **11,** 1725–9.
17. Rasmussen, K., Stoye, J., and Myers, E. W. (2006) Efficient q-Gram Filters for Finding All epsilon-Matches over a Given Length. *J. Comp. Biol.* **13,** 296–308.
18. Kurtz, S., Choudhuri, J. V., Ohlebusch, E., Schleiermacher, C., Stoye, J., and Giegerich, R. (2001) REPuter: the manifold applications of repeat analysis on a genomic scale. *Nucleic Acids Res* **29,** 4633–42.

Chapter 2

Identification of Plant microRNAs Using Expressed Sequence Tag Analysis

Taylor P. Frazier and Baohong Zhang

Abstract

microRNAs (miRNAs) are a new class of small endogenous noncoding regulatory RNAs, which play an important function in plant growth, development, phase change, and response to environmental stress. Identifying miRNAs is the first step for investigating miRNA-mediated gene regulation and miRNA function. In this chapter, we describe a comprehensive comparative genomics-based expressed sequence tag (EST) analysis for identifying miRNAs from a wide range of plant species. EST analysis is based on the conservation of miRNA sequences and the stem-loop hairpin secondary structures of miRNAs. In this method, potential miRNAs will first be identified by EST analysis followed by confirmation using TaqMan® MicroRNA qRT-PCR. This method is simple and reliable with high efficiency. This method has also been widely adopted by many scientists around the world and several hundreds of miRNAs have been identified in many plant species using this method.

Key words: microRNA, Expressed sequence tag, Comparative genomics, BLASTn, qRT-PCR, EST

1. Introduction

microRNAs (miRNAs) are a newly discovered class of noncoding endogenous small RNAs with about 20–22 nucleotides in length (1, 2). Many investigations have demonstrated that miRNAs play a fundamental role in almost all biological and metabolic processes in plants, including plant growth and development, phase change, and response to abiotic and biotic stress factors (2, 3). Multiple stages are involved in miRNA biogenesis. First, a miRNA gene is transcribed by RNA polymerase II into a long product, called primary miRNA (pri-miRNA); pri-miRNA can form into a specific stem-loop hairpin secondary structure that is sequentially processed by several enzymes, including Dicer-like enzyme 1 (DCL1) and the miRNA methyltransferase HEN, into the mature miRNA (3).

Andy Pereira (ed.), *Plant Reverse Genetics: Methods and Protocols*, Methods in Molecular Biology, vol. 678, DOI 10.1007/978-1-60761-682-5_2,

miRNAs act as post-transcriptional gene regulators by binding perfectly or near-perfectly to messenger RNAs (mRNAs) (4). miRNAs that are near-perfect complements to their target mRNA sequences bind and inhibit protein translation, whereas miRNAs that are perfect complements to the target mRNAs bind and target the mRNA for degradation (4–6).

There are two major approaches for identifying miRNAs in plants: experimental approaches and computational approaches. Experimental approaches include genetic screening, direct cloning, and recently developed next generation high throughput sequencing (2, 7). These methods are the most efficient for identifying miRNAs as they produce few false positive results and are particularly useful for discovering new and species-specific miRNAs. However, experimental methods are often extremely costly and technique dependent. Based on the features of miRNAs, several computational programs have been developed for predicting miRNAs, including miRcheck (8) and findMiRNA (9, 10). In the beginning, these programs were employed to predict many miRNAs in plants. However, a majority of computational programs are based on complete genome sequences which are available for only a limited number of model species, such as *Arabidopsis* and rice. The shortcomings limit the application of these computational programs on a wider range of species.

Studies on comparative genomics across vastly divergent taxa have demonstrated that many miRNAs are highly evolutionarily conserved from species to species, ranging from moss and gymnosperms to high flowering eudicot species (11, 12). This provides a powerful strategy to identify miRNAs from any species using a comparative genomic-based BLASTn search with already known miRNA sequences. Using this strategy, we developed an expressed sequence tag (EST) and a genome survey sequence (GSS) approach to identify conserved miRNAs (13, 14). There are several significant advantages for identifying miRNAs using comparative genomics-based EST analysis (11, 15): (1) EST analysis can be employed to identify miRNAs in any species for which there are previously determined EST sequences; (2) EST analysis not only can be used to identify conserved miRNA, but also provides direct evidence for miRNA expression because ESTs are derived from transcribed sequences (mRNA); (3) it is easy to identify miRNAs using EST analysis and no specialized software is needed; thus, this method is readily available for widespread usage. EST analysis is based on the conservation of miRNAs, and so it can only be used to identify conserved miRNAs. Since the development of this method in 2005 (14), EST analysis has been widely adopted by different laboratories to identify miRNAs from a variety of species, including several important crops such as apple (16), wheat (17, 18), tomato (19), cotton (20), soybean (15), oilseed (21), and maize (22). This method is also used to identify miRNAs in animals (23) and viruses (13). Additionally, EST analysis was used to investigate the diversity and evolution of miRNAs in the plant kingdom (11). Currently, this method has been widely adopted by scientists around the world. In this chapter, we are presenting the basic protocol for the identification of miRNAs using comparative genomics-based EST analysis.

2. Materials

2.1. RNA Collection and Extraction from Plants Using the mirVana™ miRNA Isolation Kit

1. *mir*Vana™ miRNA isolation kit (Ambion, Austin, TX)
   (a) *miRNA wash solution 1*: Add 21 mL 100% ethanol before use. This solution contains guanidinium thiocyanate which is a potential biohazard and should be handled with caution.
   (b) *Wash solution 2/3*: Add 40 mL 100% ethanol before use. This solution can be left at room temperature for up to 1 month. For longer storage periods, store at 4°C but warm to room temperature before use.
   (c) *Collection tubes*: store at room temperature.
   (d) *Filter cartridges*: store at room temperature.
   (e) *Lysis/binding buffer*: store at 4°C.
   (f) *miRNA homogenate additive*: store at 4°C.
   (g) *Acid-phenol: chloroform*: store at 4°C. Phenol is a poison and an irritant and therefore gloves or other protection should be worn when handling this reagent. Dispose of phenol waste appropriately.
   (h) *Elution solution or nuclease-free water*: preheated to 95°C when used and stored at 4°C or room temperature.
2. 100% RNase free ethanol stored at room temperature. Ethanol is flammable so handle and dispose of it accordingly.
3. Liquid Nitrogen.
4. RNase free water.

2.2. RT-PCR of RNA from Plant Tissues

1. TaqMan® MicroRNA Reverse Transcription Kit (Applied Biosystems, Foster City, CA): store at –15 to –25°C. All contents should be thawed on ice and centrifuged briefly before using
   (a) 10× RT Buffer.
   (b) dNTP mix with dTTP (100 mM).
   (c) RNase Inhibitor (20 U/μL).
   (d) Multiscribe™ RT enzyme (50 U/μL).

2. Nuclease-Free water.
3. RT Primers.

2.3. qRT-PCR Analysis of miRNA Expression in Plant Tissues

1. TaqMan 2× Universal PCR Master Mix, No AmpErase UNG (Applied Biosystems, Foster City, CA).
2. qRT Primers (Applied Biosystems, Foster City, CA).
3. Nuclease-Free water.

3. Methods

EST analysis depends on conserved plant miRNA sequences and the NCBI GenBank database in order to find potential miRNAs in other plants. Already identified and confirmed plant miRNA sequences can be obtained from the miRNA database miRBase (http://microrna.sanger.ac.uk) (24). Using a confirmed miRNA sequence, potential homologs can be found by BLASTn searching against the ESTs of other plant species. The resulting matches are further narrowed down by secondary structure analysis using mFold version 3.2 (http://frontend.bioinfo.rpi.edu/applications/mfold/cgi-bin/rna-form1.cgi) (25). Figure 1 summarizes the major steps for identifying miRNAs using EST analysis (11, 22). Figure 2 gives the general structure for a miRNA, including the pre-miRNA sequence and mature miRNA sequence.

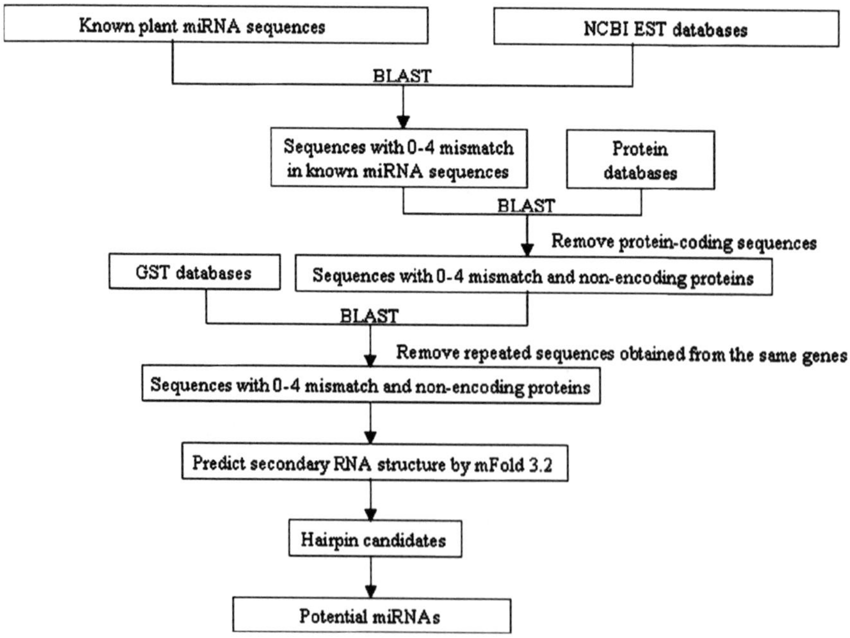

Fig. 1. Schematic representation of the miRNA gene search procedure for identifying miRNA homologs based on established miRNAs using EST analysis.

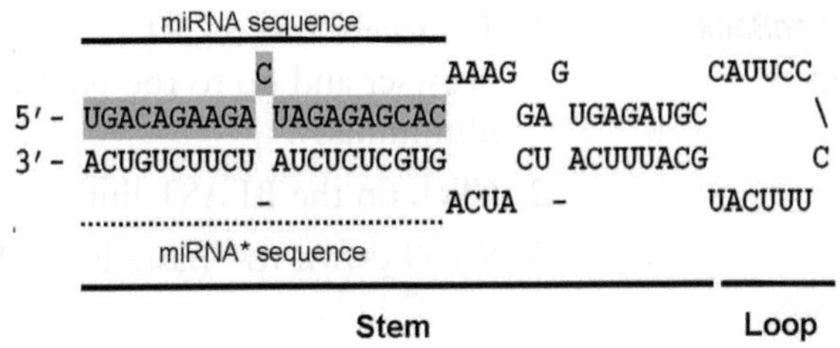

Fig. 2. An example stem-loop hairpin secondary structure of a miRNA precursor sequence.

The criteria for a potential miRNA are (11, 26): (1) predicted mature miRNAs have no more than four nucleotide substitutions compared with a known mature miRNAs; (2) an EST sequence can fold into a stem-loop hairpin secondary structure; (3) the potential mature miRNA sequence is located in one arm of the hairpin structure; (4) there are no more than six mismatches between the predicted miRNA sequence and its opposite miRNA* sequence in the secondary structure; (5) there is no loop or break in the miRNA or miRNA* sequence; and (6) the predicted secondary structure has a high negative minimal folding energy (MFE) and high MFE index (MFEI) value (27). Possible miRNAs that meet all of these criteria are then confirmed experimentally using reverse transcription-polymerase chain reaction (RT-PCR) followed by quantitative reverse transcription-polymerase chain reaction (qRT-PCR).

3.1. miRNA Sequence Acquisition from the miRNA Database

1. Open a Web Browser (Internet Explorer, Mozilla, Safari, etc.) and go to the miRNA database miRBase (http://microrna.sanger.ac.uk).
2. Click on the Circle that says, “Sequences”. This will take you to the main home page.
3. Click on the Search tab located at the top of the page.
4. At the top of the page there will be a box that says, “By miRNA identifier or keyword”. In this box, type “plants” and click on Submit Query. This will take you to a page that contains a list of all miRNAs that have been identified in different plant species.
5. Click on the link provided under the ID column for the desired miRNA sequence. For this chapter, ath-miR156a will be used as an example for demonstrative purposes.
6. Scroll down the page until the mature miRNA sequence appears and click on the link that says, “Get Sequence”.
7. Copy the miRNA name and sequence to a word document for future use.

3.2. NCBI GenBank BLAST Search

1. To access the NCBI GenBank BLAST search, open a web browser and go to the NCBI homepage at http://www.ncbi.nlm.nih.gov.
2. Click on the BLAST link located at the top of the page.
3. Scroll down to "Basic BLAST" and click on the "nucleotide blast" link.
4. In the first box where it says to "Enter Query Sequence", type in or copy and paste from the word document the desired miRNA sequence. In this case, the miRNA sequence listed for ath-miR156a was copied and pasted into the box.
5. Under "Choose Search Set", change the database to "Others". After this is done, a new tab will appear that will allow for the database to be changed. Click on the down arrow and scroll down to select "Expressed sequence tags (est)."
6. In the "Organism" box, type the scientific name of the organism that the miRNA sequence will be searched against. For the purpose of this chapter, *Nicotiana tabacum* will be used as an example organism of choice. If you wish to search for all potential miRNAs in all potential organisms, just leave the box blank.
7. Under "Programs", make sure that the circle next to "highly similar sequences (megablast)" is selected.
8. Click the BLAST button at the bottom of the page. A minute or two will be necessary for the next page to load as the sequences are being retrieved.
9. Once the page has loaded, scroll down the page to the "Alignments" section.
10. Starting with the first result, right-click on the sequence ID (such as gb|FG164766.1) and open the link in a new tab or window.

BLAST Result

```
gb|FG164766.1|  AGN_RNC012xi20r1.ab1 AGN_RNC
Nicotiana tabacum cDNA 3', mRNA
sequence.
Length=807

 Score = 32.2 bits (16),  Expect = 0.15
 Identities = 19/20 (95%), Gaps = 0/20 (0%)
 Strand=Plus/Plus

Query  1    TGACAGAAGAGAGTGAGCAC  20
            |||||||||||| ||||||
Sbjct  430  TGACAGAAGAGAGAGAGCAC  449
```

11. Scroll down the page until "Origin" appears. This is the nucleotide sequence for this particular EST.
12. Highlight and copy up to 800 nucleotides with the targeted sequence located in the middle. This is due to the fact that the mFold software can only fold 800 nucleotide sequences for an immediately folding job. Write down on a separate piece of paper the "Query as 1–20" and the "Sbjct as 430–449".

3.3. mFold to Predict miRNA Secondary Structure

1. In a separate browser window or tab, open the mFold webpage located at http://frontend.bioinfo.rpi.edu/applications/mfold/cgi-bin/rna-form1.cgi.
2. Scroll down the page and where it says to "Enter the sequence to be folded in the box", paste the nucleotide sequence copied from the BLASTn search results.
3. Scroll down to the bottom of the page and click on the button that says "Fold RNA". The software default parameters are used to predict the secondary structures of the selected sequences. It will take a few minutes for the next page to load.
4. Once the RNA has finished folding, a new page will appear with the date and time of folding. Scroll down the page and click the link for "Structure 1".
5. Search the page for the number "430" and look at the secondary structure of the EST between nucleotides 430 and 449.
6. If the secondary structure meets the criteria 1–5 listed above, then this sequence is selected and the end of 5′ and 3′ are determined. The selected EST fragment (potential miRNA precursor sequence) should go through another cycle of mFold.
7. For the potential pre-miRNA sequence, all mFold outputs, including free energy (ΔG kcal/mol), the number of nucleotides (A, G, C and U), and location of the matching regions, should be recorded in an excel data sheet. The MFEI for each sequence should be calculated as previously described (27).
8. Repeat the previous steps again to continually work on other hit sequences.
9. After inspection of all hit sequences, all selected sequences now form a dataset. Perform another BLASTn search against this dataset and remove all repeated sequences found.
10. Because plant miRNAs are unlikely to be located in protein-coding genes, the third BLASTn search should be performed by searching the potential protein-coding genes using the selected sequences from step 9.
11. After removing the protein-coding sequences, the rest of the sequences will most likely be potential miRNAs.

3.4. RNA Collection and Extraction from Plants Using the mirVana™ miRNA Isolation Kit

Young plant leaves are harvested from the greenhouse. Total RNAs are isolated using mirVana™ miRNA Isolation Kit (Ambion, Austin, TX) according to the manufacture's protocol.

1. Collect desired plant tissue using scissors and place in aluminum foil or a 1.5 mL epi-tube. Immediately freeze the tube with samples in liquid nitrogen. If not proceeding with RNA extraction, transfer tissue samples to a −80°C freezer for storage.
2. Prechill a mortar and pestle in a −80°C freezer for at least 30 min prior to RNA extraction.
3. Pipet 300 μL of Lysis/Binding buffer into a 1.5 mL epi-tube and place on ice.
4. Remove the mortar and pestle from the −80°C freezer. Take the tissue sample out of the liquid nitrogen and place in the mortar. Add liquid nitrogen slowly and using the pestle, grind the tissue sample into a fine powder. Transfer the powder, making sure it does not thaw, to the 1.5 mL tube containing the Lysis/Binding buffer making sure to keep the tube on ice.
5. Repeat the previous steps with all of the collected tissue samples.
6. Homogenize the tissue sample with a homogenizer until the tissue is thoroughly broken down.
7. Add 30 μL (1/10 the volume of Lysis/Binding buffer) miRNA Homogenate Additive to the homogenate and mix well by vortexing.
8. Keep the tube on ice for 10 min.
9. Add 300 μL (the volume equal to the Lysis/Binding buffer before miRNA Homogenate Additive addition) Acid-Phenol/Chloroform to each tube making sure to draw from the bottom phase of the bottle.
10. Mix thoroughly by inverting or vortex the tube for approximately 30–60 s to mix.
11. Centrifuge the tube at 10,000 × g at room temperature for 5 min to separate the aqueous phase from the organic phase. If the interphase is not compact after the centrifugation, repeat with a second round.
12. Carefully remove 300 μL of the aqueous upper phase, being careful not to disturb the lower phase, and transfer to a new 1.5 mL tube.
13. Add 375 μL (or 1.25 volumes of the aqueous phase) of room temperature 100% ethanol to the aqueous phase. Mix well by inverting or vortexing.
14. For each sample, place a Filter Cartridge into a new collection tube provided with the kit. Pipet the lysate/ethanol mixture

onto the filter cartridge. The maximum volume that the filter cartridge can hold is 700 μL.

15. Centrifuge at 10,000 × *g* for approximately 15 s. Discard the flow-through and place the filter cartridge back into the same tube. Repeat this procedure until all of the lysate/ethanol mixture has passed through the filter.
16. Apply 700 μL of miRNA Washing Solution 1 to the filter cartridge and centrifuge for approximately 5–10 s. Dispense of the flow-through and place the filter cartridge back into the same tube.
17. Apply 500 μL of miRNA Wash Solution 2/3 and pull the solution through the filter cartridge as detailed in the previous step.
18. Repeat step 17 with a second aliquot of equal volume of miRNA Wash Solution 2/3.
19. After discarding the flow-through from previous step, return the filter cartridge to the same collection tube and spin the assembly for 1 min at 10,000 × *g* at room temperature. This removes residual fluid from the filter.
20. Transfer the filter cartridge to a new collection tube. Apply 100 μL of preheated 95°C Elution solution or nuclease-free water to the center of the filter. Let stand for 30 s–1 min.
21. Centrifuge the tube for 20–30 s at 10,000 × *g* to recover the RNAs.
22. Remove the filter cartridge and mix the recovered RNAs by gently flicking the tube. Briefly centrifuge again to bring all contents to the bottom of the tube.
23. The quality and quantity of the total RNAs are measured using NanoDrop ND-1000 (NanoDrop Technologies, Wilmington, DE).
24. RNA samples are stored in a −80°C freezer until further use.

3.5. RT-PCR of RNA from Plant Tissues

qRT-PCR will be used to confirm the miRNAs identified by EST analysis. A two step assay is performed in TaqMan-based real-time quantification of miRNAs. The first step is a reverse transcription reaction in which a stem-loop RT primer is used to reverse transcribe mature miRNAs to cDNAs. The second step involves real-time PCR, in which the expression levels of miRNAs are monitored and quantified using qRT-PCR that includes miRNA-specific forward primer, reverse primer and FAM dye-labeled TaqMan probes (28).

1. Allow the TaqMan MicroRNA Reverse Transcription Kit reagents and Reverse Transcription Primers (RT primers) to thaw on ice. Briefly centrifuge to bring the reagents and primers to the bottom of the tubes.

2. In a PCR tube (0.2 mL tube), add the following amount of reagents for *one* reaction: 4.16 μL nuclease-Free water, 0.19 μL RNase inhibitor, 1.5 μL 10× RT Buffer, 0.15 μL dNTP mix (100 mM), and 1.00 μL Reverse Transcriptase enzyme.
3. Mix the reagents gently by flicking the tube and briefly centrifuge. Place the tube back on ice.
4. The concentration of total RNA should be 1–10 ng for every 15 μL reaction and added in a ratio such that there is 5 μL of RNA for every 7 μL of reagents. If necessary, add nuclease free water to the reaction tube to bring the volume to 12 μL.
5. Add 3 μL of RT primers to the appropriate tube bringing the total volume per tube to 15 μL. Mix the tube gently by flicking and centrifuge briefly. Incubate for 5 min on ice or until ready to load the thermal cycler.
6. Program the thermal cycler as follows:

Step type	Time (min)	Temperature (°C)
HOLD	30	16
HOLD	30	42
HOLD	5	85
HOLD	∞	4

7. If not proceeding to qRT-PCR, store the RT-PCR samples in a −20°C freezer.

3.6. qRT-PCR Analysis of miRNA Expression in Plant Tissues

1. Add 80 μL of nuclease free water to the RT-PCR products from the previous method.
2. Prepare a master mix in a new 0.2 mL PCR tube by adding the following: 22.5 μL of nuclease-free water, 37.5 μL of 2× PCR mixture, 9 μL of RT-PCR products (after addition of water), and 6 μL RT Primer. The volume of the tube should equal 75 μL.
3. Using a 96-well PCR plate, aliquot 22 μL of the master mix to three separate wells.
4. Centrifuge the plate briefly.
5. The reactions are incubated in a 96-well plate at 95°C for 10 min, followed by 40 cycles of 95°C for 15 s and 60°C for 60 s. This should take approximately 2 h.
6. Analyze the miRNA expression levels from the qRT-PCR amplification results.

7. After the completion of the real-time reactions, the threshold manually sets and the threshold cycle (Ct) will be recorded. The Ct is defined as the fractional cycle number at which the fluorescence passes the fixed threshold (28).
8. Based on the qRT-PCR results, we can conclude which miRNAs are really expressed in that plant organ.

4. Notes

1. All already known miRNA datasets can be downloaded from the miRBase (http://microrna.sanger.ac.uk/cgi-bin/sequences/browse.pl).
2. BLASTn search can be done individually or by group.
3. BLASTn search can be done online or locally by downloading the BLASTn software.
4. If the BLASTn searches reveal partial sequence similarity to a known mature miRNA sequence, the nonaligned regions should be manually inspected and compared in order to determine the number of matching nucleotides and to assess their potential as miRNA candidates.
5. If a BLASTn search hits a sequence that is (±) complementary to the known miRNA sequence, the hit sequence should be changed to the complementary sequence for final analysis.
6. If there is a greater volume of tissue sample for RNA isolation, the volume of the Lysis/Binding buffer may be increased.
7. A vacuum manifold can be used instead of a centrifuge to pull solutions through the filter cartridge.
8. When performing qRT-PCR for miRNA confirmation, the reverse transcription product needs to be diluted to avoid the potential interference of the high concentration of stem-loop primer.

Acknowledgments

This work was partially supported by East Carolina University New Faculty Research Startup Funds Program and a Science and Engineering Grant from DuPont. We would like to thank Dr. Ramsey Lewis and Mr. Ted Woodlief of North Carolina State University for kindly providing tobacco seeds.

References

1. Bartel, D. P. (2004) MicroRNAs: Genomics, biogenesis, mechanism, and function, *Cell 116*, 281–297.
2. Zhang, B. H., Pan, X. P., Cobb, G. P., and Anderson, T. A. (2006) Plant microRNA: A small regulatory molecule with big impact, *Developmental Biology 289*, 3–16.
3. Chen, X. M. (2005) microRNA biogenesis and function in plants, *FEBS Letters 579*, 5923–5931.
4. Zhang, B. H., Wang, Q. L., and Pan, X. P. (2007) MicroRNAs and their regulatory roles in animals and plants, *Journal of Cellular Physiology 210*, 279–289.
5. Eulalio, A., Huntzinger, E., and Izaurralde, E. (2008) Getting to the root of miRNA-mediated gene silencing, *Cell 132*, 9–14.
6. Pillai, R. S., Bhattacharyya, S. N., and Filipowicz, W. (2007) Repression of protein synthesis by miRNAs: How many mechanisms? *Trends in Cell Biology 17*, 118–126.
7. Zhang, B. H., Pan, X. P., Wang, Q. L., Cobb, G. P., and Anderson, T. A. (2006) Computational identification of microRNAs and their targets, *Computational Biology and Chemistry 30*, 395–407.
8. Jones-Rhoades, M. W., and Bartel, D. P. (2004) Computational identification of plant microRNAs and their targets, including a stress-induced miRNA, *Molecular Cell 14*, 787–799.
9. Adai, A., Johnson, C., Mlotshwa, S., Archer-Evans, S., Manocha, V., Vance, V., and Sundaresan, V. (2005) Computational prediction of miRNAs in *Arabidopsis thaliana*, *Genome Research 15*, 78–91.
10. Lindow, M., and Krogh, A. (2005) Computational evidence for hundreds of non-conserved plant microRNAs, *BMC Genomics 6*, 119.
11. Zhang, B. H., Pan, X. P., Cannon, C. H., Cobb, G. P., and Anderson, T. A. (2006) Conservation and divergence of plant microRNA genes, *Plant Journal 46*, 243–259.
12. Floyd, S. K., and Bowman, J. L. (2004) Gene regulation: Ancient microRNA target sequences in plants, *Nature 428*, 485–486.
13. Pan, X. P., Zhang, B. H., SanFrancisco, M., and Cobb, G. P. (2007) Characterizing viral microRNAs and its application on identifying new microRNAs in viruses, *Journal of Cellular Physiology 211*, 10–18.
14. Zhang, B. H., Pan, X. P., Wang, Q. L., Cobb, G. P., and Anderson, T. A. (2005) Identification and characterization of new plant microRNAs using EST analysis, *Cell Research 15*, 336–360.
15. Zhang, B. H., Pan, X. P., and Stellwag, E. J. (2008) Identification of soybean microRNAs and their targets, *Planta 229*, 161–182.
16. Gleave, A. P., Ampomah-Dwamena, C., Berthold, S., Dejnoprat, S., Karunairetnam, S., Nain, B., Wang, Y. -Y., Crowhurst, R. N., and MacDiarmid, R. M. (2008) Identification and characterisation of primary microRNAs from apple (*Malus domestica* cv. Royal Gala) expressed sequence tags, *Tree Genetics & Genomes 4*, 343–358.
17. Dryanova, A., Zakharov, A., and Gulick, P. J. (2008) Data mining for miRNAs and their targets in the Triticeae, *Genome 51*, 433–443.
18. Jin, W. B., Li, N. N., Zhang, B., Wu, F. L., Li, W. J., Guo, A. G., and Deng, Z. Y. (2008) Identification and verification of microRNA in wheat (*Triticum aestivum*), *Journal of Plant Research 121*, 351–355.
19. Yin, Z. J., Li, C. H., Han, M. L., and Shen, F. F. (2008) Identification of conserved microRNAs and their target genes in tomato (*Lycopersicon esculentum*), *Gene 414*, 60–66.
20. Zhang, B. H., Wang, Q. L., Wang, K. B., Pan, X. P., Liu, F., Guo, T. L., Cobb, G. P., and Anderson, T. A. (2007) Identification of cotton microRNAs and their targets, *Gene 397*, 26–37.
21. Xie, F. L., Huang, S. Q., Guo, K., Xiang, A. L., Zhu, Y. Y., Nie, L., and Yang, Z. M. (2007) Computational identification of novel microRNAs and targets in *Brassica napus*, *FEBS Letters 581*, 1464–1474.
22. Zhang, B. H., Pan, X. P., and Anderson, T. A. (2006) Identification of 188 conserved maize microRNAs and their targets, *FEBS Letters 580*, 3753–3762.
23. Weber, M. J. (2005) New human and mouse microRNA genes found by homology search, *FEBS Journal 272*, 59–73.
24. Griffiths-Jones, S., Saini, H. K., van Dongen, S., and Enright, A. J. (2008) miRBase: Tools for microRNA genomics, *Nucleic Acids Research 36*, D154–D158.
25. Zuker, M. (2003) Mfold web server for nucleic acid folding and hybridization prediction, *Nucleic Acids Research 31*, 3406–3415.
26. Ambros, V., Bartel, B., Bartel, D. P., Burge, C. B., Carrington, J. C., Chen, X. M., Dreyfuss, G., Eddy, S. R., Griffiths-Jones, S., Marshall, M., Matzke, M., Ruvkun, G., and

Tuschl, T. (2003) A uniform system for microRNA annotation, *RNA 9*, 277–279.

27. Zhang, B. H., Pan, X. P., Cox, S. B., Cobb, G. P., and Anderson, T. A. (2006) Evidence that miRNAs are different from other RNAs, *Cellular and Molecular Life Sciences 63*, 246–254.

28. Chen, C. F., Ridzon, D. A., Broomer, A. J., Zhou, Z. H., Lee, D. H., Nguyen, J. T., Barbisin, M., Xu, N. L., Mahuvakar, V. R., Andersen, M. R., Lao, K. Q., Livak, K. J., and Guegler, K. J. (2005) Real-time quantification of microRNAs by stem-loop RT-PCR, *Nucleic Acids Research 33*, e179.

Chapter 3

Microarray Data Analysis

Saroj K. Mohapatra and Arjun Krishnan

Abstract

Gene expression profiling has revolutionized functional genomics research by providing a quick handle on all the transcriptional changes that occur in the cell in response to internal or external perturbations or developmental programs. Microarrays have become the most popular technology for recording gene expression profiles. This chapter describes all the necessary steps for analyzing Affymetrix microarray data using the open-source statistical tools (R and bioconductor). The reader is walked through all the basic steps of data analysis: reading raw data, assessing quality, preprocessing/normalization, discovery of differentially expressed genes, comparison of gene lists, functional enrichment analysis, and saving results to files for future reference. Some familiarity with computer is assumed. This chapter is self-contained with installation instructions for R and bioconductor packages along with links to downloadable data and code for reproducing the examples.

Key words: Gene expression, Statistical analysis, Bioinformatics, Differential expression, Gene Ontology

1. Introduction

Transcriptional regulation is a complex process that plays a pivotal role in reprogramming cellular states in response to internal or external changes that arise because of progress through different phases of growth and recycling or perturbations. Hence, measuring and analyzing this regulation could lead to important discoveries regarding the regulatory molecules and the downstream mediators of the response. For over a decade, many high-throughput technologies have been developed for profiling gene expression by monitoring genome-wide mRNA levels, the most prominent among them being the microarray technology.

Andy Pereira (ed.), *Plant Reverse Genetics: Methods and Protocols*, Methods in Molecular Biology, vol. 678,
DOI 10.1007/978-1-60761-682-5_3, © Springer Science+Business Media, LLC 2011

1.1. Basic Principle of Microarray Technology and the Biological Data that It Offers

Microarray is a small chip that contains thousands of probes fixed to its surface, which can hybridize with fluorescently labeled RNA samples (the targets). Hybridization intensities, represented by the amount of fluorescent emission, give an estimate of the relative amounts of the different transcripts that are present in the RNA sample. Many different microarray platforms exist that differ in array fabrication and dye selection. Affymetrix Genechips (1) are high-density oligonucleotide microarrays with 25-nt probes that contain multiple probes per gene (together called a probeset) for most genes in the genome. This platform has gained enormous popularity because of reasonable accuracy and coverage.

A single microarray experimental assay, like most high-throughput technologies, records the transcript levels of all the genes in a given condition, for a given cell type or mixture, at a given time. Comparison of the transcript levels of genes (represented by virtual values) to a reference assay and between different assays is hence important for the interpretation of changes in gene expression. Here, the experimental design heavily influences the quality, quantity, and even the validity of the information obtained. Much care is therefore needed in developing a sound design taking into account several factors, including the quality of samples, the amount of replication and pooling.

Assuming that we have a completed microarray experiment and the resulting gene expression profile data, the following sections bring forth concepts underlying the various steps involved in statistical analysis of this data toward biological inference.

1.2. Statistical Analysis of Microarray Data

1.2.1. Quality Assessment and Control

Issues with RNA extraction (sample quality and amount of starting material), labeling, scanning, or even array manufacture can affect the quality of the microarrays. Visual inspection and creating diagnostic plots can help assess and possibly filter the data. Several of the following factors are taken into account when assessing the quality of microarray chips: average background (expected to be similar across all chips), scale factors (expected to be within threefold of one another), percentage of the number of genes called present (expected to be similar across all chips), and ratio of expression of 3′ probes to 5′ probes of "housekeeping" genes β-actin and GAPDH (acceptable if less than 3 and 1.25, respectively).

The reader is referred to the "Guidelines for Assessing Data Quality" section in the Affymetrix data analysis manual (available at http://www.affymetrix.com/support/downloads/manuals/data_analysis_fundamentals_manual.pdf) for further information on quality control.

1.2.2. Preprocessing

Preprocessing is the process of extracting and transforming the raw fluorescence intensities into a signal normalized for experimental errors and biological variation. The first step in preprocessing is background subtraction (removal of background noise) followed by the normalization to remove systematic sources of variation in the measured intensities due to wide variety of factors. These include early factors, such as print-tip differences (when multiple printing pins are used to print each chip), other spatial effects, and the quality of the microarray printing, and later factors such as quality of the mRNA used, separate reverse transcription and labeling, different dye labeling efficiencies, and different scanning parameters. Procedures, such as quantile normalization (2), make the assumption that the different biological samples have roughly the same distribution of RNA abundance, and transform the intensities so that the bulk of the intensity distribution is the same for all assays in an experiment, typically with some differences in the distribution tails (which might reflect actual biological differences). For Affymetrix data, this method is applied at the probe-level before the prove-level intensities are summarized into a probeset-level value. Methods such as Robust Multiarray Averaging (RMA) perform background correction, quantile normalization, and summarization (3).

1.2.3. Comparison Between Groups to Get Differentially Expressed Genes

Genes that respond to the condition under scrutiny can be discovered by comparing their expression levels across a pair of groups, treatment versus control, later- versus earlier-time-point, infected versus uninfected, etc. Genes that show consistent expression across replicates within a group and difference between groups can be uncovered by employing a statistical test. But, to deal with experimental design issues, the main one being the usual small group size (small number of replicates), modified versions of conventional statistical methods (e.g., moderated *t*-test) have to be used, as implemented in the analysis package limma (4). All the genes can then be ranked based on the significance of differential expression and genes that have a p-value < 0.05 (say) can be declared as the set of differentially expressed genes (DEGs). However, performing multiple statistical tests (one for each of the thousands of genes on the array) at the same time raises the alarm of genes having a p-value < 0.05 just by chance. Methods that control the false discovery rate (FDR) (5), the proportion of false positives among the genes declared as significantly different, have proven to be very useful in overcoming the multiple hypothesis testing problem (see ref. 6 for very good discussion of related issues).

1.2.4. Functional Analysis of Differentially Expressed Genes

Further biological knowledge can be extracted from the list of DEGs by portioning the genes into coherent groups that are known to perform biological processes or participate in a biochemical pathway. The significance of any such functional category (e.g., terms in the Gene Ontology (GO) (7)) that a subset of DEGs belong to can be tested by calculating the probability of observing as many or more genes (among the DEGs) belonging to that category given the total number of DEGs, the number of genes in the genome annotated with that category and the total number of genes in the genome. The hypergeometric test takes these quantities to calculate a *p*-value for the enrichment of genes belonging to a category among the DEGs.

Recent reviews contain more rigorous discussions of other potential analysis methods and issues (8–10).

Here, we concentrate on one simple series of steps for the analysis of gene expression data produced using a standard single-channel microarray platform (Affymetrix GeneChip). Similar steps can be extended to the data from other platforms. We will be using bioconductor tools (11) within R statistical computing environment (12) for the analysis. It is assumed that RNA extracted from some samples has been hybridized onto microarrays and the machine-output data are available for analysis. The example dataset used in this chapter contains the raw data files (CEL format; see Note 1) corresponding to three abiotic stress treatments in *Arabidopsis* along with their respective controls (13). mRNA from shoot samples harvested 30 min post-treatment was used for gene expression profiling using the *Arabidopsis* ATH1 Genome Array (see Note 2). Through a series of steps, we take you from reading in the raw data files to finding DEGs, followed by some basic functional analysis of the genes.

2. Materials

R (http://www.r-project.org/) is an excellent free software environment for statistical computing and graphics. With its wide variety of statistical and graphical techniques, and high extensibility, it has become the standard for analysis and representation of data, including biological data. Bioconductor (http://www.bioconductor.org/) is a free, open source, and open development software project for the analysis and comprehension of genomic data. It is based on R and its components are distributed as R packages, several of which are written for the analysis of microarray data. To keep with the big wave in microarray analysis, we present all the steps needed to perform the analysis entirely within R.

1. Install R on your local machine. Go to http://www.r-project.org, under "Download, Packages" section on the left, click CRAN, select the mirror site nearest to you, and then based on your operating system download the corresponding binary and install it based on the instructions provided therein.
2. In addition to the default packages in R, we need a basic set of packages from bioconductor, the *Arabidopsis* array database

and the GOstats package for functional enrichment analysis (14). Start R (see Note 3). From within R, run the following commands (without the initial ">"):

```
> source("http://www.bioconductor.org/
  biocLite.R")
> biocLite()
> biocLite("ath1121501.db")
```

For the GOstats package, go to http://bioconductor.org/packages/devel/bioc/html/GOstats.html, download the package corresponding to your operating system, and place it in any folder. If you are using Linux/UNIX/Mac do 'R CMD INSTALL GOstats_2.11.0.tar.gz' outside of the R session (in your terminal); if you are using Windows, within an R session go to the "Packages" menu, select 'Install from a local zip file', and select the downloaded file (see Note 4).
Now, load the libraries.

```
> library(affy)
> library(limma)
> library(simpleaffy)
> library(ath1121501.db)
> library(GOstats)
```

3. In your machine, create a directory for this analysis and within it, create another directory called *celfiles* for the raw data files. Then, download the CEL files from TAIR (15) into this directory. Go to http://www.arabidopsis.org/servlets/TairObject?type=hyb_descr_collection&id=ID, replacing "ID" with each of 1007966668, 1007966553, and 1007966888 for the drought, cold, and salt data, respectively. For each stress, download the four CEL files corresponding to shoot tissue: two replicates of the control samples and two replicates of 30 min post-treatment samples.
4. In the current directory (one level above "celfiles"), create a file called *Target.txt* in the format presented in Fig. 1 that

Name	**Celfile**	**Group**
Drought.Control.1	DROUGHT_CONTROL_30MIN_SHOOT_REP1.cel	Drought.Control
Drought.Control.2	DROUGHT_CONTROL_30MIN_SHOOT_REP2.cel	Drought.Control
Drought.Stress.1	DROUGHT_30MIN_SHOOT_REP1.cel	Drought.Stress
Drought.Stress.2	DROUGHT_30MIN_SHOOT_REP2.cel	Drought.Stress
Salt.Control.1	SALT_CONTROL_30MIN_SHOOT_REP1.cel	Salt.Control
Salt.Control.2	SALT_CONTROL_30MIN_SHOOT_REP2.cel	Salt.Control
Salt.Stress.1	SALT_30MIN_SHOOT_REP1.cel	Salt.Stress
Salt.Stress.2	SALT_30MIN_SHOOT_REP2.cel	Salt.Stress
Cold.Control.1	COLD_CONTROL_30MIN_SHOOT_REP1.cel	Cold.Control
Cold.Control.2	COLD_CONTROL_30MIN_SHOOT_REP2.cel	Cold.Control
Cold.Stress.1	COLD_30MIN_SHOOT_REP1.cel	Cold.Stress
Cold.Stress.2	COLD_30MIN_SHOOT_REP2.cel	Cold.Stress

Fig. 1. Sample annotation of the RNA samples used for hybridizing the microarrays. The format resembles that of the *Targets* file.

contains the annotation of the samples in 3 (tab-separated) columns: *Name* for a short sample name, *Celfile* for the name of the raw data file corresponding to the sample, and *Group* for the type of stress and treatment. The group information is especially important for creating contrasts, i.e., pairs of sample groups that are compared with each other.

Now, we are ready to begin the analysis.

3. Methods

3.1. Reading Raw Data

1. Navigate to the current directory (called *Example_dataset* here) using the following command, replacing the path here with the path of the folder in your machine.

```
> setwd("E:/Microarray_Analysis/
  Example_dataset")
```

2. Read the sample annotation from the *Targets.txt* file to an R variable called *targets.*

```
> targets = readTargets("Targets.txt")
```

3. Next, create an R object called *phenoData* that stores the metadata information about the samples. This is important for organizing the sample-level information along with gene-level expression data obtained from CEL files.

```
> rownames(targets) = targets[,1]
> nlev = as.numeric(apply(targets, 2,
  function(x)
+ nlevels(as.factor(x))))
> metadata = data.frame(labelDescription =
+ paste(colnames(targets), ": ", nlev, "
  level",
+ ifelse(nlev==1,"","s"), sep=""),
+ row.names=colnames(targets))
> phenoData = new("AnnotatedDataFrame",
+ data=targets, varMetadata=metadata)
```

4. Read the data from CEL files (along with sample metadata from *phenoData*) to create an object *dat*, which contains the raw probe-level data.

```
> dat = ReadAffy(sampleNames =
  targets$Name,
+ filenames = targets$Celfile,
+ phenoData = phenoData, celfile.path =
  "celfiles")
```

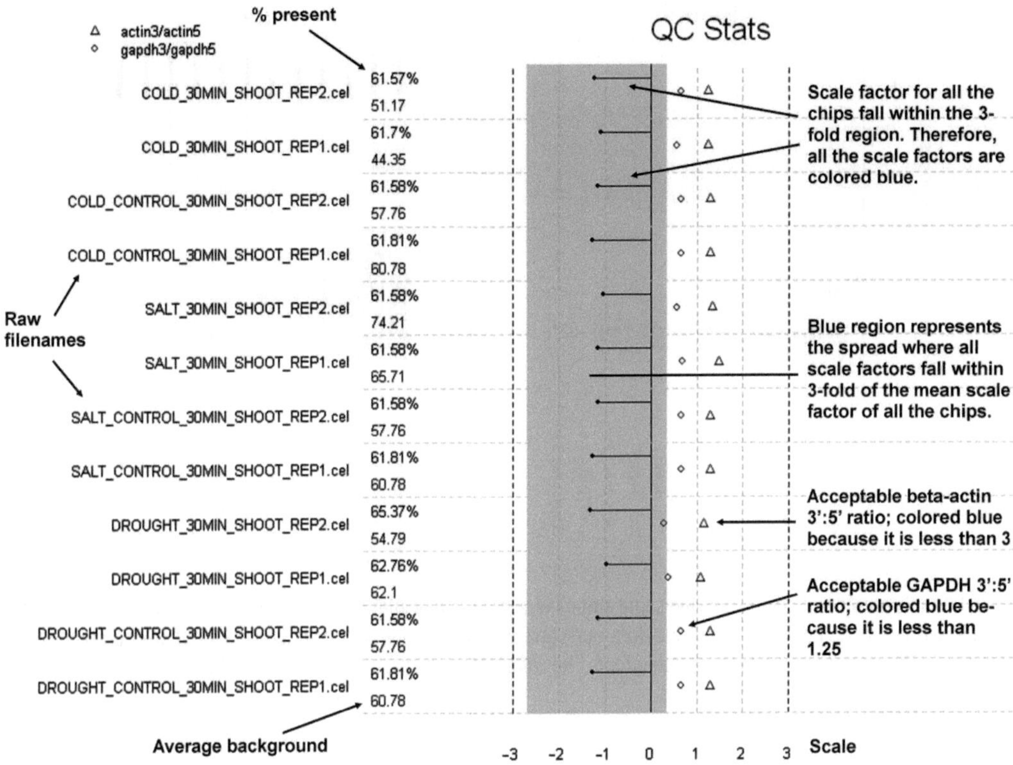

Fig. 2. QC summary statistics. All the arrays appear to be of good quality. Different metrics and their expected and observed quantitles are annotated.

3.2. Quality Assessment

1. To assess the quality of the samples, make use of the function *qc* in package *simpleaffy* (16) for this purpose (Fig. 2) which computes a number of statistics based on recommendations from the array manufacturer Affymetrix.

```
> myqc = qc(dat)
> plot(myqc)
```

3.3. Preprocessing

1. Next, probeset-level data needs to be extracted by applying a normalization algorithm. Here, RMA is used (see Note 5). The normalized data object *eset* is used in further steps below.

```
> eset = rma(dat, verbose = FALSE)
```

2. The effect of quantile normalization (part of RMA) – to bring about similar distribution of gene expression data for each individual array – can be seen by plotting a boxplot for the data before and after normalization (Fig. 3; see Note 6).

```
> par(mfrow=c(1,2), mar=c(12,5,3,3))
> mycols = rep(c("lightgreen", "orange",
  "green", "red",
```

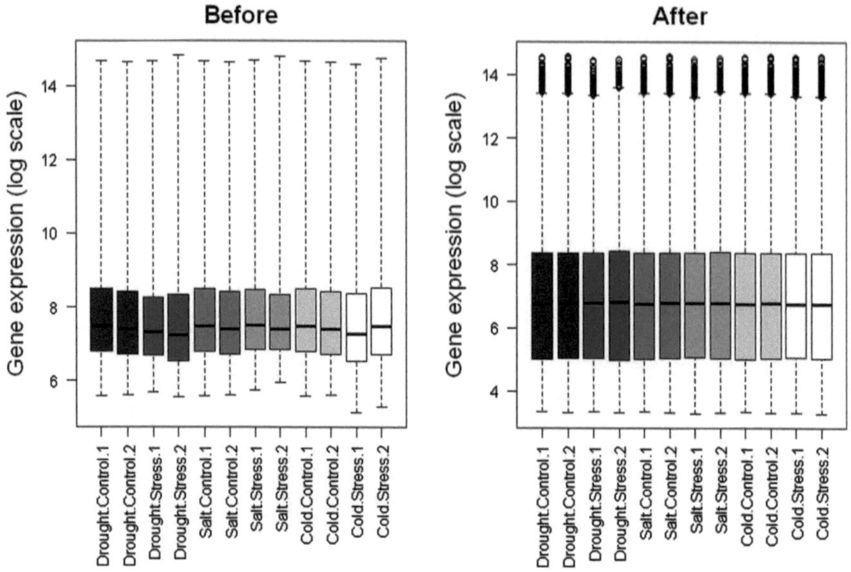

Fig. 3. Quantile normalization imposes the same empirical distribution of intensities on each array. Plots show distribution of gene expression in individual samples before (*left*) and after (*right*) normalization. Each *colored box* corresponds to one array, with same color for each sample group. The *middle line* corresponds to median. Observe that the *boxes* are more aligned at their medians on the *right*.

```
+ "blue", "pink"), each=2)
> boxplot(dat, col=mycols, main="Before",
+ ylab="Gene expression (log scale)", cex.
  lab=1.5,
+ names=dat$Name, las=2, cex.main=2)
> boxplot(data.frame(exprs(eset)),
  col=mycols,
+ main="After", ylab="Gene expression (log
  scale)",
+ cex.lab=1.5, names=eset$Name, las=2, cex.
  main=2)
```

3.4. Discovering Interesting Genes

To find interesting genes that respond to stress, we do the following four comparisons for all three types of stress: drought, salt, and cold. This can be performed using moderated *t*-test available in the package *limma*. There are four steps: (1) define the design matrix and contrast matrix, (2) fit a linear model through each gene, (3) perform empirical Bayes moderation of the standard errors, and (4) list the most interesting genes. Practically, it amounts to applying four functions from the package *limma*: *model.matrix* (+*makeContrasts*)→*lmFit*(+*contrasts.fit*)→*eBayes*→*topTable*.

1. Creating a design matrix and contrast matrix. The design matrix indicates which RNA samples have been applied to

```
> print(design)
   Cold.Control Cold.Stress Drought.Control Drought.Stress Salt.Control Salt.Stress
1             0           0               1              0            0           0
2             0           0               1              0            0           0
3             0           0               0              1            0           0
4             0           0               0              1            0           0
5             0           0               0              0            1           0
6             0           0               0              0            1           0
7             0           0               0              0            0           1
8             0           0               0              0            0           1
9             1           0               0              0            0           0
10            1           0               0              0            0           0
11            0           1               0              0            0           0
12            0           1               0              0            0           0
attr(,"assign")
[1] 1 1 1 1 1 1
attr(,"contrasts")
attr(,"contrasts")$grp
[1] "contr.treatment"
```

```
> print(cont.diff)
                  Contrasts
Levels             Cold Drought Salt
  Cold.Control       -1       0    0
  Cold.Stress         1       0    0
  Drought.Control     0      -1    0
  Drought.Stress      0       1    0
  Salt.Control        0       0   -1
  Salt.Stress         0       0    1
```

Fig. 4. *Top*: The design matrix indicates the RNA sample hybridized to each array. *Bottom*: The contrast matrix indicates the comparisons that are of interest.

each array. It is created by applying the function *model.matrix* on the *Group* information in sample metadata. Each row of the design matrix corresponds to an array and each column corresponds to a coefficient used to describe the RNA source (Fig. 4 *top*). Since we are dealing with Affymetrix arrays, the number of coefficients (columns in the design matrix) is same as the number of distinct RNA sources.

```
> grp = factor(eset$Group)
> design = model.matrix(~0+grp)
> colnames(design) = c("Cold.Control",
  "Cold.Stress",
+ "Drought.Control", "Drought.Stress",
  "Salt.Control",
+ "Salt.Stress")
> print(design)
```

We have an experiment with six groups that allows $6 \times 5 = 30$ possible pair-wise comparisons. However, we are interested in only a subset of these, i.e., three contrasts of Stress versus Control. Thus, we need to create the contrast matrix which indicates the comparisons that are of interest. This can be achieved by applying the function *makeContrasts* on the design matrix (Fig. 4 *bottom*).

```
> cont.diff = makeContrasts(
+ Cold = Cold.Stress-Cold.Control,
+ Drought = Drought.Stress-Drought.Control,
```

```
+ Salt = Salt.Stress-Salt.Control,
+ levels = design)
> print(cont.diff)
```

2. Fitting a linear model. The systematic part of the data can be fully modeled by the linear model specified by the design matrix, and the initial coefficients can be compared in selected ways (as specified in the contrast matrix).

```
> fit = lmFit(eset, design)
> fit2 = contrasts.fit(fit, cont.diff)
```

3. Empirical Bayes moderation. For assessing differential expression, the empirical Bayes method can then be used to moderate the standard errors of the estimated log ratios.

```
> fit2 = eBayes(fit2)
```

3.5. Exploring the List of DEGs

1. The list of top genes that are differentially expressed by drought stress, for e.g., (at default adjusted *p*-value cutoff of 0.05) can be obtained using the function *topTable* (Fig. 5; see Note 7).

```
> options(digits = 3)
> topTable(fit2, coef = "Drought")
```

The column *logFC* reports the fold change (log scale) in gene expression by drought stress. Because fold change is a ratio, it sometimes helps to know the absolute gene expression level: column *AveExpr* (very low values should not be trusted too much). The other important column is *adj.P.val*; a value of less than 0.05 means that the fold change is statistically significant.

2. While *topTable* does return the list of most interesting genes, it does not inform about all the genes that are differentially expressed, nor how many of such genes exist for the contrast. Another function *decideTests* from package *limma* needs to be

```
> topTable(fit2, coef = "Drought")
              ID logFC AveExpr     t  P.Value adj.P.Val     B
13239 258139_at  2.69    7.15  26.1 4.13e-09  6.40e-05 10.76
2308  247208_at  1.79    5.99  24.2 7.60e-09  6.40e-05 10.36
20166 265066_at -1.83   10.95 -23.4 1.01e-08  6.40e-05 10.17
9094  253994_at  2.08    9.07  23.0 1.12e-08  6.40e-05 10.09
17548 262448_at  2.24    4.72  21.9 1.72e-08  7.44e-05  9.78
14532 259432_at  2.42    5.44  21.5 1.96e-08  7.44e-05  9.68
6256  251156_at  2.47    4.52  20.9 2.44e-08  7.46e-05  9.52
7231  252131_at  3.76    6.10  20.7 2.62e-08  7.46e-05  9.47
2379  247279_at  1.56   10.09  19.6 4.04e-08  1.02e-04  9.13
19106 264006_at  1.35    9.89  18.8 5.57e-08  1.27e-04  8.87
```

Fig. 5. List of top genes that are differentially expressed by drought stress. The important columns are: *logFC* – fold change (log scale) in gene expression; *AveExpr* – absolute gene expression level; and, *adj.P.val* – *p*-value after adjusting for multiple hypothesis testing.

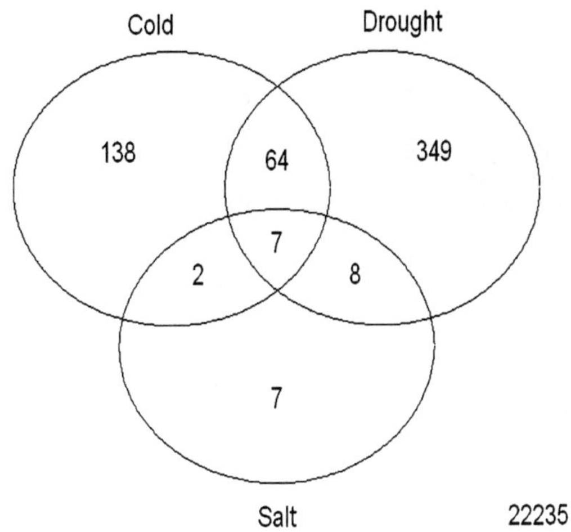

Fig. 6. Venn diagram of the number of genes differentially expressed in the three conditions. Each *circle* corresponds to a set of genes differentially expressed (either up- or downregulated) by particular stress in a statistically significant manner. The name above the *circle* refers to the type of stress. The overlapping area between *two circles* contains the number of genes that are modulated by two types of stress. The number in the *center* (7) refers to the number of genes modulated by all three types of stress. The number on the *lower right corner* indicates the genes unaffected by any stress.

used to obtain the different lists of DEGs, which can then be compared to each other using a using Venn diagram (Fig. 6).

```
> result = decideTests(fit2, lfc=1)
> vennDiagram(result)
```

3. The Venn diagram shows the distribution of stress-regulated genes ($p < 0.05$; absolute fold change >2). To identify genes of interest from here, we need to work on *result*. Let us look at the first five rows in this object (Fig. 7 *top*).

```
> print(result[1:5, ])
```

4. As we can see, each row corresponds to one gene, each column to one stress, and the numbers suggest the direction of differential expression: no change (0), upregulation (1) or downregulation (−1) by stress. Based on this, we can extract the genes in different regions of the Venn diagram and display their fold changes from the data available in the *fit2* object (Fig. 7 *bottom*):

```
> drought.genes = names(which(result
  [, "Drought"] != 0))
> cold.genes = names(which(result[, "Cold"]
  != 0))
> salt.genes = names(which(result[, "Salt"]
  != 0))
> common.genes = intersect(intersect
  (drought.genes,
```

```
> print(result[1:5, ])
            Contrasts
             Cold Drought Salt
  244901_at    0        0    0
  244902_at    0        0    0
  244903_at    0        1    0
  244904_at    0        0    0
  244905_at    0        0    0
> print(lfc)
            Contrasts
               Cold Drought  Salt
  245176_at -1.19   -1.95 -1.92
  245334_at -1.09   -1.07 -1.10
  252607_at  2.21    1.90  1.06
  257076_at -2.36   -2.85 -2.27
  261658_at -1.81   -2.06 -1.50
  263421_at -1.64   -2.65 -2.24
  264857_at -1.02   -1.13 -1.22
> print(mget(common.genes[3], ath1121501GENENAME))
$`252607_at`
[1] "xyloglucan endo-transglycosylase" "xyloglucan endo-transglycosylase"
```

Fig. 7. *Top*: A subset of the object *result* that contains the data for differential expression in the three stresses. Each row corresponds to one gene, each column to one stress, and the numbers suggest the direction of differential expression: no change (0), upregulation (1) or downregulation (−1) by stress. *Bottom*: List of seven genes commonly regulated by the three stresses, and the name of the gene consistently upregulated.

```
+ salt.genes), cold.genes)
> lfc = fit2$coef[common.genes, ]
> print(lfc)
```

The third gene is upregulated by all the stresses. The name of the gene corresponding to that probeset id can be found by referring to the *ath1121501 database* that contains mappings between several types of identifiers and annotations (see Notes 8 and 9).

```
> print(mget(common.genes[3],
  ath1121501GENENAME)
```

5. All the results from linear modeling can be saved to a tab-delimited file that contains fold changes and *p*-values for all probe set ids thus:

```
> write.fit(fit2, file = "limmaResult.txt",
  adjust = "BH")
```

The file *limmaResults.txt* will be created in your current folder and can be explored using a text editor or programs like Excel. The file contains the log fold changes (*Coef.Cold*, *Coef.Drought*, *Coef.Salt*), adjusted *p*-values (*p.value.adj.Cold*, *p.value.adj.Drought*, *p.value.adj.Salt*) as well as other estimates. This file is suitable for manipulating the results outside of R environment and doing any further analysis.

3.6. GO Enrichment Analysis

1. For this analysis, let us concentrate on a particular stress, drought, and identify the genes upregulated by this stress (see Note 10).

```
> drought.genes.up = names(which(result[,
  "Drought"] > 0))
```

2. We can use the genes from this object as our selected list of genes to analyze the enrichment of functional categories.

```
> selectedGenes = unlist(mget(drought.
  genes.up,
+ ath1121501ACCNUM))
```

3. For the enrichment analysis, we also need to define the list of all genes that are in the array as the universe (total number of genes in the genome).

```
> allGenes = featureNames(eset)
> geneUniverse = unlist(mget(allGenes,
  ath1121501ACCNUM))
```

4. Now, set the parameters for the hypergeometric function in the object *params* (see Note 11) and perform the enrichment (overrepresentation) analysis of GO biological process (BP) categories using the function *hyperGTest* (see Note 12).

```
> params = new("GOHyperGParams", geneIds =
  selected Genes,
+ universeGeneIds = geneUniverse,
+ annotation = "ath1121501.db", ontology =
  "BP",
+ pvalueCutoff = 0.001, conditional =
  FALSE,
+ testDirection = "over")
> hgOver = hyperGTest(params)
```

The results of the analysis can be exported as an HTML report that contains the various statistics of each enriched GO term, which are hyperlinked to the GO database (http://www.geneontology.org/).

```
> htmlReport(hgOver, file = "report_hgOver.
  html")
```

Although you could go to your current directory and click this report to open it in your browser, R provides a more elegant way of doing this:

```
> browseURL("file:///E:/Microarray_Analysis/
  Example_dataset/report_hgOver.html",
+ browser="C:/Program Files/Mozilla Firefox/
  firefox.exe")
```

As always, replace the file path and Firefox's (or your favorite browser's) location with the ones in your machine.

5. The report contains the GO Biological Process identifiers (*GOBPID*), their *p*-values (*Pvalue*) that indicate the level of significance, odds ratio (*OddsRatio*) that is an indicator

of the level of enrichment of genes within the list as against the universe, expected number of genes annotated with that term (*ExpCount*) in the list based on the sizes of the list and the universe, actual number of genes in the list annotated with that term (*Count*), number of genes in the universe annotated with that term (*Size*), and the GO term (*Term*).

One of the significantly enriched terms that is of interest is "transcription, DNA-dependent" (GO:0006351) that annotates transcription factors (TFs). There are 25 drought-regulated TFs in the list. These genes can be obtained for scrutiny thus:

```
> drought.tf.genes.up = intersect
  (drought.genes.up,
+ unlist(get("GO:0006355",
  ath1121501GO2ALLPROBES)))
```

Finally, any set (or all) of these genes can be listed with their gene ids and annotations using the following command (Fig. 8).

```
> for(i in 23:25) { y=unlist(mget(drought.
  tf.genes.up[i],
+ ath1121501GENENAME));
+ x=unlist(mget(drought.tf.genes.up[i],
+ ath1121501ACCNUM));
+ cat(strwrap(unlist(c(x[1],y[1])),width=70),
  sep="\t");
+ cat(sep="\n")}
```

```
> for(i in 23:25) { y=unlist(mget(drought.tf.genes.up[i], ath1121501GENENAME));
+      x=unlist(mget(drought.tf.genes.up[i], ath1121501ACCNUM));
+      cat(strwrap(unlist(c(x[1],y[1])), width=70),sep="\n"); cat(sep="\n") }
AT2G46680
encodes a putative transcription factor that contains a homeodomain
closely linked to a leucine zipper motif. Transcript is detected in
all tissues examined.  Is transcriptionally regulated in an
ABA-dependent manner and may act in a signal transduction pathway
which mediates a drought response.

AT2G44840
encodes a member of the ERF (ethylene response factor) subfamily B-3
of ERF/AP2 transcription factor family. The protein contains one AP2
domain. There are 18 members in this subfamily including ATERF-1,
ATERF-2, AND ATERF-5.

AT2G38470
Member of the plant WRKY transcription factor family.  Regulates the
antagonistic relationship between defense pathways mediating
responses to P. syringae and necrotrophic fungal pathogens.
```

Fig. 8. Three of the 25 TFs upregulated by drought stress.

4. Notes

1. CEL file format description: The CEL file includes for each probe on the array an intensity value, standard deviation of the intensity, the number of pixels used to calculate the intensity value, a flag to indicate an outlier as calculated by the algorithm, and a user defined flag indicating the feature should be excluded from future analysis. Generally, this file is used as an input for further analysis and supporting annotation files are either already present as part of the system or automatically downloaded.
2. The Arabidopsis ATH1 Genome Array contains >22,500 probe sets representing ~21,000 genes.
3. R typically starts in a console with a prompt (">"). Therefore, R code presented here starts with the prompt symbol (">") and is written in a different font to distinguish it from the rest of the text. In some cases, the command extends beyond the line and the symbol "+" is used to indicate continuation from the previous line. To reproduce the analysis presented here, please strip off any ">" or trailing "+" and issue the same command in your R console. Alternatively, you could use the R code provided along with the raw data. You should be able to get the same (or similar) result as presented here.
4. GOstats (version 2.10.0) that you could get from within R (using biocLite("GOstats")) does not support GO enrichment analysis for *Arabidopsis*, while this functionality has been added to the development version of the package, which is GOstats_2.11.0. However, please pay attention to the fact that these development version packages might be unsupported, not fully tested and even buggy. So if you then encounter problems, be warned.
5. There are other options for the normalization algorithm such as mas5 (17) and gcrma (18) that you can try in the place of rma. Also, try the command with option `verbose=TRUE` to see the steps printed out as the method progresses.
6. Use the menu "Windows" to alternate between the window displaying the figure and the R console.
7. If you wish to look into more genes and filter them for (adjusted) *p*-value and fold-change, try the following command: `topTable(fit2, coef = "Drought", number = 100, p.value = 0.01)`. *coef* here refers to the contrast name. For finding out the top genes for other contrasts, just change the parameter from *Drought* to *Cold* or *Salt*. Now, it is a good time to show the command that opens up a help/manual

page on any function: `?topTable`. Simply prefix the function by a question mark. If you simply wish to find out all the functions that have anything to do with "table" within R from all the loaded packages, type: `??table`.

8. You can find out about the other types of identifier and annotation mappings available in this database using the following command: `help(package="ath1121501.db")`.
9. This gene is a xyloglucan endo-transglycosylase that plays a role in plant cell wall biogenesis. Members of this family have been reported to be upregulated and assume a protective role in cell wall modification in response to diverse environmental stresses and hormone stimuli (see ref. 7 and the references therein).
10. Type `length(drought.genes.up)` to find the number of genes.
11. Duplicate identifiers, which come about because of multiple probesets mapping to the same gene, will be removed from both the selected and the universe lists.
12. Type `print(hgOver)` to find out the summary of the enrichment test.

References

1. Lockhart, D., Dong, H., Byrne, M., Follettie, M., Gallo, M., Chee, M., et al. (1996) Expression monitoring by hybridization to high-density oligonucleotide arrays. *Nat Biotechnol.* **14**: 1675–1680.
2. Bolstad, B. M., Irizarry, R. A., Åstrand, M., and Speed, T. P. (2003) A comparison of normalization methods for high-density oligonucleotide array data based on variance and bias. *Bioinformatics.* **19**: 185–193.
3. Irizarry, R. A., Hobbs, B., Collin, F., Beazer-Barclay, Y. D., Antonellis, K. J., Scherf, U., et al. (2003) Exploration, normalization, and summaries of high density oligonucleotide array probe level data. *Biostatistics.* **4**: 249–264.
4. Smyth, G. (2004) Linear models and empirical bayes methods for assessing differential expression in microarray experiments. *Stat Appl Genet Mol Biol.* **3**: Article3.
5. Benjamini, Y., and Hochberg, Y. (1995) Controlling false discovery rate: a practical and powerful approach to multiple testing. *J R Stat Soc Series B.* **57**: 289–300.
6. Nettleton, D. (2006) A discussion of statistical methods for design and analysis of microarray experiments for plant scientists. *Plant Cell.* **18**: 2112–2121.
7. Ashburner, M., Ball, C. A., Blake, J. A., Botstein, D., Butler, H., Cherry, J. M., et al. (2000) Gene ontology: tool for the unification of biology. The Gene Ontology Consortium. *Nat Genet.* **25**: 25–29.
8. Clarke, J. D., and Zhu, T. (2006) Microarray analysis of the transcriptome as a stepping stone towards understanding biological systems: practical considerations and perspectives. *Plant J.* **45**: 630–650.
9. Allison, D. B., Cui, X., Page, G. P., and Sabripour, M. (2006) Microarray data analysis: from disarray to consolidation and consensus. *Nat Rev Genet.* **7**: 55–65.
10. Cordero, F., Botta, M., and Calogero, R. A. (2007) Microarray data analysis and mining approaches. *Brief Funct Genomic Proteomic.* **6**: 265–281.
11. Gentleman, R. C., Carey, V. J., Bates, D. M., Bolstad, B., Dettling, M., Dudoit, S., et al. (2004) Bioconductor: open software development for computational biology and bioinformatics. *Genome Biol.* **5**: R80.
12. R Development Core Team. (2008) R: a language and environment for statistical computing. R Foundation for Statistical Computing, Vienna, Austria. ISBN 3-900051-07-0. URL http://www.R-project.org.

13. Kilian, J., Whitehead, D., Horak, J., Wanke, D., Weinl, S., Batistic, O., et al. (2007) The AtGenExpress global stress expression data set: protocols, evaluation and model data analysis of UV-B light, drought and cold stress responses. *Plant J.* **50**: 347–363.

14. Falcon, S., and Gentleman, R. (2007) Using GOstats to test gene lists for GO term association. *Bioinformatics.* **23**: 257–258.

15. Swarbreck, D., Wilks, C., Lamesch, P., Berardini, T. Z., Garcia-Hernandez, M., Foerster, H., et al. (2008) The Arabidopsis Information Resource (TAIR): gene structure and function annotation. *Nucleic Acids Res.* **36**: D1009–D1014.

16. Wilson, C. L., and Miller, C. J. (2005) Simpleaffy: a BioConductor package for Affymetrix quality control and data analysis. *Bioinformatics.* **21**: 3683–3685.

17. Gautier, L., Cope, L., Bolstad, B. M., and Irizarry, R. A. (2004) affy – analysis of Affymetrix GeneChip data at the probe level. *Bioinformatics.* **20**: 307–315.

18. Wu, Z., Irizarry, R. A., Gentleman, R., Murillo, F. M., and Spencer, F. (2004) A model based background adjustment for oligonucleotide expression arrays. *J Am Stat Assoc.* **99**: 909–917.

19. Iliev, E. A., Xu, W., Polisensky, D. H., Oh, M. H., Torisky, R. S., Clouse, S. D., et al. (2002) Transcriptional and posttranscriptional regulation of Arabidopsis TCH4 expression by diverse stimuli. Roles of cis regions and brassinosteroids. *Plant Physiol.* **130**: 770–783.

Chapter 4

Setting Up Reverse Transcription Quantitative-PCR Experiments

Madana M.R. Ambavaram and Andy Pereira

Abstract

Quantitative real-time PCR (qRT-PCR), in conjunction with reverse transcriptase, has been used for the systematic measurement of plant physiological changes in gene expression. In the present paper, we describe a qRT-PCR protocol that illustrates the essential technical steps required to generate quantitative data that are reliable and reproducible. To demonstrate the methods used, we evaluated the expression stability of five [*actin* (*ACT*), *actin1* (*ACT1*), *β-glyceraldehyde-3-phosphate dehydrogenase* (*GAPDH*), *cyclophilin* (*CYC*), and *elongation factor 1α* (*EF-1α*)] frequently used housekeeping genes in rice. The expression stability of the five selected housekeeping genes varied considerably in different tissues (seedlings, vegetative and reproductive stages) in a given stress condition. The analysis allowed us to choose a set of two candidates (*ACT1* and *EF-1α*) that showed more uniform expression and are also suitable for the validation of weakly expressed genes (≥0.5 fold), identified through microarray analysis.

Key words: qRT-PCR, Housekeeping genes, SYBR Green, Normalization, Drought

1. Introduction

The state-of-the-art technology for confirmation and quantitative analysis of relative changes in gene expression levels is "quantitative real-time RT-PCR" (qRT-PCR). Nevertheless, the invention of real-time PCR has revolutionized the field of gene expression analysis and has become routine in many of today's research laboratories. In conventional PCR, the amplified product or amplicon is detected by an end-point analysis by running DNA on an agarose gel after the reaction has finished. In contrast, real-time PCR monitors the amount of amplicon generated as the reaction occurs. Usually, the amount of product is directly related to the fluorescence of a reporter dye and the measured fluorescence

Andy Pereira (ed.), *Plant Reverse Genetics: Methods and Protocols*, Methods in Molecular Biology, vol. 678,
DOI 10.1007/978-1-60761-682-5_4, © Springer Science+Business Media, LLC 2011

reflects the amount of amplified product in each cycle. The fluorescent chemistries employed for this purpose include DNA-binding dyes (SYBR Green I) and fluorescently labeled sequence specific primers or probes (TaqMan probes). In a TaqMan assay, the probe is labeled at the 5′ end with a fluorescent reporter molecule and at the 3′ end with another fluorescent molecule, which act as a quencher for the reporter. However, the most commonly used DNA-binding dye for real-time PCR is SYBR Green I, which specifically binds double-stranded DNA (dsDNA) by intercalating between base pairs, and fluoresces only when bound to dsDNA. Therefore, the overall fluorescent signal from a reaction is proportional to the amount of dsDNA present, and will increase as the target is amplified. Furthermore, the advantages of using dsDNA-binding dyes include simple assay design, the ability to test multiple genes quickly without designing multiple probes (for example, validation of gene expression data from many genes in a microarray experiment), and the ability to perform a melt-curve analysis to check the specificity of the amplification reaction.

When performing a qRT-PCR analysis, several parameters need to be controlled to obtain reliable quantitative expression measures. These include variations in initial sample amount, RNA recovery, RNA integrity, efficiency of cDNA synthesis, and differences in the overall transcriptional activity of the tissues or cells analyzed (1, 2). Besides being extremely a powerful technique, real-time PCR suffers from certain pitfalls, most importantly selection of gene specific primer pairs and the normalization with a reference or housekeeping gene (3).The expression of a reference gene used for the normalization of real-time PCR analysis should remain constant between the cells of different tissues/organs and under different experimental conditions. However, in recent years, it has become clear that no single gene is constitutively expressed in all cell types and under all experimental conditions, implying that the expression stability of the intended control gene has to be verified before each experiment.

One of the fastest expanding applications of real-time RT-PCR is the confirmation of data obtained from microarray analysis. Indeed, the reliability of microarray experiments may sometimes be questioned. Since plants display a high number of multigene families, cross-hybridization between cDNA representatives of members of gene families on cDNA-based chips may lead to false interpretations (4). On the other hand, microarray experiments can analyze thousands of genes in one step, whereas real-time PCR is often limited to far fewer genes. Real-time PCR requires the design of specific oligonucleotides for each gene to be analyzed, and because of the limited number of both fluorophores and light spectra detected by real-time PCR machines, this allows the detection of fewer than five genes per multiplex PCR run. However, a maximum of two genes are analyzed routinely in the same tube.

Therefore, a widely used strategy is to point out a handful of potentially interesting genes with microarray experiments and to confirm those candidates by real-time RT-PCR analysis (5).

Keeping in view, the importance of control gene(s) in the normalization of real-time PCR, various housekeeping genes have been evaluated for stable expression under gradual drought stress conditions in rice. We found that the potential internal control genes differed widely in their expression stability over the different tissues/or developmental stages and environmental conditions studied. Therefore, it is necessary to validate the expression stability of a control gene under specific experimental conditions prior to its use for the normalization.

2. Materials

1. We use RNeasy Plant Mini Kit (Qiagen) and the Trizol Reagent (Invitrogen) for the microarray and GS-FLX transcriptome analysis, respectively (see Note 1).
2. Diethylpyrocarbonate (DEPC) (Sigma) [DEPC-Treated H_2O: Add 1 mL of DEPC to 1 L of MilliQ H_2O and incubate overnight with stirring after that autoclave and cool to room temp before use).
3. DNase I (Qiagen).
4. i- Script™ cDNA Synthesis Kit (Bio-Rad) (see Note 2).
5. RNasin® Plus RNase Inhibitor (Promega).
6. iQ™ SYBR® Green Super Mix (Bio-Rad) (see Note 3).
7. 96-well optically clear plate and Optical plate cover (Bio-Rad).
8. Oligo (dT)15 Primer (Promega).
9. SuperScript™ III Reverse Transcriptase (Invitrogen).

3. Methods

3.1. Primer Design Considerations

A successful real-time PCR reaction requires efficient and specific amplification of the product (see Note 4). There are numerous Web-based programs available for PCR primer design, both free and commercial. We used Becon Designer (http://www.premierbiosoft.com) for the current study, and it is probably the most comprehensive commercial program. It also facilitates direct link to NCBI databases to enable sequence retrieval using accession numbers as well as BLAST searches. However, there are some general considerations to design the primers (see Note 5).

3.2. Quantification of RNA

The Nanodrop spectrophotometer (Nanodrop ND-1000) is used to quantify the total RNA. The NanoDrop ND-1000 spectrophotometer (260/280 nm) can measure 1 μl samples with concentrations between 2 ng/μl and 3,000 ng/μl without dilution. Briefly, after blanking and setting the system to zero with 2 μl of distilled water, place 1–2 μl of RNA onto the sensor and measure the RNA concentration; the instrument automatically calculates the RNA concentration.

3.3. On-Column DNase Digestion

An RNase-free DNase set (Qiagen) is used for efficient on-column digestion of DNA during RNA purification as per manufactures instructions. Further, the DNase is efficiently removed in subsequent wash steps.

3.4. First-Strand cDNA Synthesis

cDNA synthesis has been carried out as "two-step RT-qPCR". In the two-step method, RNA is first transcribed into cDNA in a reaction using reverse transcriptase by using oligo (dT) or random primes as mentioned below. An aliquot of the resulting cDNA can then be used as a template source for multiple qPCR reactions.

1. Add the following components to a nuclease-free microcentrifuge tube. 1 μg of total RNA; 1 μl of Oligo $(dT)_{15}$ (50 μM) primer; 1 μl 10 mM dNTP mix and adjust reaction volume to 13 μl with sterile distilled water.
2. Heat mixture to 65°C for 5 min and incubate on ice for at least 1 min.
3. Collect the contents of the tube by brief centrifugation and add 4 μl of 5× First-Strand Buffer; 1 μl of 0.1 M DTT; 1 μl RNase Inhibitor and 1 μl of SuperScript™ III Reverse Transcriptase (200 units/μl). Ensure that the final reaction volume is 25 μl.
4. Mix by pipetting gently up and down. If using random primers, incubate tube at 25°C for 5 min.
5. Then, incubate at 50°C for 60 min.
6. Terminate the reaction by heating at 70°C for 15 min, and then place on ice until required. The cDNA now can be used as a template for amplification in PCR, the rest can be stored at –20°C for up to 6 months.

3.5. Performing qPCR Reactions

Preformulated real-time PCR master mixes containing buffer, DNA polymerase, dNTPs, and SYBR Green dye are available from several vendors (see Note 6). We have used iQ™ SYBR® Green Super Mix from Bio-Rad. We generally do 25 μl reactions in each well of the 96-well plates by setting up a qPCR mastermix by adding the reagents in the order shown as below

1. iQ SYBR Green Supermix for the final concentration to 1×.
2. 200 nM of each forward and reverse primer.

3. 1:10 dilution of the cDNA generated from the above RT reaction.
4. Finally, adjust the reaction volume to 25 μl with sterile distilled water. After mixing the reaction components gently by pipetting up and down also brief spin, followed a qPCR step as mentioned in step 5.
5. Run in qPCR instrument (Bio-Rad, IQ™5) using a three-step protocol according to the thermal profile as below.

1.	1 cycle	Activation	95°C for 3 min
2.	40 cycles	Denaturation	95°C for 10 s
		Annealing	55°C for 30 s (see Note 7)
3.	40 cycles	Melt curve	Between 55°C and 95°C for 30 s

3.6. Melt-Curve Analysis

Melt curves are a powerful means of providing accurate identification of amplified products and distinguishing them from primer dimers and other small amplification artifacts (5). This is due to the fact that SYBR Green will detect any double-stranded DNA, including primer dimers, contaminating DNA, and PCR product from misannealed primer. Melting analysis is often conveniently performed immediately after PCR in the same reaction tube. An optimized SYBR Green I qPCR reaction should have a single peak in the melt curve (Fig. 1a), where as nonspecific products that may have been coamplified with the specific product can be identified by melt-curve analysis as shown in Fig. 1b.

3.7. qRT-PCR Data Analysis

We use the "relative quantification" method to validate microarray results. In relative quantification, normalizers (β-actin or other RNAs as an endogenous control) are used to ensure that the target quantities from equivalent amounts of samples are compared (see Note 8). The $2^{-\Delta\Delta CT}$ method is a convenient and widely used to analyze the relative changes in gene expression from real-time quantitative PCR experiments (6, 7). Before using the $2^{-\Delta\Delta CT}$ method, it is essential to verify the amplification efficiencies of the target and the reference genes. Once we establish that the target and reference genes have similar and nearly 100% amplification efficiencies, the relative difference in expression level of target gene(s) in different samples can be determined as mentioned below.

1. Normalize the C_T (the endpoint of real-time PCR analysis is the threshold cycle or C_T) of the target gene to that of the reference gene.

$$\Delta C_{T(Test)} = C_{T(target,test)} - C_{T(ref,test)}$$

2. Normalize the ΔC_T of the test sample to the ΔC_T of the calibrator.

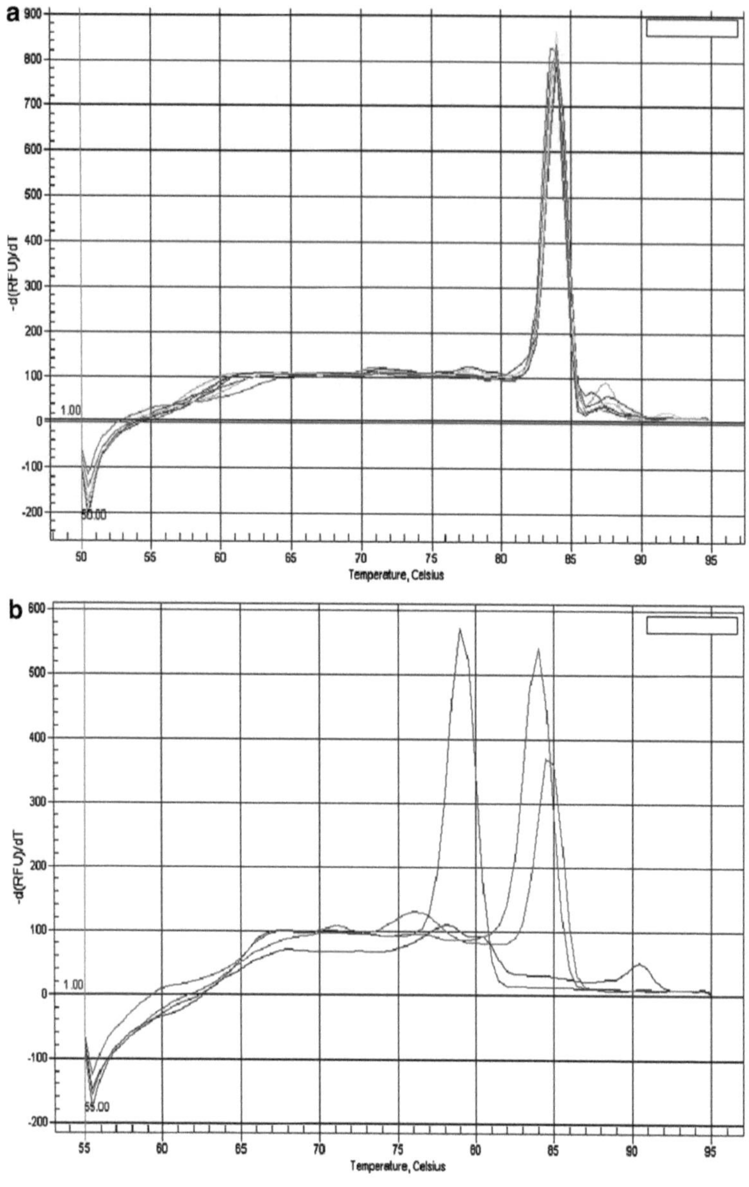

Fig. 1. Validation of SYBR Green I reaction using melt-curve analysis. (**a**) An optimized SYBR Green I qPCR reaction showing a single peak at the end of reaction. (**b**) The presence of a nonspecific product, in case of primer dimers seen as additional peaks.

$$\Delta\Delta C_{\mathrm{T}} = \Delta C_{\mathrm{T(Test)}} - \Delta C_{\mathrm{T(calibrator)}}$$

3. Finally, calculate the expression ratio by $2^{-\Delta\Delta C_{\mathrm{T}}}$ = Normalized expression ratio.

The result obtained here is the fold increase or decrease of the target gene in the test sample relative to the calibrator sample and is normalized to the expression of a reference gene.

3.8. Results and Discussion

The analysis of gene expression requires sensitive, precise, and reproducible measurements for specific mRNA sequences. Real-time RT-PCR is at present the most sensitive method for the detection of low abundance mRNA and is increasingly becoming the method of choice for high-throughput gene expression analysis. However, to get reliable results from real-time PCR, specific PCR conditions and an appropriate internal control must be determined.

RNA preparations are usually contaminated with low amounts of genomic DNA and protein, which can result in a nonspecific amplification and also affect the efficacy of qRT reactions. Therefore, the quality of RNA is assessed prior to cDNA synthesis by various quality controls (Formaldehyde agarose gel electrophoresis, Nanodrop, and Bioanalyzer). A high quality total RNA from various developmental stages of rice was reverse transcribed from three independent biological replicates of each sample and used for real-time PCR analysis. To evaluate the stability of RNA transcription levels, the accurate normalization of gene expression against a control gene is required. Generally, in qRT-PCR, transcripts of stably expressed genes, also called reference genes, are employed for the data normalization. The expression of reference gene used for the normalization of real-time PCR analysis should remain constant between the cells of different tissues and also under different experimental conditions.

In plants, most commonly used housekeeping genes are 18S ribosomal RNA (18S r RNA), actin (ACT), actin1 (ACT1), ubiquitin5 (UB5), β-tubulin (TUB), glyceraldehyde 3-phosphate dehydrogenase (GAPDH), ubiquitin5 (UBQ5), elongation factor 1α (EF-1α), expressed protein (EP), and TIP41-like protein (TIP41). In rice, several reports demonstrate that the transcript levels of these genes also vary considerably under different experimental conditions (8–10), and are consequently unsuitable for gene expression studies. This is true for other plants as well (11, 12). The reason for this expression variability may be that the housekeeping genes not only take part in basic cell metabolism, but also participate in other cellular process (13, 14). Therefore, it is necessary to select a suitable housekeeping gene(s), which has a constant expression level in certain experimental conditions for getting accurate results in gene expression studies.

To identify the most suitable reference gene in rice under drought stress conditions and also stable expression across various developmental stages, we initially selected five candidates (*actin* (*ACT*), *actin1* (*ACT1*), *β-glyceraldehyde-3-phosphate dehydrogenase* (*GAPDH*), *cyclophilin* (*CYC*), and *elongation factor 1α* (*EF-1α*) based on earlier reports (8–10), and moreover which are commonly used housekeeping genes in other plants as well. Though the *18SrRNA* was found to be stable and remained constant in different rice cultivars (8, 10), we did not consider it further for our

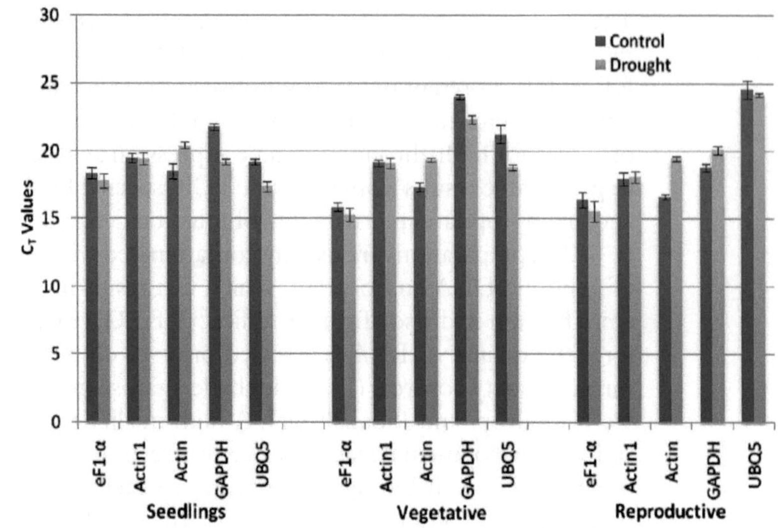

Fig. 2. RNA transcription levels of selected housekeeping genes in rice, presented as C_T mean value in the different developmental stages and environmental conditions. *Error bars* indicate the standard deviation of the three independent biological replicates.

analysis because of high expression, and it requires the use of random hexamers instead of oligo (dT) as primers for the reverse transcriptase. Figure 2 describes the validation of five selected housekeeping genes in rice across different developmental stages (seedlings, vegetative, and reproductive) in response to drought stress. The results indicate that *ACT1* is the most stably expressed and followed by *EF-1α*. Further, we also tested the sensitivity and accuracy of the real-time PCR by choosing some of the highly and weakly expressed genes in rice (Ambavaram et al.; unpublished).The results suggested that *ACT1* was sufficiently accurate for the normalization of rice even for low-abundant transcripts such as transcription factors. In conclusion, our data suggests that housekeeping genes are expressed variably in different developmental stages. Based on the results from ΔC_T analysis, *ACT1* and or *EF-1α* could be used as most stably expressed internal controls to normalize gene expression studies in response to drought stress.

4. Notes

Several parameters must be evaluated and optimized independently to achieve the maximum potential of real-time PCR. These parameters fall into three categories: general laboratory practices, template and primer design, and reaction conditions.

1. For all procedures, use DNase/RNase-free consumables.
2. Defrost all reagents on ice and mix well prior to making up reaction mixes, use calibrated pipets dedicated to PCR.
3. Avoid exposing fluorescent probes and fluorescent nucleic acid binding dyes to light.
4. Dilute the template so that between 3 μl and 10 μl are added to each reaction. This reduces inaccuracies because of very low volumes.
5. Amplify a template region of 75–150 bp (shorter amplicons are typically amplified with higher efficiency), and also try to choose a primer pair that straddles an intron to avoid amplification of genomic sequence.

Optimization conditions can vary with assay type. Therefore, these conditions should be considered when establishing a new assay:

6. Commercial master mixes are available in 2× concentration, but the $MgCl_2$ concentration should be adjusted according to dNTP concentration (increasing the dNTP will require an increase in $MgCl_2$).
7. Annealing temperature (50–60°C) and the extension time (dependent on primer T_m and product length).
8. Perform reactions in duplicate (ideally as triplicates). If the reproducibility is consistently low, the assay should be reoptimized.

References

1. Andersen, C.L., Jensen, J.L., and Orntoft, T.F. (2004) Normalization of real-time quantitative reverse transcription-PCR data: a model-based variance estimation approach to identify genes suited for normalization, applied to bladder and colon cancer data sets. *Cancer Research* **64**, 5245–5250.
2. Udavardi, M., Czechowski, T., and Scheible, W.-R. (2008) Eleven golden rules of quantitative RT-PCR. *The Plant Cell* **20**, 1736–1737.
3. Bustin, S.A., and Nolan, T. (2004) Pitfalls of quantitative real-time reverse-transcription polymerase chain reaction. *Journal of Biomolecular Techniques* **15**, 155–66.
4. Gachon, C., Mingam, A., and Charrier, B. (2004) Real-time PCR: what relevance to plant studies? *Journal of Experimental Botany* **55**, 1115–1454.
5. Klok, E.J., Wilson, I.W., Wilson. D., Chapman, S.C., Ewing, R.M., Somerville, S.C., Peacock, W.J., Dolferus, R., and Dennis, E.S. (2002) Expression profile analysis of the low-oxygen response in *Arabidopsis* root cultures. *The Plant Cell* **14**, 2481–2494.
6. Nolan, T., Hands, R.E., and Bustin, S.A. (2006) Quantification of mRNA using real-time RT-PCR. *Nature Protocols* **1**, 1559–1582.
7. Livak, K.J., and Schmittgen, T.D. (2001) Analysis of relative gene expression data using real-time quantitative PCR and the $2^{-\Delta\Delta CT}$ method. *Methods* **25**, 402–408.
8. Kim, B.R., Nam, H.Y., Kim, S.U., Kim, S.I., and Chang, Y.J. (2003) Normalization of reverse transcription quantitative-PCR with housekeeping genes in rice. *Biotechnology Letters* **25**, 1869–1872.
9. Jain, M., Nijhawan, A., Tyagi, A.K., and Khurana, J.P. (2006) Validation of housekeeping genes as internal control for studying gene expression in rice by quantitative real-time PCR. *Biochemical and Biophysical Research Communications* **345**, 646–651.
10. Cladana, C., Scheible, W.-F., Muller-Robert, B., and Ruzicic, S. (2007) A quantitative

RT-PCR platform for high-throughput expression profiling of 2500 rice transcription factors. *Plant Methods* **3**, 7.

11. Czechowski, T., Bari, R.P., Stitt, M., Schebile, W.R., and Udvardi, M.L. (2004) *Arabidopsis* transcription factors: unprecedented sensitivity reveals novel root and shoot-specific genes. *Plant Journal* **38**, 366–379.
12. Jian, B., Liu, B., Bi, Y., Hou, W., Wu, C., and Han, T. (2008) Validation of internal control for gene expression study in soybean by quantitative real-time PCR. *BMC Molecular Biology* **9**, 59.
13. Singh, R., and Green, M.R. (1993) Sequence-specific binding of transfer RNA by glyceraldehydes-3-phosphate dehydrogenase. *Science* **259**, 365–368.
14. Ishitan, R., Sunaga, K., Hirano, A., Saunders, P., Katsube, N., and Chuang, D.M. (1996) Evidence that glyceraldehydes-3-phosphate dehydrogenase is involved in age-induced apoptosis in mature cerebellar neurons in culture. *Journal of Neurochemistry* **66**, 928–935.

Chapter 5

Virus-Induced Gene Silencing in *Nicotiana benthamiana* and Other Plant Species

Andrew Hayward, Meenu Padmanabhan, and S.P. Dinesh-Kumar

Abstract

Virus-induced gene silencing (VIGS) is an efficient tool for high throughput reverse genetic screens. VIGS engages the endogenous RNA-silencing machinery of the plant host, and can yield an 85–95% reduction of target transcripts. Gene silencing is rapid, target-specific, and does not require the creation of stable transformants. The technique has been used successfully in numerous Solanaceae species as well as in *Arabidopsis*, maize, and rice. Here we describe a protocol for conducting a VIGS screen in *Nicotiana benthamiana* using *Tobacco Rattle Virus* (TRV) based silencing vectors. This protocol can readily be adapted to many other model plant species.

Key words: Virus-induced gene silencing, Reverse genetic screen, TRV-VIGS vector, RNA-silencing

1. Introduction

Virus-induced gene silencing (VIGS) is an efficient tool for silencing endogenous transcripts in plants. It has been successfully applied in both forward and reverse genetic experiments, and is particularly convenient for high-throughput reverse genetic screens (1, 2). VIGS relies on the endogenous activity of the host cell RNA-silencing machinery. Recombinant viral vectors trigger this machinery, and these viral vectors can be modified to target any host transcript of interest (3, 4).

VIGS has several advantages over other available gene-silencing methods. The technique is fast, typically requires only 4–5 weeks to complete an experiment in *Nicotiana benthamiana*, and VIGS does not require the creation of stable plant lines. VIGS also silences redundant copies of target genes, a welcome convenience for those

Andy Pereira (ed.), *Plant Reverse Genetics: Methods and Protocols*, Methods in Molecular Biology, vol. 678,
DOI 10.1007/978-1-60761-682-5_5, © Springer Science+Business Media, LLC 2011

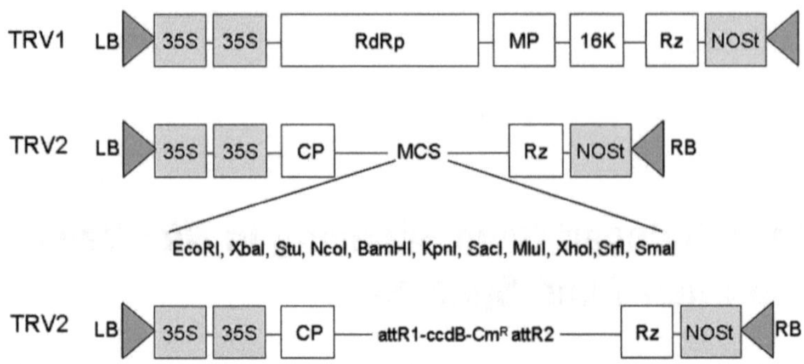

Fig. 1. Map of TRV based VIGS vectors. TRV cDNA was inserted between duplicated 35 S promoters and the Nopaline synthase terminator (NOSt) within the *Agrobacterium* TDNA vector. TRV1 vector contains the viral RNA dependent RNA polymerase (RdRp), Movement protein (MP), a 16 kDa Cysteine rich protein (16 K) along with a self cleaving ribozyme (Rz). TRV2 contains the viral Coat protein (CP) and either a multiple cloning site (MCS) or Gateway based recombination sites for the incorporation of target sequences. LB and RB represent the *left and right borders* of the TDNA.

working with polyploid plant species, or non-sequenced genomes. VIGS has been used to successfully silence host genes in many Solanaceae species including *Nicotiana* (5–7), tomato (8–10), pepper (11), potato (12), petunia (13), and poppy (14), as well as in the Brassicaceae family – *Arabidopsis thaliana* (15–17); and mon-cotyledonous plants including barley, maize, and rice (18, 19).

The protocol herein relies upon *Tobacco Rattle Virus* (TRV) based silencing vectors (6, 20). TRV is a bipartite positive sense RNA virus that provides numerous advantages over other VIGS vectors. These include a fairly broad host range, uniform cell invasion, and relatively mild disease symptoms. TRV RNA1 contains the viral replicase, RNA-dependent RNA polymerase, and movement protein, while RNA2 contains the coat protein and other non-essential elements. RNA2 has been modified to incorporate a fragment of the host gene to be silenced and both constructs have been cloned into T-DNA cassettes for *Agrobacterium tumefaciens* mediated delivery (Fig. 1) (21). While the protocol below primarily describes VIGS in *N. benthamiana*, it can be easily modified for other systems.

2. Materials

2.1. Preparing Target Silencing Constructs

1. Target gene of interest.
2. TRV2 cloning vector – either pYL156 for restriction-based cloning or pYL279 for Gateway-based cloning (available from ABRC, Ohio State University).
3. Reagents for restriction digestion and ligation or for Gateway cloning (including pDONR vector; Invitrogen).
4. Equipment for PCR.

2.2. Preparing N. benthamiana Plants

1. *N. benthamiana* seeds.
2. Conical tubes (15 ml).
3. Agarose solution (0.1% w/v; Sigma).
4. Soil (Professional Growth Medium No. 2 recommended; Conrad Farard, Inc.).
5. Pots (1-pint, 4 in. sq.).
6. Pot trays.
7. Clear plastic domes.
8. Light source (40-W Gro-Lux fluorescent light bulb, Sylvania).

2.3. Preparing Agrobacterium

1. TRV1 and TRV2-empty vectors, TRV2-*PDS* (available from ABRC, Ohio State University).
2. TRV2-Target construct, as prepared in Subheading 3.1.
3. *Agrobacterium* strain GV2260.
4. LB plates containing 50 mg/L kanamycin, 25 mg/L rifampicin, 50 mg/L streptomycin and 50 mg/L carbenicillin.
5. LB liquid medium containing 50 mg/L kanamycin, 25 mg/L rifampicin, 50 mg/L streptomycin and 50 mg/L carbenicillin.
6. Conical tubes (50 ml).
7. Agroinfiltration medium containing 200 μM acetosyringone (3′,5′-dimethoxy-4′-hydroxyacetophenone; prepared from a 0.2 M stock in dimethylformamide[1]), 10 mM MES (2-[N-morpholine]ethanesfonic acid), and 10 mM $MgCl_2$.

2.4. Agroinfiltration

1. *N. benthamiana* plants, approximately 3 weeks post germination, as prepared in Subheading 3.2.
2. Needleless syringes (1 ml).
3. Razor blades.
4. *Agrobacterium* transformant mixture, as prepared in Subheading 3.3.

2.5. Quantification of Silencing

1. Silencing target-specific and internal control-specific RT-PCR primers (for example, EF-1α, or actin).
2. RNAeasy Plant Mini Kit and RNAse free DNAse (Qiagen).
3. Equipment for PCR and reagents for RT-PCR (including reverse transcriptase such as SuperScript II and oligo-d(T) primer; Invitrogen).

2.6. Materials for Other Plant Systems

1. *Agrobacterium* strain GV3101.
2. Artist airbrush (model V180; Paasche).
3. Carborundum.

---

[1]Dimethylformamide is an eye and skin irritant, and is toxic to the liver and kidney. Wear appropriate safety equipment and dispense in a chemical fume hood.

3. Methods

Successful targeting of the host RNA-silencing machinery to your gene of interest is of primary importance for successful VIGS. While a minimal homologous sequence of 23nt should theoretically be sufficient for gene silencing, in practice larger fragments are recommended to ensure a high efficiency of silencing. The optimal size is between 500 and 700 bp, and any coding region, or 5′/3′ UTR can be used. Note that when choosing your target sequence, at least 30 bp of sequence should remain untargeted for confirmation of silencing by RT-PCR. It is also advisable to create multiple silencing constructs per target, as silencing efficiency can vary based on the region of homology. When the target gene is thought to have other family members with high sequence homology, and potentially redundant function, the silencing constructs can be designed using highly conserved regions with the aim to silence all of these family members simultaneously. Alternatively, the constructs can be designed using the more variable UTR sequences in order to minimize off-target silencing.

Changes in temperature or photoperiod can have dramatic effects on silencing efficiency, so it is important to determine the optimal plant growth conditions for your silencing experiment. Optimization can be performed by silencing a positive control such as *Phytoene desaturase* (*PDS*), which causes a visible white photobleaching phenotype in silenced leaves (Fig. 2). For quantification of silencing, *PDS* mRNA levels can be measured by RT-PCR. Strong and reproducible *PDS* silencing efficiency suggests that conditions are optimal for silencing of other target genes.

Fig. 2. Virus Induced Gene Silencing of the *PDS (Phytoene desaturase)* gene in *N. benthamiana*. TRV-*NbPDS* was agro-infiltrated in *N. benthamiana* plants and visualized 14 days post infiltration. *PDS* silenced plants show characteristic photo-bleaching on the upper un-inoculated leaves in comparison to the control plants that received the empty TRV vector.

It is also good practice to include the empty TRV2 vector as a negative control to monitor TRV-infection phenotypes unrelated to gene silencing. Although silencing efficiency varies by target, successful VIGS can reduce *PDS* transcript level by 85–95% in *N. benthamiana*.

Modifications to the standard silencing protocol for silencing in tomato and *A. thaliana* are included, briefly, in Subheading 3.6.

3.1. Preparing Silencing Constructs

1. Design primers to amplify a 500–700 bp fragment of the target coding sequence by PCR (see Note 1). Primers should contain either the appropriate restriction sites for cloning into the TRV2 vector pYL156, or the att-B1 forward (GGGG-ACA-AGT-TTG-TAC-AAA-AAA-GCA-GGC-TNN) and att-B2 reverse (GGG-AC-CAC-TTT-GTA-CAA-GAA-AGC-TGG-GTN) sites for gateway cloning into the TRV2 vector pYL279.
2. For pYL156, clone the PCR-fragment by restriction digest followed by ligation. For pYL279, the PCR-fragment must first be integrated into an entry clone such as pDONR 207 by BP Gateway reaction. After confirming the pDONR construct by sequencing, transfer the fragment into pYL279 by LR Gateway reaction (see Invitrogen Gateway Cloning manual for details).

3.2. Preparing N. benthamiana Plants

1. Place 40–50 seeds (or as appropriate for 3–5 plants per silencing construct) into a 15 ml conical tube and add 7 ml 0.1% agarose. Incubate seeds at room temperature for 3–5 days. This initiates seed germination and eases distribution of seeds into pots for growth.
2. Place soil in ten 1-pint pots in a pot tray. Distribute seeds evenly among the pots. Cover the tray with a clear plastic dome.
3. Remove plastic dome immediately after seeds germinate, and continue growing seedlings under continuous light at 24–26°C for 10–12 days. Plants should be watered as necessary by adding 0.5–1 in. of water to the pot trays.
4. Transplant seedlings individually into new 1-pint pots with soil. Grow plants until they reach the four-leaf stage (approximately 3 weeks) (see Note 2).

3.3. Preparing Agrobacterium

1. Transform the TRV1, TRV2-empty vector, TRV2-Target, and TRV2-*NbPDS* vectors separately into *Agrobacterium* strain GV2260 and plate transformants on LB plates containing 50 mg/L kanamycin, 25 mg/L rifampicin, 50 mg/L streptomycin, and 50 mg/L carbenicillin.
2. Incubate plates for 2 days at 28°C.
3. Inoculate TRV1, TRV2-empty vector, TRV2-Target, and TRV2-*NbPDS* clones separately into 10 ml LB liquid

containing 50 mg/L kanamycin, 25 mg/L rifampicin, 50 mg/L streptomycin, and 50 mg/L carbenicillin.

4. Grow cultures overnight at 28°C.
5. Pellet cells at room temperature by centrifugation at 3,000 × *g* for 15 min and resuspend to OD600 of 1.0 in agroinfiltration medium.
6. Incubate cultures for 3–6 h at room temperature.
7. Mix the cultures containing TRV1 and either TRV2-empty vector, TRV2-Target, or TRV2-*NbPDS* in a 1:1 ratio prior to agroinfiltration

3.4. Agroinfiltration

1. Load a 1 ml needless syringe with the appropriate mixture of *Agrobacterium* cultures.
2. Using a corner of a fresh razor blade, nick the third and fourth leaves of the *Nicotiana* plant.
3. Place the opening of the needleless syringe against a nick on the underside of the leaf. Seal the nick on the opposite side of the syringe using a finger from the other hand. Inject the *Agrobacterium* mixture into the leaf until the infiltrate will proceed no further into the tissue surrounding the nick.
4. Repeat steps 1–3 until the entire two leaves have been infiltrated with the *Agrobacterium* mixture.
5. Repeat this process for all constructs, including the TRV1 + TRV2-*NbPDS* silencing control and the TRV1 + TRV2-empty vector (see Note 3). To avoid cross contamination, fresh gloves, blades, and syringes should be used for each new construct (see Note 4).

3.5. Quantification of Gene Silencing

1. After agroinfiltration, *N. benthamiana* should be maintained on growth carts at 25°C and under continuous light (see Note 5). Allow 6–10 days post-infiltration for gene silencing to progress. Successful execution of the silencing protocol can be qualitatively assessed by visual inspection of plants infiltrated with the TRV2-*NbPDS* positive control. PDS silencing will cause white bleaching of the upper leaves of the silenced plant.
2. Silencing efficiency can be assessed by semi-quantitative RT-PCR. Briefly – Extract total RNA from Target-silenced, *NbPDS*-silenced, and vector control-silenced plants and treat with RNAse-free DNAse. Synthesize first strand cDNA using 2 μg total RNA, oligo-d(T) primer, and SuperScript II reverse transcriptase. RT-PCR should then be performed using primers that anneal outside of the region targeted for gene silencing. Collect samples at 15, 20, 25, 30, and 35 PCR cycles, and quantify using appropriate imaging software. RT-PCR analysis of EF-1α or actin serves as an appropriate internal control.

3. Alternative quantification techniques include real-time RT-PCR and Northern blot analysis.

3.6. Modifications for Other Plant Systems

1. *Arabidopsis* – The primary modification necessary for gene silencing in *Arabidopsis* is the use of *Agrobacterium* strain GV3101 as opposed to GV2260. GV3101 growth media should contain 50 mg/L kanamycin and 15 mg/L gentamycin. Silencing in *Arabidopsis* is most efficient in seedlings inoculated with *Agrobacterium* at OD600 of 1.5 at the two- to three-leaf stage (approximately 15–18 days post germination). Plants should be grown under a long day (16/8 h) photoperiod (see Note 5). Silencing efficiency of 80–95% has been reported using this modified protocol (15).
2. *Tomato* – Silencing in tomato plants also requires the GV3101 vector. However, silencing by syringe infiltration as described in Subheading 3.4, above, is less efficient in tomato plants (yielding ~20–30% reduction in target transcript). Alternatively, highly efficient silencing (>90%) can be achieved in tomato plants by spray inoculation (see below) (8), or vacuum infiltration (see ref. 9). For spray inoculation, *Agrobacterium* cultures should be resuspended to OD600 of 2.0 with agroinfiltration medium. After 3–6 h of incubation, TRV1 and TRV2-Target cultures should be mixed in a 1:1 ratio. Add 75–100 mg carborundum to the mixed culture. Next, using an artist's airbrush set to 80 psi, spray the underside of the two lower leaves of the tomato plant with the *Agrobacterium* mixture for 1–5 s from a distance of approximately 8 in. Tomato plants should be maintained between 18°C and 21°C during silencing under a long day (16/8 h) photoperiod.

4. Notes

1. It may be expedient to conduct an initial screen using a single VIGS construct per target gene. However, for any further characterization of potential genes-of-interest, the preparation of multiple silencing constructs can be crucial to optimizing your silencing efficiency. For targets of 1–2 kb, we have found it convenient to start by targeting the full length mRNA, and two 700 bp regions at either the 5′ or 3′ end of the gene. If necessary, target sequences as short as 300 bp can still give optimal silencing. Special consideration must also be given to silencing of genes in gene families, as discussed in the introduction to Subheading 3.
2. The age of plants targeted for silencing can be critically important to silencing efficiency. Younger plants invariably perform better during silencing experiments, and it is often wiser to wait

a week for new plants to reach the proper size than to perform a less efficient silencing experiment on plants that are too old.

3. Great care should be taken to prevent cross-contamination during a VIGS experiment. RNA inhibition is a catalytic process, and it can take only a single transformation event to initiate the TRV-mediated spread of an siRNA catalyst throughout the infiltrated plant. Serial dilution experiments performed in our lab revealed that 1 ml of a 1:1,000 dilution of TRV2-*NbPDS* was sufficient to induce photobleaching. Notably, for this experiment the *Agrobacterium* mixture was not infiltrated into the plant leaves, but rather ***placed on the soil above the plant roots.*** It is then easy to envision accidental contamination of experimental constructs with a positive control, resulting in loss of time pursuing false leads.
4. The use of fresh gloves, blades, and syringes during infiltration of silencing constructs should be considered a minimal level of care to prevent cross-contamination. Among other precautions, infiltrations targeting a given construct should be conducted in spatial isolation from other plants to prevent contamination by stray infiltration media. Plants silenced for different constructs should also be maintained in separate trays to avoid soil- and water-borne *Agrobacterium* contamination.
5. As with age considerations, photoperiod and humidity can also have dramatic effects on silencing efficiency. For example, in Burch-Smith et al., (15) PDS silencing was successful in only 10% of *Arabidopsis* grown under a short day (8/16 h) photoperiod, while plants grown under a long day (16/8 h) photoperiod showed 90–100% successful silencing. Growth conditions should be optimized prior to performing any large-scale silencing experiment.

Acknowledgements

We thank past NSF funding in support of VIGS work in S, P. D-K lab.

References

1. Lu, R., Malcuit, I., Moffett, P., Ruiz, M. T., Peart, J., Wu, A. J., Rathjen, J. P., Bendahmane, A., Day, L., and Baulcombe, D. C. (2003) High throughput virus-induced gene silencing implicates heat shock protein 90 in plant disease resistance. *EMBO J.* 22:5690–5699.
2. Liu, Y., Schiff, M., Czymmek, K., Talloczy, Z., Levine, B., and Dinesh-Kumar, S. P. (2005) Autophagy regulates programmed cell death during the plant innate immune response. *Cell.* 121:567–577.
3. Baulcombe, D. C. (1999) Fast forward genetics based on virus-induced gene silencing. *Curr Opinion Plant Biol.* 2:109–113.
4. Dinesh-Kumar, S. P., Anandalakshmi, R., Marathe, R., Schiff, M., and Liu, Y. (2003)

Virus-induced gene silencing. *Methods Mol Biol.* 236:287–294.

5. Ratcliff, F., Martin-Hernandez, A. M., and Baulcombe, D. C. (2001) Tobacco rattle virus as a vector for analysis of gene function by silencing. *Plant J.* 25:237–245.
6. Liu, Y., Jin, H., Yang, K. Y., Kim, C. Y., Baker, B., and Zhang, S. (2003) Interaction between two mitogen-activated protein kinases during tobacco defense signaling. *Plant J.* 34:149–160.
7. Caplan, J. L., Mamillapalli, P., Burch-Smith, T. M., Czymmek, K., and Dinesh-Kumar, S. P. (2008) Chloroplastic protein NRIP1 mediates innate immune receptor recognition of a viral effector. *Cell.* 132:449–462.
8. Liu, Y., Schiff, M., and Dinesh-Kumar, S. P. (2002) Virus-induced gene silencing in tomato. *Plant J.* 31:777–786.
9. Ekengren, S. K., Liu, Y., Schiff, M., Dinesh-Kumar, S. P., and Martin, G. B. (2003) Two MAPK cascades, NPR1, and TGA transcription factors play a role in Pto-mediated disease resistance in tomato. *Plant J.* 36:905–917.
10. Fu, D. Q., Zhu, B. Z., Zhu, H. L., Jiang, W. B., and Luo, Y. B. (2005) Virus-induced gene silencing in tomato fruit. *Plant J.* 43:299–308.
11. Chung, E., Seong, E., Kim, Y. C., Chung, E. J., Oh, S. K., Lee, S., Park, J. M., Joung, Y. H., and Choi, D. (2004) A method of high frequency virus-induced gene silencing in chili pepper (*Capsicum annuum* L. cv. Bukang). *Mol Cells.* 17:377–380.
12. Brigneti, G., Martin-Hernandez, A. M., Jin, H., Chen, J., Baulcombe, D. C., Baker, B., and Jones, J. D. (2004) Virus-induced gene silencing in *Solanum* species. *Plant J.* 39:264–272.
13. Chen, J. C., Jiang, C. Z., Gookin, T. E., Hunter, D. A., Clark, D. G., and Reid, M. S. (2004) Chalcone synthase as a reporter in virus-induced gene silencing studies of flower senescence. *Plant Mol Biol.* 55:521–530.
14. Hileman, L. C., Drea, S., Martino, G., Litt, A., and Irish, V. F. (2005) Virus-induced gene silencing is an effective tool for assaying gene function in the basal eudicot species *Papaver somniferum* (opium poppy). *Plant J.* 44:334–341.
15. Burch-Smith, T. M., Schiff, M., Liu, Y., and Dinesh-Kumar, S. P. (2006) Efficient virus-induced gene silencing in *Arabidopsis. Plant Physiol.* 142:21–27.
16. Cai, X. Z., Xu, Q. F., Wang, C. C., and Zheng, Z. (2006) Development of a virus-induced gene-silencing system for functional analysis of the RPS2-dependent resistance signalling pathways in *Arabidopsis. Plant Mol Biol.* 62:223–232.
17. Pflieger, S., Blanchet, S., Camborde, L., Drugeon, G., Rousseau, A., Noizet, M., Planchais, S., and Jupin, I. (2008) Efficient virus-induced gene silencing in *Arabidopsis* using a 'one-step' TYMV-derived vector. *Plant J.* 56:678–690.
18. Ding, X. S., Schneider, W. L., Chaluvadi, S. R., Mian, M. A., and Nelson, R. S. (2006) Characterization of a Brome mosaic virus strain and its use as a vector for gene silencing in monocotyledonous hosts. *Mol Plant Microbe Interact.* 19:1229–1239.
19. Scofield, S. R., Huang, L., Brandt, A. S., and Gill, B. S. (2005) Development of a virus-induced gene-silencing system for hexaploid wheat and its use in functional analysis of the Lr21-mediated leaf rust resistance pathway. *Plant Physiol.* 138:2165–2173.
20. Burch-Smith, T. M., Anderson, J. C., Martin, G. B., and Dinesh-Kumar, S. P. (2004) Applications and advantages of virus-induced gene silencing for gene function studies in plants. *Plant J.* 39:734–746.
21. Liu, Y., Schiff, M., Marathe, R., and Dinesh-Kumar, S. P. (2002) Tobacco Rar1, EDS1 and NPR1/NIM1 like genes are required for N-mediated resistance to tobacco mosaic virus. *Plant J.* 30:415–429.

Chapter 6

Agroinoculation and Agroinfiltration: Simple Tools for Complex Gene Function Analyses

Zarir Vaghchhipawala, Clemencia M. Rojas, Muthappa Senthil-Kumar, and Kirankumar S. Mysore

Abstract

Agroinoculation, first developed as a simple tool to study plant–virus interactions, is a popular method of choice for functional gene analysis of viral genomes. With the explosive growth of genomic information and the development of advanced vectors to dissect plant gene function, this reliable method of viral gene delivery in plants, has been recruited and morphed into a technique popularly known as agroinfiltration. This technique was developed to examine the effects of transient gene expression, with applications ranging from studies of plant–pathogen interactions, abiotic stresses, a variety of transient expression assays to study protein localization, and protein–protein interactions. We present a brief overview of literature which document both these applications, and then provide simple agroinoculation and agroinfiltration methods being used in our laboratory for functional gene analysis, as well as for fast-forward and reverse genetic screens using virus-induced gene silencing (VIGS).

Key words: Agroinoculation, Agroinfiltration, Plant–pathogen interactions, Virus-induced gene silencing, Abiotic stress, Tobacco Rattle Virus vector

1. Introduction

Gene function analysis in plants has become fairly more amenable with the advent of advanced binary vectors. Agroinoculation has become a preferred delivery tool for a variety of viral genomes of interest *via* expression in plants through *Agrobacterium* binary vectors. The earliest usage of agroinoculation in gene function analyses has been in the studies of plant–virus interactions wherein whole viral genomes, or open reading frames (ORFs) were cloned into binary vectors and delivered into plants *via* agroinoculation (1). Agroinoculation is also being used to deliver *Tobacco mosaic*

Andy Pereira (ed.), *Plant Reverse Genetics: Methods and Protocols*, Methods in Molecular Biology, vol. 678, DOI 10.1007/978-1-60761-682-5_6, © Springer Science+Business Media, LLC 2011

virus (TMV)- and *Potato virus X* (PVX)-based expression vectors to transiently express a protein in desired leaf segment or whole plant. Agroinoculation results in strong transient gene expression which is several folds higher than in stably transformed plants due to the high number of viral transcripts expressing the gene of interest. Gene expression is stable for several days after agroinoculation allowing for high levels of proteins for analysis without the need to produce stable transgenic plants (2).

With the discovery of visual reporter proteins like GUS and GFP, which could be expressed in plants as gene fusions (3), agroinoculation was soon modified to a technique termed "agroinfiltration", which today is primarily used to transiently express transgene(s) or fragments thereof, in plants at high levels for functional analysis. In one of the earliest reports, Rossi and colleagues (4) adapted a vacuum-aided agroinfiltration technique, using GUS activity assays to measure efficiency of *Agrobacterium* T-DNA transfer in tobacco. The most popular method of agroinfiltration involves introduction of *Agrobacterium* into plant leaves using a needless syringe. Simple *Agrobacterium*-mediated gene overexpression protocols (agroinfiltration) have now been optimized for several plant species (5). High level transgene expression can be used for several downstream applications such as biotic and abiotic stress analyses. Co-expression of silencing suppressors will prevent possible post-transcriptional gene silencing (PTGS) during transgene overexpression (6).

In addition to transient gene overexpression, *Agrobacterium*-mediated transient silencing assay in intact tissues has also emerged as a rapid and useful method to analyze gene function in plants (7). Infiltration of *Agrobacterium* cultures harboring hairpin RNAi construct, carrying fragment of endogenous gene to be silenced, induces silencing of corresponding endogenous gene. Agroinfiltration-mediated silencing may also produce systemic silencing signals that confer silencing in infiltrated cells and also non-infiltrated cells adjacent to infiltration zone (8). Agroinfiltration based transient gene silencing method has also been particularly used to identify genes with RNAi suppressor function.

1.1. Applications of Agroinfiltration and Agroinoculation

A variety of new assays coupled with agroinoculation or agroinfiltration have become handy tools to study gene function. Primary applications of these techniques include gene-for-gene interaction studies, analysis of bacterial/viral gene expression in planta, and functional analysis of genes in response to biotic and abiotic stresses. This chapter provides a brief research summary of the variety of applications available for gene function analyses and follows up with a simple protocol for each technique. Helpful tips on down-stream procedures, post-agroinoculation, are provided in Subheading 4.

1.2. Study of Plant–Pathogen Interactions

This section takes a wide-ranging look at the plethora of applications of these two closely related protocols and for ease of understanding, it has been divided into well studied subclasses, which include *R–Avr* interactions, using either transient assays or virus-induced gene silencing (VIGS) and plant–virus interactions.

1.2.1. Study of Gene-for-Gene Interactions

One important event in the evolution of plant immune response is the recognition of a potential pathogen by the surveillance system present in plants. This system is able to recognize general features in the potential pathogen, called Pathogen Associated Molecular Patterns (PAMPs), or specific molecules delivered by the pathogen into the plant cell, called effectors (9). The more specific recognition of effectors is mediated by the products of resistance genes (*R*) present in the host plant. The direct or indirect interaction between a pathogen effector and its matching R protein triggers a defense signal transduction cascade that results in a rapid localized cell death at the site of infection called hypersensitive response (HR) that restricts pathogen growth. Agroinoculation and agroinfiltration have been used to understand the mechanism(s) of disease resistance in plant–pathogen interactions.

1.2.2. Agroinfiltration-Mediated Transient Assays to Identify and Study Interactions Between Pathogen Effectors and Plant Resistance Proteins

Transient assays involve the delivery of *Agrobacterium* containing candidate avirulence (*Avr*) gene(s) within the T-DNA of the binary vector, *via* agroinfiltration into the apoplast of a plant harboring the matching *R* gene. Evaluation of the interaction is based on the timing, occurrence, and severity of the HR. Similarly, *Agrobacterium* – mediated transient assay has been used in screens to identify resistance genes by co-expressing a candidate *R* gene with its matching *Avr* gene in plants (10, 11). Furthermore, components of the disease resistance signal transduction pathway have been identified and characterized by co-expressing *R* and *Avr* genes in a specific genetic background and observing whether resistance or susceptibility ensues (12). Agroinfiltration has been effectively used to study the signal transduction pathway during direct or indirect interactions of Avr proteins such as Avr4, Avr9, AvrPto, and tvEIX with their respective matching plant R proteins Cf-4, Cf-9, Pto, and LeEix2 in tobacco (12, 13). HR-mediated defense response induced by these interactions has since been exploited in fast-forward genetic screens using virus-induced gene silencing (VIGS), as described below, to identify the signaling intermediates of HR. This has improved our understanding of plant disease resistance cascades.

1.2.3. Agroinoculation-Mediated Virus-Induced Gene Silencing to Identify Components of the Defense Signaling Transduction Pathway

VIGS is a post-transcriptional gene silencing mechanism (PTGS) to transiently suppress endogenous expression of a target gene by infecting plants with a recombinant virus vector carrying host-derived sequence (14). Infection and systemic spreading of the virus causes targeted degradation of the gene transcripts. The choice of a virus vector depends on whether it can naturally infect and propagate within the plant to be silenced. The objective of rapid genome scale analysis of gene function using VIGS is achievable using agroinoculation. Our laboratory employs the bipartite *Tobacco rattle virus* (TRV) based vector (developed in Dr. Dinesh Kumar's laboratory at Yale University) for silencing endogenous plant genes. A normalized plant cDNA library cloned into the viral RNA2 genome within an *Agrobacterium* binary vector and delivered *via* agroinoculation has lead to the identification of plant genes necessary for both type I and type II nonhost disease resistance ((15); Senthil kumar Muthappa, unpublished results). VIGS has been successfully used to identify and characterize many plant genes involved in defense against pathogens (16).

1.2.4. Plant–Virus Interactions Using Agroinoculation

Agroinoculation offers a simple, efficient, and powerful approach for delivery of plant viral genomes for understanding viral replication, assembly, and movement. Since the first application of agroinoculation to study *Cauliflower mosaic virus* (CaMV) and *Maize streak virus* (1, 17), this simple protocol has revolutionized the study of plant–virus interactions by facilitating the expression of both DNA/RNA and mono/bipartite virus genomes. Applications discussed below include, the validation of a viral isolate as causal agent of disease, characterization of viral ORFs through expression and mutagenesis, recombinatorial analyses between related viruses and lastly, identification of resistant germplasms.

Insect transmission and infectivity of *Tobacco golden mosaic virus* (TGMV) (18), *Tomato yellow leaf curl virus* (TYLCV) and *Potato yellow mosaic geminivirus* (PYMV) were shown for the first time using agroinoculation to deliver the viruses. Mutational analyses of various viral genomes has revealed the role of various genes/ORFs in determining symptom development, severity, and viral accumulation (19). Functional studies of viral ORFs allowed for the distinction between narrow and broad host range strains, and also helped determine genetic interchangeability of ORFs (20). Germplasm screening for virus resistance using agroinoculation has identified resistant germplasms in tomato (21), rice, and maize. The 126 kDa protein of *Tobacco mosaic virus* (TMV) was shown to be a suppressor of gene silencing using the RNA interference (RNAi) approach and agroinoculation (22). A combination of agroinoculation (VIGS) with RNAi technology is one of the main reasons for an immense progress in the area of plant functional genomics.

1.3. Study of Plant Abiotic Stress Tolerance

Genes involved in imparting abiotic stress tolerance can be analyzed through either a virus-based vector (agroinoculation), or a binary vector (agroinfiltration) depending on the type of vector in use. The latter method employs delivering the construct carrying gene of interest into plant leaves, and subjecting the leaf segments collected from the infiltrated area to stress analyses, post-inoculation. A recent study using tobacco leaves transiently expressing a tomato Phospholipid Hydroperoxide Glutathione Peroxidase (LePHGPx) gene cloned in a PVX-based vector, delivered by agroinoculation, showed enhanced salinity and heat tolerance (23). Furthermore, rapid analyses to determine the role of plant promoters and transcription factors in abiotic stresses is possible by simple agroinfiltration of the appropriate plasmid constructs into tobacco leaves (24). Compared to several biotic stress inducible promoters, salt-, drought-, cold-, and heat-inducible promoters are least affected by the presence of *Agrobacterium*, hence, facilitating the identification and characterization of abiotic stress responsive promoters in the intact plants (24). Multiple transient expression assays can be performed on a single leaf, thereby enabling the analyses of large number of abiotic stress responsive candidate genes in a shorter time frame. The use of agroinoculation to introduce VIGS vectors for silencing and characterizing genes and cellular processes involved in abiotic-stress tolerance, in particular drought and oxidative stress, have been demonstrated. Agroinoculation, has allowed for the analysis of abiotic stress-induced genes from heterologous species, in *Nicotiana benthamiana,* thus extending its applicability to analyze genes of VIGS recalcitrant plant species (25). VIGS-based fast-forward genetic screen is an emerging option to initially analyze large number of genes in order to narrow down the list to a few promising genes that might have a role in abiotic stress tolerance.

1.4. Other Applications

Silencing of *N. benthamiana* cDNA library using agroinoculation based VIGS followed by infection of silenced tissue with a tumorigenic *Agrobacterium tumefaciens* strain lead to the identification of novel plant genes involved in *Agrobacterium*-mediated plant transformation (26). *Agrobacterium*-mediated transient expression has also been used for producing recombinant proteins in plants (27) and to make marker free transgenic plants by delivering a TMV vector harboring a CRE recombinase (28). The field of protein–protein interactions studies has been revolutionized with the design of novel vectors that are being used to deliver recombinant genes into plants *via* agroinfiltration for facilitating immuno/co-precipitation and Bimolecular florescence complementation (BiFC) studies (29). To summarize, agroinoculation and agroinfiltration have become very important tools to study silencing mechanisms of resident genes, to characterize promoter and transcription factors *in vivo*, to investigate sub-cellular localization and intracellular trafficking of gene products, to test gene expression constructs in non-transgenic plants, to analyze gene function *via* overexpression, to engineer a whole pathway *via* expression of single or multiple genes, and as a tool for rapid bio-physical analysis of plant ion channels. Figure 1 depicts some of the applications mentioned above that are currently used in our laboratory.

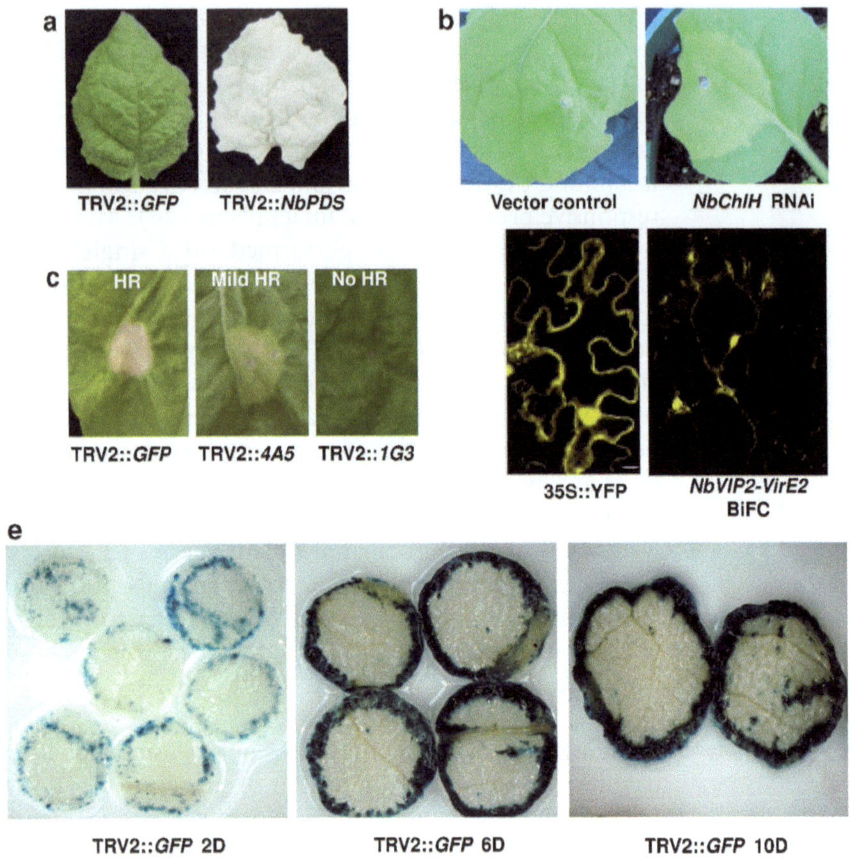

Fig. 1. Applications of agroinoculation and agroinfiltration methods for transgene expression and gene silencing. (**a**) Virus-induced gene silencing (VIGS). *Nicotiana benthamiana* plants silenced for *NbPDS* gene (encoding phytoene desaturase) using *Tobacco rattle virus* (TRV)-based gene silencing vector delivered *via* agroinoculation, showed photo-bleaching phenotype. (**b**) On-the-spot gene silencing. Transient silencing of endogenous gene was done by infiltration of the *Agrobacterium* strain carrying an RNAi construct. Photographs shows *ChlH* gene (encoding H subunit of Mg-chelatase) silencing in *N. benthamiana* leaf. (**c**) Study of R-Avr interactions in silenced plants using agroinfiltration. Silencing of two unique genes from a plant cDNA library leads to variable intensities of the hypersensitive response during a R–Avr interaction mediated by ethylene inducing xylanase (EIX) and its cognate receptor LeEix2 in the gene silenced leaves. Silencing as well as inoculation of the *R*-gene and *Avr*-gene constructs was mediated *via* agroinoculation. (**d**) Bimolecular Flourescence Complementation Assay (BiFC). In planta interaction of *N. benthamiana* VIP2 and *A. tumefaciens* VirE2 protein using the BiFC vectors delivered *via* agroifiltration (29). (**e**) Transient GUS expression assay. Leaf disks from *TRV2::GFP*–silenced plants were co-cultivated with non-tumorigenic strain *A. tumefaciens* GV2260 carrying pBISN1 (has the *uidA*-intron gene within the T-DNA). Silenced leaf disks were periodically collected at 2 days, 6 days and 10 days post inoculation and stained with X-Gluc for GUS expression.

2. Materials

Routinely employed methods in our laboratory include: (a) Agroinfiltration-mediated transient assays to study interactions of pathogen and its effectors with plant proteins and (b) Agroinoculation based VIGS for fast-forward and reverse genetic screens to study nonhost resistance, abiotic stress tolerance, and *Agrobacterium*-mediated plant transformation. Detailed methods are presented for each application.

2.1. Agroinfiltration-Mediated Transient Assays to Identify Gene-for-Gene Interactions

1. Plants: 5-week-old *N. benthamiana*.
2. *A. tumefaciens* strain GV2260 harboring genes (*Avr* gene or *R* gene) in a binary vector (see Note 1).
3. AB minimal medium plates and AB liquid medium supplemented with appropriate antibiotic (selection marker for the binary vector) and rifampicin (10 μg/mL; *Agrobacterium* chromosomal marker) to grow *Agrobacterium* strains.
4. Induction medium (adjusted to pH 5.5 with 1M KOH): 30 mM MES, 1.7 mM NaH_2PO_4, 1% mannitol, and 200 μM acetosyringone.
5. Infiltration medium: 10 mM MES adjusted to pH 5.5.
6. 1-mL tuberculin syringes.

2.2. VIGS

1. Plants: 3-week-old *N. benthamiana* grown in the greenhouse at 25°C.
2. *A. tumefaciens* GV2260 harboring *pTRV1*.
3. *A. tumefaciens* GV2260 harboring *pTRV2::GFP* (mock control) (see Note 5a).
4. *A. tumefaciens* GV2260 harboring *pTRV2::PDS* (positive control) (see Note 5b).
5. *A. tumefaciens* GV2260 harboring *pTRV2* with cloned sequence of interest (see Note 5c).
6. Luria Bertani (LB) plates and liquid medium supplemented with kanamycin (50 μg/mL) and rifampicin (10 μg/mL) to grow *Agrobacterium* strains.
7. Induction medium (adjusted to pH 5.5 with 1M KOH): 10 mM $MgCl_2$, 10 mM MES buffer – pH5.6, and 200 μM acetosyringone.
8. Infiltration medium: 10 mM MES adjusted to pH 5.5.
9. 1-mL tuberculin syringes.

3. Methods

3.1. Agroinfiltration-Mediated Transient Assays to Identify Gene-for-Gene Interactions

1. Germinate and transfer 3-week-old *N. benthamiana* seedlings to greenhouse and grow for 2 more weeks. Plants can be grown at 22 ± 2°C with occasional mist to maintain 55–65% relative humidity (RH).
2. Select fully expanded, newly formed third and fourth leaf from the top for inoculation (see Note 2).
3. For transient expression assay explained here, the *tvEIX* (*Avr*) – *LeEix2* (*R*) gene interaction mediated HR assay is taken as an example. However, this protocol can be applied to induce HR for most *Avr–R* gene interactions. Inoculate a single colony of disarmed *A. tumefaciens* strain GV2260 carrying *tvEIX* and *LeEix2* constructs in AB minimal medium and grow overnight at 28°C in an incubator shaker. Grow the cultures till $OD_{600} = 0.7$. Spin down the cultures and wash twice with induction medium. Incubate the cultures in induction medium for additional 6–8 h at 22 ± 2°C with gentle shaking.
4. Spin down the culture and re-suspend in infiltration medium. Dilute the cultures to $OD_{600} = 0.5–0.7$ using infiltration medium.
5. Spot infiltrate the suspensions expressing *35S:tvEIX* and *35S:LeEix2* in a mixture (1:1) using needleless tuberculin syringe into the leaves. Maintain the infiltrated plants at 22 ± 2°C to facilitate expression of transgenes in plants.
6. After 4–5 days post infiltration (see Note 3), HR manifested as cell death in the infiltrated region will be evident when both the constructs are expressed in the same spot. This indicates recognition between effector tvEIX and R protein LeEix2, and subsequent signaling to induce HR. For extended applications of this assay, see Note 4.

3.2. VIGS

1. Streak *A. tumefaciens* strains containing *pTRV1* and *pTRV2* plasmids on LB plates supplemented with antibiotics, and incubate at 28°C for 3 days.
2. Inoculate single colony into 20-mL LB broth containing kanamycin (50 μg/mL) and rifampicin (10μg/mL). Incubate overnight at 28°C in a shaker.
3. Harvest cultures by centrifugation at 3,200 × *g* for 10 min. Discard supernatant.
4. Re-suspend pellets in induction buffer supplemented with acetosyringone and shake at room temperature for 6 h (see Note 6).

5. Harvest induced cultures by centrifugation at 3,200 × *g* for 10 min. Discard supernatant.
6. Re-suspend pellets in 10 mM MES – pH5.5 and dilute to a final $OD_{600} = 1.0$ (see Note 7).
7. Mix *Agrobacterium* cultures containing *pTRV1* and *pTRV2* construct in 1:1 V/V ratio.
8. Infiltrate two lower leaves of plants using a 1-mL needleless syringe (see Note 8). Maintain plants in greenhouse at 23–25°C.
9. Observe silencing phenotypes after 3 weeks and confirm transcript down-regulation by RT-PCR (see Notes 9 and 10).
10. Silenced plants are now ready for further downstream analyses.

4. Notes

1. Co-expression of cloned *tvEIX* and *LeEix2* from binary vectors is required to observe HR since *N. benthamiana* genome does not encode the *Eix2* gene (13). Our laboratory uses pBINPLUS based binary vectors for expressing *tvEIX*, an *Avr* gene product and *LeEix2*, an *R* gene product. In addition, *pBTEX* vectors can also be used (30). We found that *Agrobacterium* strain GV2260 is most efficient for transient expression studies in *N. benthamiana*, although other strains namely GV3101, EHA105, and LBA4404 have also been used in other studies. Efficiency of *Agrobacterium* strains varies with plant species. Hence using the most efficient strain for the species under study is important.
2. Selection of appropriate leaf is very important to achieve efficient transient expression and induction of HR. Select only fully expanded, newly formed leaves, and of same developmental stage for better comparison among different treatments or plants. Maintaining buffer and vector control inoculations are highly recommended. For *Avr–R* interaction mediated HR assays in *N. benthamiana*, the AvrPto–Pto interaction can be used as positive control. Studying several different *Avr–R* gene interactions in the same leaf minimizes variation and makes the assay amenable for large scale experiments.
3. Time taken for HR development varies with type of *Avr–R* interaction. For example, co-expression of *35S::AvrPto* and *35S::Pto* produces HR in 48–96 h post inoculation while tv-EIX and LeEix2 interaction produces HR after 3–4 days.

4. Application of the agroinfiltration based *Avr–R* gene interaction assay can be extended to identify signaling components involved in pathogen defense responses. This has been used in mutant and fast-forward genetic VIGS-based screening to identify genes involved in defense-related signaling cascades.
5. (a) *Vector control*: A 451-bp *GFP* fragment was amplified using primers gfpattB1: 5′-ggggacaagtttgtacaaaaaagcaggct CTTTTCACTGGAGTTGTCCC-3′ and gfpattB2: 5′-ggggaccactttgtacaagaaagctgggtGCTTGTCGGCCATGATGTA-3′ from *GFP* gene, and cloned into *pTRV2*. Since there is no *GFP* homolog in plants, this vector when inoculated, will not cause any gene silencing effect in plants.
   (b) *Positive control*: As a visual marker for gene silencing, a 409 bp *NbPDS* (phytoene desaturase) fragment was cloned in *pTRV2* using the primers 5′-GGGGACAAGTTTTGTACAAAAAAGCAGGCCGGTCTAGAGG-CACTCAACTTTATAAACC-3′ and 5′-GGGGACCACTTTGTACAAGAAAGCTGGGCGGGGATCCCTTCAGTTTTCTGTCAAACC-3′.
   (c) For analyzing gene function, introduce a 200–500 bp PCR product from the gene of interest into *pTRV2* vector *via* gateway cloning. Choose sequences from the 3′-UTR region whenever possible, to prevent off-target silencing.
6. Inducing *Agrobacterium Vir* genes with acetosyringone is important for efficient infection. Induction can be carried out from a minimum of 4–6 h up to 24 h.
7. Inoculation of higher *Agrobacterium* OD_{600} induces localized cell death in the infiltration zone. Hence, this should be optimized before experimentation. Bacterial culture suspensions less dense than $OD_{600} = 0.1$ results in weak transgene expression, and above $OD_{600} = 1.0$ results in tissue yellowing or necrosis.
8. Several methods of Agroinoculation viz. spot inoculation using needleless syringe, vacuum inoculation of entire leaf and agrodrench (31) at the root zone are available. Users have choice to select any one of these methods.
9. Confirmation of down-regulation is done by reverse transcription of RNA from silenced leaves, using oligo-dT primer (random primer may lead to amplification of viral transcripts, giving erroneous results).
10. For large-scale analysis of plant cDNA library clones, we inoculate two leaves per plant and two pots per construct. For validation purposes, we inoculate at least 12 pots per construct and repeat twice to validate the results.

Acknowledgements

Projects involving *Agrobacterium*-mediated transient assays in the KSM laboratory are supported by the Samuel Roberts Noble Foundation, The National Science Foundation (Grant # IOB 0445799), and U.S.-Israel Binational Agricultural Research & Development Fund (BARD; Project # IS-3922-06).

References

1. Grimsley, N., Hohn, B., Hohn, T., and Walden, R. (1986) "Agroinfection," an alternative route for viral infection of plants by using the Ti plasmid *Proc Natl Acad Sci USA* **83**, 3282–86.
2. Zottini, M., Barizza, E., Costa, A., Formentin, E., Ruberti, C., Carimi, F., and Lo Schiavo, F. (2008) Agroinfiltration of grapevine leaves for fast transient assays of gene expression and for long-term production of stable transformed cells *Plant Cell Rep* **27**, 845–53.
3. Jefferson, R. A., Kavanagh, T. A., and Bevan, M. W. (1987) GUS fusions: beta-glucuronidase as a sensitive and versatile gene fusion marker in higher plants *EMBO J* **6**, 3901–7.
4. Rossi, L., Escudero, J., Hohn, B., and Tinland, B. (1993) Efficient and sensitive assay for T-DNA-dependent transient gene expression *Plant Mol Biol Rep* **11**, 220–29.
5. Wroblewski, T., Tomczak, A., and Michelmore, R. (2005) Optimization of *Agrobacterium*-mediated transient assays of gene expression in lettuce, tomato and *Arabidopsis Plant Biotechnol J* **3**, 259–73.
6. Johansen, L. K., and Carrington, J. C. (2001) Silencing on the spot. Induction and suppression of RNA silencing in the *Agrobacterium*-mediated transient expression system *Plant Physiol* **126**, 930–38.
7. Schob, H., Kunz, C., and Meins, F., Jr. (1997) Silencing of transgenes introduced into leaves by agroinfiltration: a simple, rapid method for investigating sequence requirements for gene silencing *Mol Gen Genet* **256**, 581–5.
8. Silhavy, D. (2005) *in* "Gene silencing by RNA interference: Technology and application" (Sohail, M., Ed.), pp. 357–63, CRC press, Oxford.
9. Chisholm, S. T., Coaker, G., Day, B., and Staskawicz, B. J. (2006) Host-microbe interactions: shaping the evolution of the plant immune response *Cell* **124**, 803–14.
10. Tai, T. H., Dahlbeck, D., Clark, E. T., Gajiwala, P., Pasion, R., Whalen, M. C., Stall, R. E., and Staskawicz, B. J. (1999) Expression of the *Bs2* pepper gene confers resistance to bacterial spot disease in tomato *Proc Natl Acad Sci USA* **96**, 14153–8.
11. Bendahmane, A., Querci, M., Kanyuka, K., and Baulcombe, D. C. (2000) *Agrobacterium* transient expression system as a tool for the isolation of disease resistance genes: application to the Rx2 locus in potato *Plant J* **21**, 73–81.
12. Van der Hoorn, R. A. L., Laurent, F., Roth, R., and De Wit, P. J. G. M. (2000) Agroinfiltration is a versatile tool that facilitates comparative analyses of Avr9/Cf-9-induced and Avr4/Cf-4-induced necrosis. *Mol Plant-Microbe Interact* **13**, 439–46.
13. Ron, M., and Avni, A. (2004) The receptor for the fungal elicitor ethylene-inducing xylanase is a member of a resistance-like gene family in tomato *Plant Cell* **16**, 1604–15.
14. Dinesh-Kumar, S. P., Anandalakshmi, R., Marathe, R., Schiff, M., and Liu, Y. (2003) *in* "Plant Functional Genomics: Methods and Protocols" (Grotewolk, E., Ed.), Vol. 236, pp. 287–93, Humana Press, Inc, Totowa.
15. Mysore, K. S., and Ryu, C. M. (2004) Nonhost resistance: how much do we know? *Trends Plant Sci* **9**, 97–104.
16. Burch-Smith, T. M., Anderson, J. C., Martin, G. B., and Dinesh-Kumar, S. P. (2004) Applications and advantages of virus-induced gene silencing for gene function studies in plants *Plant J* **39**, 734–46.
17. Grimsley, N., Hohn, T., Davies, J. W., and Hohn, B. (1987) *Agrobacterium*-mediated delivery of infectious maize streak virus into maize plants *Nature* **325**, 177–79.
18. Elmer, J. S., Sunter, G., Gardiner, W. E., Brand, L., Browning, C. K., Bisaro, D. M., and Rogers, S. G. (1988) *Agrobacterium*-mediated inoculation of plants with tomato golden mosaic virus DNAs *Plant Mol Biol* **10**, 225–34.
19. Sung, Y., and Coutts, R. (1995) Mutational analysis of potato yellow mosaic geminivirus *J Gen Virol* **76**, 1773–80.

20. Boulton, M. I., King, D. I., Markham, P. G., Pinner, M. S., and Davies, J. W. (1991) Host range and symptoms are determined by specific domains of the maize streak virus genome *Virology* **181**, 312–18.
21. Kheyr-Pour, A., Gronenborn, B., and Czosnek, H. (1994) Agroinoculation of tomato yellow leaf curl virus (TYLCV) overcomes the virus resistance of wild *Lycopersicon* species *Plant Breed* **112**, 228–33.
22. Ding, X. S., Liu, J., Chen, N. -H., Folimonov, A., Hou, Y. -M., Bao, Y., Katagi, C., Carter, S. A., and Nelson, R. S. (2004) The *Tobacco mosaic virus* 126-kDa protein associated with virus replication and movement suppresses RNA silencing *Mol Plant Microbe Interact* **17**, 583–92.
23. Chen, S., Vaghchhipawala, Z., Li, W., Asard, H., and Dickman, M. B. (2004) Tomato phospholipid hydroperoxide glutathione peroxidase inhibits cell death induced by Bax and oxidative stresses in yeast and plants *Plant Physiol* **135**, 1630–41.
24. Yang, Y., Li, R., and Qi, M. (2000) *In vivo* analysis of plant promoters and transcription factors by agroinfiltration of tobacco leaves *Plant J* **22**, 543–51.
25. Senthil-Kumar, M., Rame Gowda, H. V., Hema, R., Mysore, K. S., and Udayakumar, M. (2008) Virus-induced gene silencing and its application in characterizing genes involved in water-deficit-stress tolerance *J Plant Physiol* **165**, 1404–21.
26. Anand, A., Vaghchhipawala, Z., Ryu, C. M., Kang, L., Wang, K., del-Pozo, O., Martin, G. B., and Mysore, K. S. (2007) Identification and characterization of plant genes involved in *Agrobacterium*-mediated plant transformation by virus-induced gene silencing, *Mol Plant Microbe Interact* **20**, 41–52.
27. Sheludko, Y. V. (2008) *Agrobacterium*-mediated transient expression as an approach to production of recombinant proteins in plants *Recent Pat Biotechnol* **2**, 198–208.
28. Jia, H., Pang, Y., Chen, X., and Fang, R. (2006) Removal of the selectable marker gene from transgenic tobacco plants by expression of Cre recombinase from a tobacco mosaic virus vector through agroinfection *Transgenic Res* **15**, 375–84.
29. Anand, A., Krichevsky, A., Schornack, S., Lahaye, T., Tzfira, T., Tang, Y., Citovsky, V., and Mysore, K. S. (2007) *Arabidopsis* VIRE2 INTERACTING PROTEIN2 is required for *Agrobacterium* T-DNA integration in plants *Plant Cell* **19**, 1695–708.
30. Frederick, R. D., Thilmony, R. L., Sessa, G., and Martin, G. B. (1998) Recognition specificity for the bacterial avirulence protein AvrPto is determined by Thr-204 in the activation loop of the tomato Pto kinase *Mol Cell* **2**, 241–5.
31. Ryu, C. M., Anand, A., Kang, L., and Mysore, K. S. (2004) Agrodrench: a novel and effective agroinoculation method for virus-induced gene silencing in roots and diverse Solanaceous species *Plant J* **40**, 322–31.

Chapter 7

Full-Length cDNA Overexpressor Gene Hunting System (FOX Hunting System)

Mieko Higuchi, Youichi Kondou, Takanari Ichikawa, and Minami Matsui

Abstract

Full-length cDNAs (fl-cDNAs) are important resources for the characterization of gene function, since they contain all the information required for the production of functional RNAs and proteins. Large sets of fl-cDNA clones have been collected from several plant species and have become available for functional genomic analysis. We have developed a system for the identification of gene function by screening for transgenic plants ectopically expressing fl-cDNAs and named it the FOX (*fl*-cDNA *o*ver*ex*pressor gene) hunting system. This system can be applied to almost all plant species without prior knowledge of their genome sequences because only fl-cDNAs are required. For utilization of the FOX hunting system, *Agrobacterium* libraries and *Arabidopsis* seeds carrying rice and *Arabidopsis* fl-cDNAs are available. Here, we will describe the procedure followed in the FOX hunting system from the generation of expression vectors carrying fl-cDNAs to the confirmation of phenotype in retransformed plants.

Key words: Full-length cDNA, *Arabidopsis*, FOX hunting system, Gain-of-function, Heterologous gene expression, Transgenic plants

1. Introduction

Classical forward genetics using gene tags is performed by phenotypic screening of loss-of-function mutant populations obtained by T-DNA and transposon insertional mutagenesis. High-throughput screens using these mutant populations provide a means to analyze plant gene function. Gain-of-function mutants are additional fundamental resources for studying gene function. Overexpression may offer a useful route for analyzing gene families when the gene of interest has functionally redundant homologs in the genome, because the function of these

Andy Pereira (ed.), *Plant Reverse Genetics: Methods and Protocols*, Methods in Molecular Biology, vol. 678, DOI 10.1007/978-1-60761-682-5_7, © Springer Science+Business Media, LLC 2011

genes would be difficult or impossible to uncover using a knockout approach. Moreover, overexpression is also useful for genes whose loss of function leads to lethality.

Activation tagging is a powerful method for generating gain-of-function mutants in plants (1). Activation-tagging lines are generated by random insertion of a T-DNA carrying the enhancer elements from the cauliflower mosaic virus (*CaMV*). However, the *CaMV 35S* enhancer can influence the expression of genes up to several kilo bases from the insertion site, thereby causing difficulties in identifying the genes responsible for the observed mutant phenotypes. We have, therefore, developed a different approach to generate gain-of-function mutants systematically. We introduced mixed and normalized full-length cDNAs (fl-cDNAs) into an expression vector under the control of the *CaMV 35S* promoter (Fig. 1). This cDNA library was introduced into *Agrobacterium*, which was then used to transform *Arabidopsis* in planta. The introduced cDNA can be cloned easily using T-DNA-specific primers. Thus, the cDNA that caused the observed phenotype should be directly linked to the function. We named this system as the FOX hunting system (*F*ull-length cDNA *O*vere*X*pressing gene hunting system), and a scheme of this is presented in Fig. 2 (2). We first generated about 15,000 FOX *Arabidopsis* lines that expressed *Arabidopsis* fl-cDNAs under the *CaMV 35S* promoter and isolated morphologic mutants with their corresponding genes (2).

The FOX hunting system is unique in that only fl-cDNAs are required for the functional analysis of genes. Thus, any fl-cDNA library can be applied to the FOX hunting system if the vector carrying the cDNAs has *Sfi*I sites for cloning. This is advantageous for some plant species having large or unsequenced genomes, because no genome sequence information is required. *Arabidopsis* is one of the best organisms for the host plant because it has a very efficient, fast, and high-throughput transformation system when compared with those available for other plant species. As a model case, we generated more than 23,000 independent *Arabidopsis* transgenic lines (rice FOX *Arabidopsis* lines) that expressed rice fl-cDNAs, to verify heterologous gene expression (3). We demonstrated that it is possible to investigate the

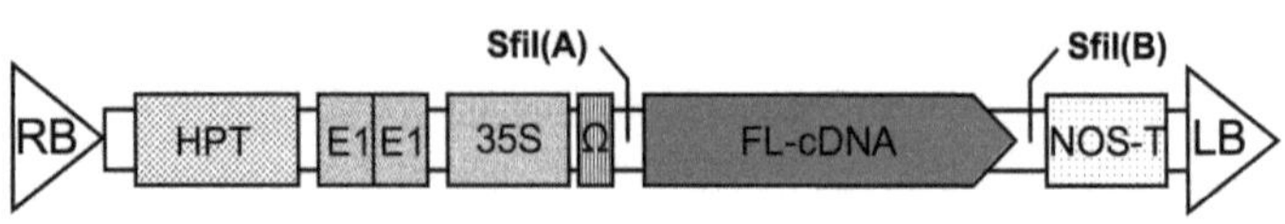

Fig. 1. The structure of the T-DNA region in expression vector for the FOX hunting system. Rice full-length cDNAs were cloned into pBIG2113SF. HPT, hygromycin resistance gene; E1, 5′-upstream sequence of the *CaMV 35S* promoter (–419 to –90); 35S, *CaMV 35S* promoter (–90 to –1); Ω, 5′-upstream sequence of tobacco mosaic virus (TMV); NOS-T, polyadenylation signal of the gene for nopaline synthase in the Ti plasmid; RB, T-DNA *right* border; LB, T-DNA *left* border.

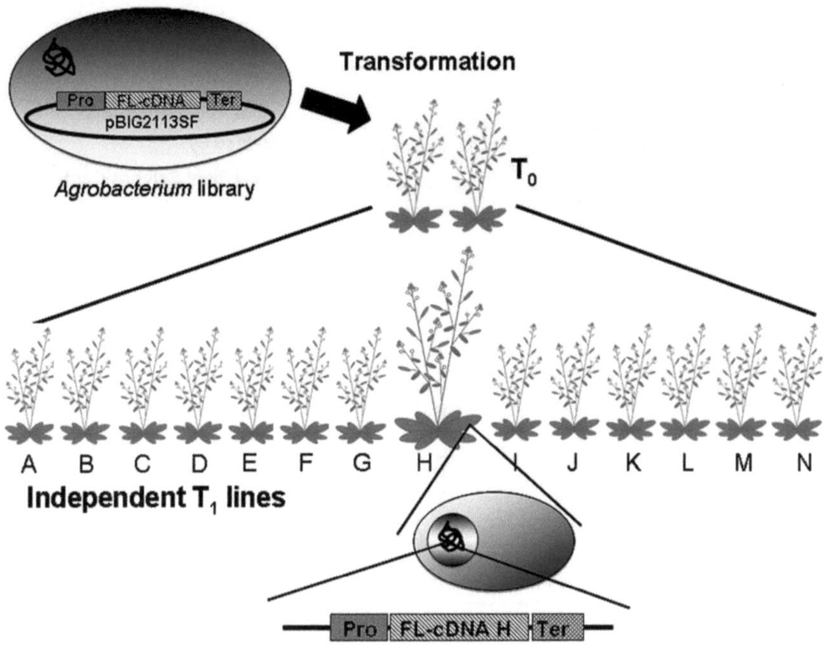

Fig. 2. Scheme of the FOX hunting system. *Arabidopsis* plants are transformed with the FOX *Agrobacterium* library carrying fl-cDNAs in the expression vector (pBIG2113SF). Generated T_0 FOX plants are self-pollinated, and then many independent T_1 FOX seed libraries are obtained. From the T_1 FOX lines, a phenotypic mutant line (in this case, the "H" line) is identified. The corresponding gene of the "H" mutation is easily and immediately identified by PCR using T-DNA-specific primers and sequencing.

function of genes in different plant species by utilizing a heterologous expression system. In this protocol, we will describe how to create an expression library carrying rice fl-cDNAs, generate rice FOX *Arabidopsis* lines using an *Agrobacterium* library, and determine the rice cDNA responsible for an observed phenotype as an illustration of the procedure used in the FOX hunting system.

1.1. Outline

The outline of the procedure is shown in Fig. 3. The cDNA library is cloned into the expression vector and then an *Agrobacterium* library carrying fl-cDNAs is generated. We used the *Agrobacterium* in planta transformation method (4) to introduce the fl-cDNAs into *Arabidopsis* and generated T_1 seeds. These T_1 seeds of the FOX *Arabidopsis* lines can be used for screening in parallel with antibiotic selection. Candidate plants are self-pollinated to generate T_2 seeds. The T_2 seeds undergo a secondary screening to confirm the observed phenotype. Alternatively, T_1 plants selected for antibiotic resistance are self-pollinated and the T_2 seeds can then be used for screening. After confirmation of the phenotype in the T_2 generation plants, the introduced cDNA is isolated from the candidate FOX line. The isolated cDNA is reintroduced into *Arabidopsis* independently to confirm that its expression is responsible for the observed phenotype.

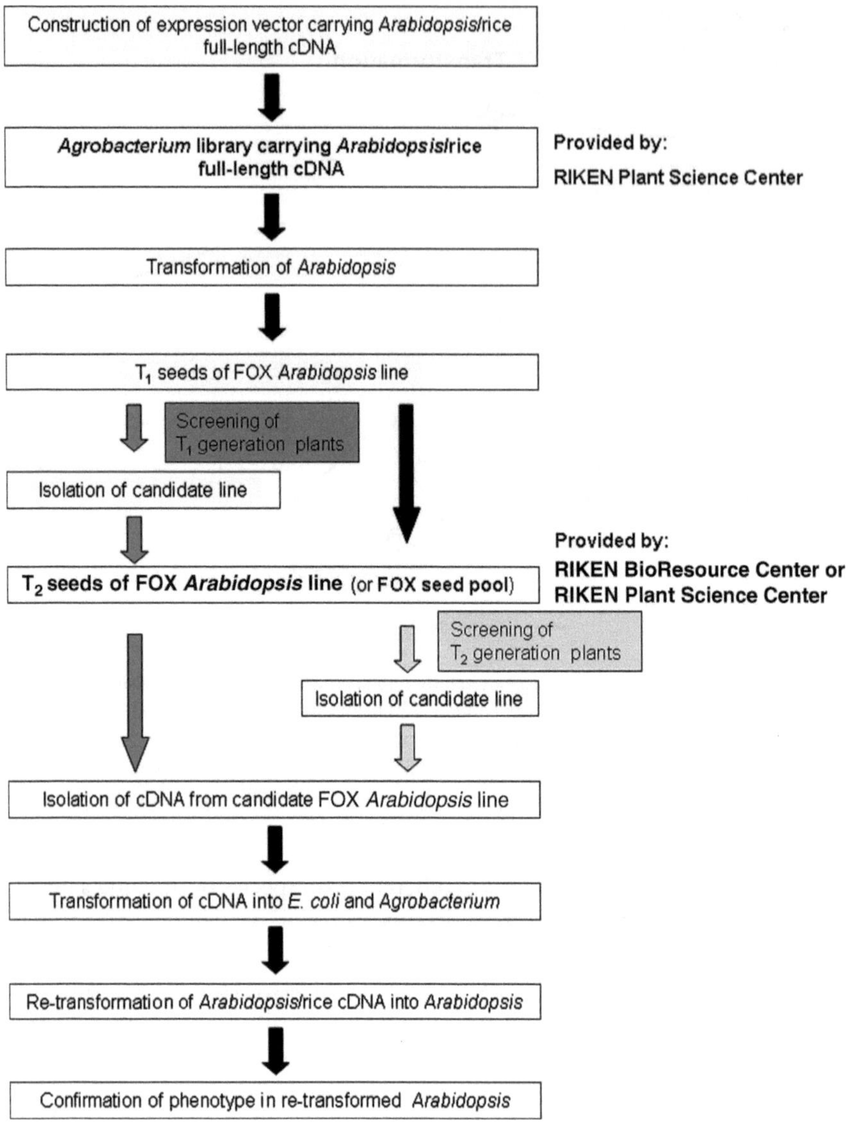

Fig. 3. Flow chart illustrating the FOX hunting system. Screening can be carried out using either own-generated transgenic plants of the T_1 (*dark gray arrows*) or T_2 (*light gray arrows*) or the FOX seed pools (*light gray arrows*) provided by the RIKEN BioResource Center. The sequence data of integrated cDNAs in the rice FOX *Arabidopsis* lines are available from the public website. Independent rice FOX lines are also available from the RIKEN BioResource Center. The *Agrobacterium* library is available from RIKEN Plant Science Center.

We provide several types of resources for the FOX hunting system for rapid identification of gene function (Fig. 3). *Agrobacterium* libraries carrying rice and *Arabidopsis* fl-cDNAs are available from RIKEN Plant Science Center (http://www.psc.riken.jp/english/index.html). We have determined the sequences of introduced cDNAs in FOX *Arabidopsis* and rice FOX *Arabidopsis* lines. The sequence data are available from our website (FOX *Arabidopsis*; http://nazunafox.psc.database.riken.jp, rice FOX *Arabidopsis*; http://ricefox.psc.riken.jp/). Seeds of independent FOX *Arabidopsis* and rice FOX *Arabidopsis* lines are available from the RIKEN BioResource Center (http://www.brc.riken.go.jp/lab/epd/Eng/) or RIKEN Plant Science Center. BioResource Center also provides seed pool sets of FOX *Arabidopsis* and rice FOX *Arabidopsis* lines (a seed pool contains approximately 400 seeds from 50 lines, eight seeds per line; one seed pool set contains 20 seed pools, equivalent to 1,000 lines).

2. Materials

2.1. Cloning of Rice Fl-cDNA Library into Expression Vector

1. Expression vector pBIG2113SF is used in this protocol.
2. 3 M sodium acetate, pH 4.8: dissolve 40.8 g of $NaOAc{\cdot}3H_2O$ in 100 mL of water. Adjust to pH 4.8 with glacial acetic acid before autoclaving.
3. *Sfi*I, 10× M buffer, and bovine serum albumin (BSA) solution were supplied by Takara Bio Inc.
4. Ligation solution and T4 DNA ligase: add 1 μL of 10× ligation buffer to 10 μL of distilled water for the ligation solution. 10× ligation buffer and T4 DNA ligase (400 U/μL) were supplied by New England Biolabs, Inc.

2.2. Plant Material and Growth Conditions

1. *Arabidopsis thaliana* plants are grown at 22°C in long-day conditions (16 h light and 8 h dark). The ecotype used in this protocol is Columbia-0.
2. For cultivation of *Arabidopsis* plants, 3,000 mL of a 1,000-fold dilution of HYPONeX (HYPONeX Japan Corp., Ltd.) is added to a mixture of 1.5 kg of PRO-MIX (Premier Tech Ltd.) and 0.9 kg of vermiculite (Fukushima VERMI KK). The soil is autoclaved at 120°C for 30 min before use (see Note 1).

2.3. Preparation of Agrobacterium and E. coli Cultures

1. LB medium: dissolve 10 g of tryptone peptone (Becton, Dickinson and Company), 5 g of yeast extract (Becton, Dickinson and Company), and 5 g of NaCl in 1,000 mL of water and autoclave at 120°C for 20 min. For solid medium, add 16 g of agar powder (Nacalai Tesque, Inc.) to 1,000 mL of LB medium.
2. SOC medium: autoclave 1,000 mL of water containing 20 g of tryptone peptone, 5 g of yeast extract, 0.19 g of KCl, 2.03 g of $MgCl_2{\cdot}6H_2O$, 2.46 g of $MgSO_4{\cdot}7H_2O$, 0.58 g of NaCl, and 3.6 g of glucose.
3. Kanamycin (Sigma-Aldrich Corp.) and gentamycin (Nacalai Tesque, Inc.): dissolve 50 mg/mL and 10 mg/mL in water, respectively, sterilize by filtering through a 0.22-μm membrane, and store at −20°C.
4. *Agrobacterium* strain: GV3101 pMP90 is used in this protocol.
5. *E. coli* strain: DH10B is used in this protocol.

2.4. Transformation of Arabidopsis Plants

1. Infiltration medium: dissolve half a packet of Murashige and Skoog (MS) inorganic salts (Wako Pure Chemical Industries, Ltd.) and 50 g of sucrose in 1,000 mL of water. Add 112 μL of Gamborg's 1,000× vitamin solution (Sigma-Aldrich Corp.), 10 μL of benzylaminopurine stock solution, and 200 μL of Silwet L-77 (Agri-Turf Supplies, Inc.) (see Note 2).
2. Benzylaminopurine stock solution: dissolve 1 mg of benzylaminopurine (Wako Pure Chemical Industries, Ltd.) in 1 mL of dimethyl sulfoxide (DMSO).
3. Solid Basic Agar Medium (BAM) (5): autoclave 1,000 mL of water containing 101 mg of KNO_3 and 8 g of Bacto Agar (Becton, Dickinson and Company).
4. 0.2% Water agar: autoclave 1,000 mL of water containing 2 g of Bacto Agar.
5. Bleaching solution: mix 10 mL of sodium hypochlorite solution (Nacalai Tesque, Inc.) and 100 μL of Triton X-100 (Nacalai Tesque, Inc.) in 100 mL of water.
6. Hygromycin B (Sigma-Aldrich Corp.) and cefotaxime (Sigma-Aldrich Corp.): dissolve 20 mg/mL and 100 mg/mL in water, respectively, sterilize by filtering through a 0.22-μm membrane, and store at −20°C.

2.5. Recloning of Rice Fl-cDNA into Expression Vector

1. Enzyme for polymerase chain reaction (PCR): PrimeSTAR HS DNA polymerase with GC buffer (Takara Bio Inc.) is used in this protocol. 2× PrimeSTAR GC buffer I and a solution containing 2.5 mM of each dNTP are included in the package.
2. Reaction solution for PCR and colony PCR: add 50 μL of 2× PrimeSTAR GC buffer I, 8 μL of the dNTP mixture, 20 pmol of each primer, an appropriate volume of DNA template, and 2.5 U of PrimeSTAR HS DNA polymerase to 100 μL of distilled water.

3. Methods

3.1. Making of Agrobacterium Library Transformed with Rice Fl-cDNA Expression Library

3.1.1. Cloning of Rice fl-cDNA Library into Expression Vector

1. Mix 20 μL of expression vector (20 ng/μL), 40 μL of normalized rice fl-cDNA library (30 ng/μL), 10 μL of 10× M buffer, 1 μL of BSA solution, 24 μL of distilled water, and 5 μL of *Sfi*I, and incubate overnight at 37°C (see Note 3).
2. Add another 5 μL of *Sfi*I to the reaction mixture and incubate for at least 3 h at 50°C spinning down every hour.
3. Precipitate the DNA by adding 0.1 vol of 3 M sodium acetate, pH 4.8, and 1 vol of isopropanol.
4. Wash the DNA pellets with 70% ethanol twice and dry.

5. Resuspend the DNA pellets in 9 μL of ligation solution and add 1 μL of T4 DNA ligase.
6. Incubate the ligation mixture overnight at 16°C.

3.1.2. Transformation of Expression Vector into E. coli Cells by Electroporation

1. Thaw electrocompetent *E. coli* strain DH10B cells on ice and at the same time chill a cuvette with 1-mm electrode gap ready for electroporation (see Note 4).
2. Mix 40 μL of competent cells with 2 μL of ligated DNA on ice and transfer to the ice-cold cuvette.
3. Apply a pulse at 4 ms, 1.5 kV, 200 Ohm, and 25 μF.
4. Immediately add 500 μL of SOC medium warmed beforehand to 37°C and mix the cells.
5. Transfer the cells to a 1.5-mL tube and incubate for 1 h at 37°C.
6. Dilute the cells with SOC medium to obtain approximately 5,000–10,000 colonies per 10-cm Petri dish. Ultimately, about 150,000 colonies are required to make an *Agrobacterium* library. Plate the diluted cells on solid LB medium supplemented with 50 μg/mL kanamycin.
7. Incubate the Petri dishes overnight at 37°C.
8. Pour 1 mL of LB medium into each Petri dish and scrape the colonies in LB medium using a spreader.
9. Collect the cell cultures from all Petri dishes including equivalent to 150,000 colonies and isolate plasmid DNA.

3.1.3. Transformation of Expression Vector into Agrobacterium Cells by Electroporation

1. Thaw electrocompetent *Agrobacterium* strain GV3101 pMP90 cells on ice and at the same time chill a cuvette with 2-mm electrode gap ready for electroporation.
2. Mix 40 μL of *Agrobacterium* competent cells with 2 μL of plasmid DNA on ice and transfer to the ice-cold cuvette.
3. Apply a pulse at 4 ms, 2.5 kV, 200 Ohm, and 25 μF.
4. Immediately add 500 μL of SOC medium warmed beforehand to 28°C and mix with the cells.
5. Transfer the cells to a 1.5-mL tube and incubate for 1–3 h at 28°C.
6. Dilute the cells with SOC medium to obtain approximately 5,000–10,000 colonies per 10-cm Petri dish. Ultimately, about 150,000 colonies are required to make an *Agrobacterium* library. Plate the diluted cells on solid LB medium supplemented with 50 μg/mL kanamycin and 10 μg/mL gentamycin.
7. Incubate the Petri dishes for 2 days at 28°C.

8. Pour 1 mL of LB medium to each Petri dish and scrape the colonies with LB medium using a spreader. Collect the cell cultures from all Petri dishes, including equivalent to 150,000 colonies. These cell cultures were used as the *Agrobacterium* library.

3.2. Transformation of Arabidopsis Plants with Rice FOX Library Using Agrobacterium

3.2.1. Preparation of Agrobacterium Culture

1. Inoculate 2 μL of *Agrobacterium* cells transformed with the rice fl-cDNA expression library into 200 mL of liquid LB medium supplemented with 50 μg/mL kanamycin and 10 μg/mL gentamycin, and grow to an OD_{600} of 1.2–1.5 on a shaker at 28°C (see Note 5).
2. Transfer the culture to a centrifuge tube and centrifuge at 6,000 × *g* for 13 min.
3. Remove the supernatant and resuspend the bacterial pellet in infiltration medium to an OD_{600} of 0.8.

3.2.2. Transformation of Arabidopsis Plants

1. *Arabidopsis* plants are grown in pots until the flowering stage (see Note 6).
2. Invert the pot over the infiltration medium containing the *Agrobacterium* (described above) and dip the plants in the medium ensuring that they get soaked.
3. Remove and put the pot into a plastic bag and seal it.
4. Keep the plastic bag overnight under long-day conditions and then open it.
5. Keep it overnight again under the same conditions and finally remove the pot from the plastic bag.
6. Grow the *Arabidopsis* plants in the pot until it is ready to harvest.
7. Harvest all the T_1 seeds from these T_0 plants.

3.2.3. Selection of Rice FOX Arabidopsis Lines

1. Leave 0.25 g of T_1 seeds in 70% ethanol for 1 min for surface sterilization.
2. Remove the 70% ethanol and then treat the seeds with bleaching solution for 10 min.
3. Remove the bleaching solution and rinse the seeds with sterile water three times.
4. Remove the water and suspend the seeds in 0.2% water agar.
5. Plate the seeds in solid BAM supplemented with 20 μg/mL of hygromycin B and 100 μg/mL of cefotaxime sodium salt (10 cm Petri dishes).
6. Keep for at least 2 days at 4°C under dark conditions to induce germination.
7. Grow the seedlings on the medium under long-day conditions for 5–10 days, or under test conditions (for example,

using BAM with high levels of salt, growing under high-light conditions, and so on), if you use the T_1 plants for screening (see Notes 7–9).

8. Pick seedlings that develop true leaves, and transfer to soil.
9. Grow these T_1 plants, rice FOX *Arabidopsis* lines, until harvest.
10. Harvest the T_2 seed.

3.3. Screening and Isolation of Rice Fl-cDNA from Rice FOX Arabidopsis Lines

3.3.1. Screening and DNA Isolation from rice FOX Arabidopsis Lines

1. Screen the rice FOX *Arabidopsis* lines in the T_2 generation under appropriate conditions to isolate rice FOX *Arabidopsis* mutants that show interesting phenotypes (see Note 10).
2. Grow these mutants and take samples of true leaf or other organ.
3. Isolate chromosomal DNA from these samples (see Note 11).

3.3.2. Amplification of Rice Fl-cDNA Fragments Using PCR and Determination of Sequences

1. Use <200 ng of chromosomal DNA isolated from the mutants as the DNA template in a 50-μL reaction solution for PCR. Use the primer sets (F1 and R1) in Table 1 (see Notes 12 and 13).
2. Using a thermocycler perform an initial denaturation at 94°C and then 30 cycles at 98°C for 30 s, 52°C for 30 s, 72°C for 3 min, finishing with a 10-min 72°C extension cycle.
3. Electrophorese the PCR products on a 0.8% agarose gel and extract each band of amplified PCR fragments (see Notes 14–16).

Table 1
Primers used for amplification of rice fl-cDNAs from genomic DNA of rice FOX *Arabidopsis* lines

Forward primer		Reverse primer	
Name	Sequence	Name	Sequence
F1	GGAAGTTCATTTATTCGGAGAG	R1	GGCAACAGGATTCAATCTTAAG
F2	CATTTATTCGGAGAGGTACGTAT	R2	GGATTCAATCTTAAGAAACTTTATT GCCAA
F3	GTACGTATTTTTACAACAATTA CCAACAAC	R3	CAAATGTTTGAACGATCGGGGAAAT
F4	ATTACATTTTACATTCTACAACT ACATCT	R4	GATCCTCTAGAGGCCCTTAT
F5	CCCCCCCCCCCCD (A or G or T)	R5	AAAAAAAAAAAAB (C or G or T)

The primers (F1–F4, R1–R4) are designed outside of the *Sfi*I sites in pBIG2113SF. F2-F4 and R1-R4 primers were designed as nested primer sets and listed in order of outer to inner sequences. The F5 and R5 are adaptor sequences used for *Arabidopsis* cDNA cloning and can be used for *Arabidopsis* fl-cDNA sequencing. We usually use the F1 and R1 primer set to amplify the cDNA

3.4. Retransformation of Arabidopsis Plants with Expression Vectors Containing Isolated Rice Fl-cDNAs Using Agrobacterium

3.4.1. Recloning of Rice Fl-cDNA into Expression Vector

1. Mix 0.9 μL of expression vector (5–20 ng/μL), 7 μL of the amplified PCR fragment extracted from the agarose gel, 1 μL of the 10× H buffer, 0.1 μL of the BSA solution, and 0.5 μL of *Sfi*I, and incubate overnight at 37°C (see Note 17).
2. Add another 0.5 μL of *Sfi*I to the reaction mixture and incubate for at least 3 h at 50°C spinning down every hour.
3. Precipitate the DNA by adding 0.1 vol of 3 M sodium acetate, pH 4.8, and 1 vol of isopropanol.
4. Wash the DNA pellets with 70% ethanol twice and dry.
5. Resuspend the pellets in 4.5 μL of ligation solution and add 0.5 μL of T4 DNA ligase.
6. Incubate the ligation mixture overnight at 16°C.
7. Precipitate the ligated DNA by adding 0.1 vol of 3 M sodium acetate, pH 4.8, and 2 vol of ethanol; wash the DNA pellets with 70% ethanol; dry; and resuspend in 5 μL of distilled water (see Note 18).

3.4.2. Transformation of Expression Vector into E. coli Cells

1. Transform *E. coli* cells with 5 μL of the ligated DNA as described in Subheading 3.1.2 of this protocol without diluting the transformed cells with SOC medium.
2. Pick each antibiotic-resistant colony into a 10-μL reaction solution for PCR without any DNA template.
3. Run the same PCR program as that for the amplification of rice fl-cDNA from chromosomal DNA.
4. Electrophorese 10 μL of the PCR products on a 0.8% agarose gel to check the size of the DNA fragment.
5. If the size of the DNA fragment amplified by colony PCR is appropriate, repick the antibiotic-resistant colony into fresh liquid LB medium containing 50 μg/mL of kanamycin. Grow overnight on a shaker at 37°C.
6. Isolate plasmid DNA from the culture.
7. Sequence the plasmid DNA using the primer sets presented in Table 1 to determine the fl-cDNAs introduced in the expression vector (see Note 19).

3.4.3. Transformation of Expression Vector into Agrobacterium Cells

1. Transform *Agrobacterium* cells with 1 μL of plasmid DNA as described in Subheading 3.1.3 of this protocol without diluting the transformed cells with SOC medium.
2. Pick antibiotic-resistant colonies into 2 mL of liquid LB medium supplemented with 50 μg/mL of kanamycin and 10 μg/mL of gentamycin, and grow overnight on a shaker at 28°C for subculturing.

3.4.4. Checking the Phenotypes of Retransformants

1. Use the *Agrobacterium* cultures from above for retransformation of *Arabidopsis* plants. Refer to the transformation procedure described in Subheading 3.2.2 of this protocol.
2. Check for the desired phenotypes in the retransformants (see Notes 20 and 21).

4. Notes

1. For proper sterilization, it is necessary to wet the soil completely with water before autoclaving.
2. The transformation efficiency of *Arabidopsis* is little affected by the absence of benzylaminopurine and Gamborg's vitamin solution from the infiltration medium.
3. When nonnormalized fl-cDNAs are used in the FOX hunting system, the cDNA species may be biased in accordance with the abundance of cDNAs in the library.
4. Electrocompetent *E. coli* cells with appropriate competence should give at least 1×10^8 transformants/μg of pUC19 plasmid. Improved results can be obtained using cells with higher competence.
5. The density of the culture is not critical; it is possible to transform *Arabidopsis* irrespective of the growth of the *Agrobacterium* cells.
6. It is very important to use healthy *Arabidopsis* plants for floral dipping. They should be grown in optimal conditions and have numerous unopened floral buds and few siliques at the time of transformation.
7. It has been noted that transformants may show growth defects as a result of the integrated transgene in some cases. Since it may be difficult to distinguish these from hygromycin-sensitive individuals, it may take a while before the transformants have grown sufficiently on the selection medium to distinguish between the two.
8. About 30–40 transformants can be obtained when 150 mg of seeds (approximately 5,500 seeds) are sown.
9. The generated T_1 seed library can be used for the desired screening in parallel with antibiotic selection. For example, Du et al. (6) isolated the salt stress-resistant plants in half-strength MS medium containing 200 mM of NaCl.
10. The candidate mutants should show the dominant phenotype because the FOX hunting system is a gain-of-function resource. It is better to isolate candidate lines that show a dominant phenotype in the T_2 generation.

11. There are a lot of protocols for the isolation of plant genomic DNA. For the FOX hunting system, a method capable of isolating pure and high molecular weight DNA is required to amplify the introduced cDNA by PCR.
12. We usually used PrimeSTAR HS DNA polymerase with GC buffer for amplification of rice fl-cDNAs due to the high GC content in rice. Some experimentation may be needed to determine which PCR enzyme is the best for amplification from different plant species. We recommend LA-Taq (Takara BioInc.) or KOD-FX (Toyobo Co., Ltd) when it is difficult to amplify the introduced cDNA.
13. We usually use the F1 and R1 primer set to amplify the cDNA and obtain good results, but other primers shown in Table 1 can also be used if the PCR fragment cannot be obtained using F1 and R1 primers. In some cases, the fragment could only be obtained by nested PCR using the first PCR product and an internal primer set.
14. There is no need to proceed if the size of the PCR fragment is less than 200 bp, as a fragment of approximately 200 bp is obtained when the empty vector is used as the template.
15. Multiple PCR fragments were obtained in some FOX lines because there were multiple T-DNA insertions in the *Arabidopsis* genome. If this is the case, all fragments can be recovered and then each one can be used to generate retransformed plants to confirm which cDNA caused the phenotype.
16. We found that tandem different cDNAs can insert into the expression vector in some FOX *Arabidopsis* lines. Thus, direct sequencing of PCR fragments can be carried out using primers at both ends to identify the transgene.
17. If cloning of fl-cDNAs into the expression vector is not efficient, an increase in the concentration of the amplified PCR fragment added to the reaction mixture for *Sfi*I digestion is recommended.
18. Ethanol precipitation of the ligated DNA is carried out to eliminate salts in ligation mixture for high transformation efficiency by electroporation. Therefore, this step may not be necessary. If the ligation mixture is directly added to competent cells, mix 1 μL of the ligation mixture with 30 μL of *E. coli* competent cells.
19. The fl-cDNA insertion must be checked by sequencing and the fl-cDNA inserted into the expression vector should also be checked for PCR errors at this step.

20. The incidence rate of retransformants that show the objective phenotypes varies depending on the fl-cDNA inserted in the expression vector used for transformation of *Arabidopsis*.
21. If a phenotype of a rice FOX *Arabidopsis* mutant is recaptured in a retransformed plant, the expression level of the transgene should be checked.

Acknowledgments

This work is supported by a Special Coordination Fund for Promoting Science and Technology awarded to M.M., K.O., and H.H. This study is also supported by a Grant-in-Aid for Young Scientists (B) from the Ministry of Education, Culture, Sports and Technology of Japan (19710055) to Y.K. We thank Dr. Hirofumi Kuroda, Ms. Yoko Horii, and Dr. Yuko Tsumoto (RIKEN Plant Science Center) for their technical support. We appreciate the helpful discussions with Dr. Takeshi Yoshizumi (RIKEN Plant Science Center).

References

1. Tani, H., Chen, X., Nurmberg, P., Grant, J.J., SantaMaria, M., Chini, A. et al. (2004) Activation tagging in plants: a tool for gene discovery. *Funct Integr Genomics*, **4**, 258–266.
2. Ichikawa, T., Nakazawa, M., Kawashima, M., Iizumi, H., Kuroda, H., Kondou, Y. et al. (2006) The FOX hunting system: an alternative gain-of-function gene hunting technique. *Plant J*, **48**, 974–985.
3. Kondou, Y., Higuchi, M., Takahashi, S., Sakurai, T., Ichikawa, T., Kuroda, H. et al. (2009) Systematic approaches to using the FOX hunting system to identify useful rice genes. *Plant J*, **57**, 883–894.
4. Clough, S.J. and Bent, A.F. (1998) Floral dip: a simplified method for *Agrobacterium*-mediated transformation of *Arabidopsis thaliana*. *Plant J*, **16**, 735–743.
5. Nakazawa, M. and Matsui, M. (2003) Selection of hygromycin-resistant *Arabidopsis* seedlings. *Biotechniques*, **34**, 28–30.
6. Du, J., Huang, Y.P., Xi, J., Cao, M.J., Ni, W.S., Chen, X. et al. (2008) Functional gene-mining for salt-tolerance genes with the power of *Arabidopsis*. *Plant J*, **56**, 653–664.

Chapter 8

Activation Tagging with *En/Spm-I/dSpm* Transposons in *Arabidopsis*

Nayelli Marsch-Martínez and Andy Pereira

Abstract

Activation tagging is a powerful strategy to find new gene functions, especially from genes that are redundant or show lethal knock-out phenotypes. It has been applied using T-DNA or transposons. *En/Spm-I/dSpm* engineered transposons are efficient Activation tags in *Arabidopsis*. An immobilized transposase source and an enhancer-bearing non-autonomous element are used in combination with positive and negative selectable markers to generate a population of single or low copy, stable insertions. This method describes the steps required to select the best parental lines, generate a population of stable insertions, and gene identification.

Key words: Activation tagging, Transposons, *Arabidopsis*, *En/Spm-I/dSpm* system, Dominant mutants

1. Introduction

1.1. Activation Tagging

For many genes, conventional knock-out insertional mutagenesis does not provide indications of their functions, mainly due to functional redundancy, lethality, or because the particular mutant phenotype can only be visualized in specific conditions (1–3). "Activation tagging", consists in using enhancers inside an insertion tag, and can overcome some of these problems (4). Enhancers positively influence gene expression, even when located at a considerable distance to the target promoter, and can increase endogenous gene expression (5). The major advantages of this strategy over simple knock-out strategies are: (a) it produces dominant in place of recessive mutations; (b) it may produce viable individuals in cases where gene knock-outs lead to lethal phenotypes; (c) it may produce evident phenotypes for genes with overlapping functions; (d) well suited to perform positive selection screens

Andy Pereira (ed.), *Plant Reverse Genetics: Methods and Protocols*, Methods in Molecular Biology, vol. 678,
DOI 10.1007/978-1-60761-682-5_8, © Springer Science+Business Media, LLC 2011

(i.e. resistance or tolerance to chemical, physical, or biological stresses), and to study metabolic pathways; (e) where the goal is to produce biotechnological applications, this strategy can lead to the direct application of discovered genes.

First proposed for plants in the 1990s (4), the activation tagging approach has been successfully applied since then. In most examples, the CaMV 35S promoter or enhancer has been used (6). In many cases, tags containing the enhancer cause quantitative increases in expression, following the original expression pattern of a gene (7–9). This is an advantage when studying those genes for which true ectopic or constitutive overexpression may affect early development leading to lethal phenotypes.

1.2. Considerations About Transposons Versus T-DNA

The activation tagging populations described to date for *Arabidopsis* and other plants (i.e. poplar, rice, tomato, barley, and petunia) employ T-DNA or transposons containing the promoter or enhancer sequences (i.e. (8–22)). Modified stabilized transposons can be more effective for activation tagging purposes than T-DNA insertions (13). This might be due to a reduction in the frequency of integration configurations that can lead to epigenetic effects affecting the enhancer (23), or hinder the isolation of adjacent sequences.

Moreover, re-introduction of the proper transposase in the plant, promoting transposon mobilization allows following reversion and targeted tagging strategies. Reversion, or the excision of the transposable element from its original locus, can be used to prove that a mutant phenotype is caused by the transposable element. With targeted tagging, a transposon can be remobilized from a known position to a new linked position. The frequency in which this occurs varies among transposon families, and has been estimated for the two main maize families used – *Arabidopsis* (24). The transposition frequency to linked loci has been estimated to be 30% for the *En/Spm-I/dSpm* system (25), and 68% for the *Ac-Ds* system, with distribution variations depending on the donor site (26).

1.3. Considerations About the Construct

A CaMV 35S enhancer tetramer has been commonly used in Activation tag populations (4, 9, 13). However, other enhancers, or inducible systems can also be used (27, 28). For the tag, a modified *En/Spm-I/dSpm* system works effectively in *Arabidopsis* (25).

The native transposon system generally consists of:

(a) An "autonomous", active element, which has specific terminal inverted repeats, and codes for a functional transposase that recognizes the termini of the transposon, cuts and introduces it into a new position.
(b) A "non-autonomous", inactive element, which has specific terminal inverted repeats, but no longer codes for a functional transposase. These elements can have a mutated transposase

or lack it, and they require the transposase coded by an active element to transpose (5).

The modified system allows to obtain stable insertions and to reduce transposition variability (i.e. frequency and time of transposition). For this, the transposase is "immobilized" by removing the transposon ends and is regulated by the CaMV 35S promoter. On the other hand, the non-autonomous element consists exclusively of transposon terminal inverted repeats, with suitable markers between them in order to follow transposition and/or other markers or regulatory sequences for trapping (i.e. promoter or gene), or activation tagging purposes. Transposon borders should be kept at the minimum length in order to avoid natural methylation sites present before them and can affect transposition (i.e. (29)), or lower enhancer activity (14). The transposon used here contains 200 and 400 bp long ends.

Selectable markers facilitate the high-throughput identification of stable transpositions, especially when the markers allow selection in the greenhouse. The immobilized transposase is linked to a negative selectable marker, and the mobile non-autonomous element bears a positive selectable marker. It is convenient that the negative marker can also be easily recognized in the plant, for example, by producing a visible phenotype.

A useful double selection system (30) employs the *bar* gene (basta resistance) as the positive marker (31, 32), placed between the transposon ends, and the *SU1* gene as the negative marker. *SU1* converts the otherwise innocuous compound R7402 to sulfonylurea that inhibits or reduces plant growth (DuPont; (33)). Conveniently, *SU1* produces a dwarf, dark green, reduced apical dominance phenotype in *Arabidopsis* plants. This facilitates the identification of progeny plants (T2) containing the *SU1* gene (and hence the transposase) to be used as parentals for the generation of stable insertions.

In principle, the two components (transposase-negative marker & transposon-positive marker) can be placed in separate vectors and plants (*in trans*), or together in a single vector and finally plant (*in cis*). The *in cis* strategy is better suited to obtain large numbers of plants with stable insertions. However, the construct used for the *in cis* strategy is large, which may be a disadvantage in plants that are difficult to transform. In that case, and when plants can be crossed, the *in trans* strategy may be advantageous, since the constructs are smaller.

The construct used for *Arabidopsis* in this chapter, named WAT, is depicted in Fig. 1 (13). For some other plants or specific strategies, different transposon systems or markers may be better.

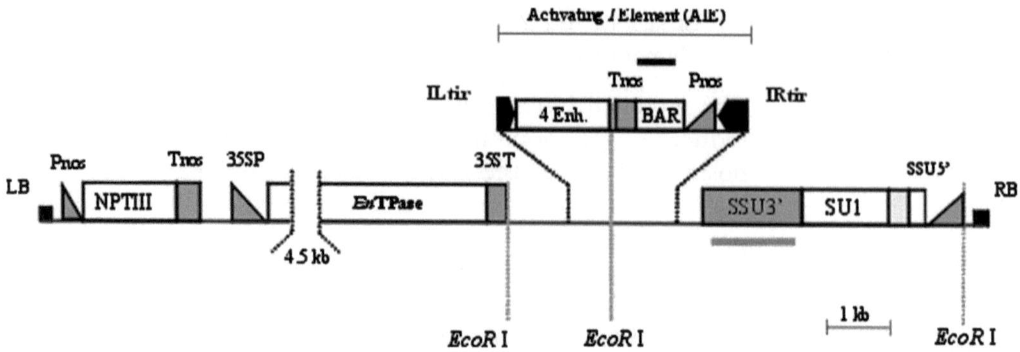

Fig. 1. Schematic representation of the WAT construct used for plant transformation. Relevant *Eco*RI sites used for Southern analysis are indicated. *LB* left border, *RB* right border; 35SP, 35ST, CaMV 35S promoter and terminator, respectively; *En*TPase, *En* immobile transposase source; ILtir, IRtir, *I*-element left and right terminal-inverted repeat, respectively; 4 Enh., tetramer of the CaMV 35S enhancer; Pnos, Tnos, promoter and terminator sequences from the nopalinesynthase gene, respectively; SSU5′, SSU3′, promoter and transit signal peptide to the chloroplast and terminator of the small subunit of *Rubisco* gene, respectively. The gene specific probes (*bar* and SSU3′) used for blot hybridization are indicated as bars above or below the figure. Reprinted from PLANT PHYSIOLOGY ONLINE by Marsch-Martinez, N. et al. Copyright 2002 by American Society of Plant Biologists. Reproduced with permission of American Society of Plant Biologists in the format Textbook via Copyright Clearance Center.

1.4. Procedure Outline

This procedure describes the steps required from generating a population of stable insertions to the identification of genes, and includes:

1. Selecting the best genotypes using a double selection assay and calculating stable transposition and independent transposition frequencies.
2. Generating collections of stable insertions using T2, T3, or F2 seed (from a cross to WT).
3. Identifying mutants, analyzing mutation segregation, and locating transposon insertions.
4. Gene identification, first assessing gene overexpression and then testing using recapitulation assays.

2. Materials

1. First transformants containing the WAT construct.
2. 0.7 mL/L Finale (commercial formulation – Aventis – that contains 150 g/L glufosinate ammonium) (see Note 1).
3. 100 μg/L R7402 (see Note 2), with Silwet L-77 50 μL/L.
4. Materials for Genomic DNA isolation (see Note 3).
5. Materials for Southern blot hybridization.

6. *Eco*RI restriction enzyme and buffer (see Note 4).
7. Agarose.
8. TAE buffer.
9. Standard gel electrophoresis materials.
10. Hybond N + membrane.
11. Labeled *bar* fragment (see Note 5).
12. TAIL-PCR Materials (conventional PCR components and TAIL-PCR degenerate oligos) (34–36).

The specific nested primers for the *I* element are:

First TAIL-PCR: **Int2** 5′-CAG GGT AGC TTA CTG ATG TGC G-3′,

Second TAIL-PCR: **Irj-201** 5′-CAT AAG AGT GTC GGT TGC TTG TTG-3′, and

Third TAIL-PCR: **DSpm1** 5′-CTT ATT TCA GTA AGA GTG TGG GGT TTT GG-3′ (30).

Sequencing primer: **Itir 3** 5′-CTT ACC TTT TTT CTT GTA GTG-3′ (see Note 6).

13. Materials for RNA isolation (see Note 7).
14. DNAse I.
15. Materials for RT-PCR.
16. Specific oligos for candidate genes and a control gene (i.e. actin, or tubulin).
17. Standard molecular biology materials to make recapitulation constructs.
18. Standard materials for *E. coli* and *A. tumefaciens* transformation.
19. Materials for plant transformation & transformant selection.

3. Methods

3.1. Selecting Best Genotypes (Stable Transposition vs. Independent Transposition Frequencies)

3.1.1. Double Selection Assay and Calculation of Stable Transposition

1. Start from 20 to 30 first transformants (T1) containing the WAT construct. Let them grow in the best conditions to maximize seed set (see Note 8).
2. Sow a known quantity of progeny seed (T2 seed) in undivided trays filled with soil. Sow by dispersing the dry seed uniformly over the soil (see Note 9).
3. Give a 3-night cold treatment to the seed before transferring to the greenhouse (see Note 10). Keep trays covered with a transparent lid or plastic for 3 days, and remove the cover afterwards.
4. Transfer to the greenhouse.

5. Eight days after transference, spray seedlings with a mix of 0.7 mL/L Finale and 100 μg/L R7402. For the next 8 days, keep on spraying R7402 every day, and only spray Finale three times in this period (see Note 11).
6. Five to seven days after the last spray, carefully transfer resistant plants to new soil (pots or trays) with individual separations. Some morphological phenotypes can already be identified at this stage (see Note 12).
7. Count the number of doubly resistant progeny plants from each genotype and calculate the stable transposition frequency (STF) as:

$$\text{Stable transposition frequency} = \frac{\text{Number of double resistant plants} \times 100}{\text{Number of sown seed}}$$

3.1.2. Calculation of Independent Transposition Frequency

1. Isolate genomic DNA either from single or pooled double resistant plants.
2. Digest at least 500 ng (for single plants), or 1 μg (for up to ten pooled plants) genomic DNA with *Eco*RI, run in an electrophoresis gel and transfer to a Hybond N+membrane (see Note 13).
3. Perform a Southern hybridization using a bar fragment as labeled probe (shown above the bar gene in Fig. 1).
4. The *bar* probe will allow visualizing different *I/dSpm* transpositions in the genome (Fig. 2 shows an example of pooled progeny plants having different insertions with varying frequencies). Use the number of independent insertions to calculate the frequency of independent insertions from each new transformant (see Notes 20 and 21). The independent insertion frequency (ITF) is calculated as:

$$\text{Independent transposition frequency} = \frac{\text{Number of different visualized instructions} \times 100}{\text{Number of progency plants assayed}}$$

Lines showing the lowest stable transposition frequency (STF, number of plants that survive the double selection assay), together with the highest independent transposition frequency (ITF, many different insertions in a group of plants), are the best ones to build a population. STF should be less than 20 plants in 1,000 seeds. ITF should be at least 50%, but there are lines that reach more than 100% (meaning as many different insertions as plants in the pool), and even higher when some plants in the pools contain more than one insertion, which is different from any insertion in other plants.

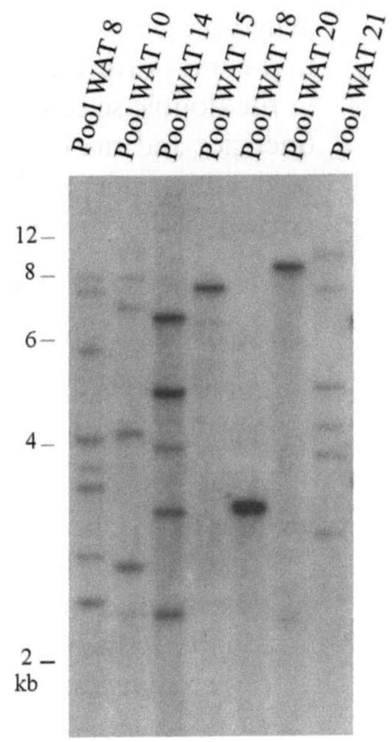

Fig. 2. Main steps to be followed to generate a population of stable insertions. The process starts from the first transformants (T_1) containing the construct shown in Fig. 1. Several rounds of stable insertion selections can be carried out using either self or F_2 seed as parental lines. Reprinted from PLANT PHYSIOLOGY ONLINE by Marsch-Martinez, N. et al. Copyright 2002 by American Society of Plant Biologists. Reproduced with permission of American Society of Plant Biologists in the format Textbook via Copyright Clearance Center.

3.2. Generating Large Collections of Stable Insertions (New Double Selection Rounds)

If enough T2 seed from the best genotypes is available, new rounds of selection can be performed as indicated in Subheading 3.1.1. However, to generate large collections of stable insertions using pre-selected best genotypes, it is necessary to use T3 or F2 seed. All the parental plants should be checked for presence of positive and negative selectable markers. Moreover, the double selection assay will be more efficient if heterozygous *SU1* parentals are pre-selected.

3.2.1. Using T3 Seed Directly (Exemplified by Step 3 in Fig. 3) (see Note 14)

1. Handle T2 seeds as in Subheading 3.1.1, but sow seeds separately (not as a bulk).
2. Spray 0.7 mL/L Finale over the seedlings at days 8 and 11, after transference to the greenhouse.
3. Screen basta resistant plants visually for the presence of the *SU1* marker, which confers a dwarf, dark green, reduced apical dominance phenotype.
4. Grow basta resistant, *SU1* positive plants in the best conditions to achieve maximum seed set.
5. To generate new stable transpositions, use T3 seeds from single parentals to perform the double selection (basta and R7402) treatments as indicated in Subheading 3.1.1 (see Notes 15 and 16). Only transfer double resistant plants from parentals with low stable transposition frequencies (lower than 5%).

3.2.2. Using F2 Seed

Selfed progeny can harbor "fixed" insertions that occurred early in the transformed lines or initial generations, and that will appear in the double selection assay, leading to high stable insertion frequencies and masking new transposition events. The use of F2 seed reduces this problem. Moreover, all F1 parentals are heterozygous, avoiding the need of checking for *SU1* heterozygosity.

1. Handle as in Subheading 3.2.1, steps 1–3.
2. Cross dwarf, basta resistant T2 plants to wild type.
3. Grow F1 seeds and select dwarf, basta resistant plants again.
4. Grow these plants in the best conditions to achieve maximum seed set.
5. Use F2 seed from single parentals to perform the double selection (basta and R7402) treatments as indicated in Subheading 3.1.1 [exemplified by step 4 in Fig. 3 (see Note 16)]. Only transfer double resistant plants from parentals with low stable transposition frequencies (lower than 5%).

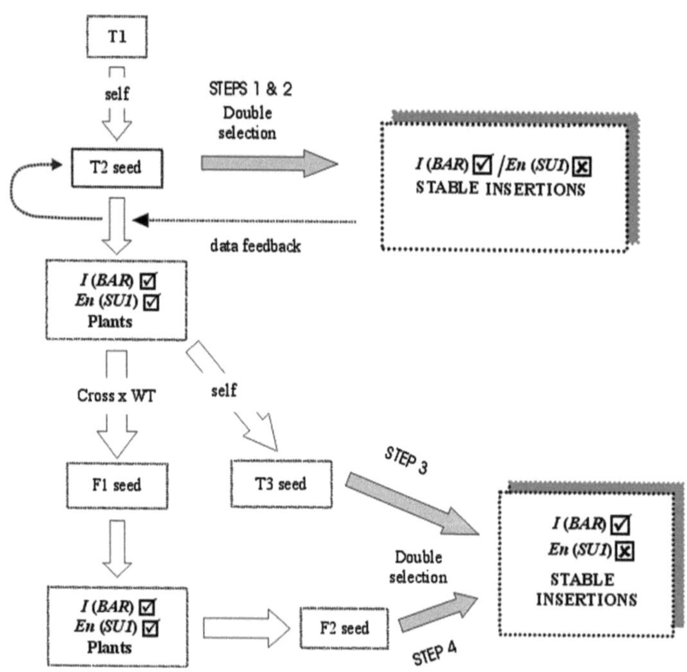

Fig. 3. Southern blot hybridization of double-resistant plants revealing stable inserts with a *bar* gene probe. Double-resistant progeny pools from different first transformants (WATs) are displayed. Each band shows an independent insertion. A ladder showing the size in kilobase pairs is indicated on the *left* side. The numbers of plants per pool are: WAT 8, seven plants; WAT 10, eight plants; WAT 14, six plants; WAT 15, eight plants; WAT 18, ten plants; WAT 20, eight plants; and WAT 21, nine plants. Reprinted from PLANT PHYSIOLOGY ONLINE by Marsch-Martinez, N. et al. Copyright 2002 by American Society of Plant Biologists. Reproduced with permission of American Society of Plant Biologists in the format Textbook via Copyright Clearance Center.

3.3. Mutant Identification and Establishment of Transposon Position

A variety of mutant phenotypes can be identified at different stages. Many leaf phenotypes can already be identified when transferring double resistant plantlets to new soil, and already be kept apart from the collection, while other phenotypes (i.e. stem length, flowering time, floral organ development) can be identified later (see Note 17). Activation tagging mutations are dominant. Therefore, have always wild type plants available to cross sterile plants when necessary and to perform genetic analysis. Moreover, collect tissue of plants with compromised or sterile phenotypes that cannot be crossed, so candidate genes causing the phenotype can be identified and assayed with recapitulation constructs (inducible when necessary) to recover the original phenotype.

3.3.1. Genetic Analysis

1. Once a mutant has been identified, backcross it to its wild type ecotype.
2. Perform genetic analyses to check phenotype segregation using both selfed seed and F1 from crosses to wild type plants.

3.3.2. Insertion Number Analysis

Mutants have single stable insertions in most cases, but this should be assessed (see Note 18). To evaluate insertion number:

1. Collect mutant tissue.
2. Isolate DNA and perform a Southern assay as in Subheading 3.1.2, steps 1–4.

3.3.3. Identification of Adjacent Sequences

Use the isolated DNA to perform a TAIL-PCR to find the position of the insert in the genome (see Notes). Use Int2 for the first, Irj-201 for the second, and DSpm1 for the third TAIL-PCR as specific nested primers (see Notes 19–21). For sequencing, use a fourth nested primer [Itir3 (see Note 6)].

3.3.4. Genomic Context Analysis

BLAST (37) the sequences to the *Arabidopsis* genome to find the position of the insert, and identify adjacent genes both up and downstream (see Note 22).

3.4. Gene Identification

3.4.1. Evaluation of Transcript Overaccumulation

1. Isolate total RNA from mutant and wild type plants (see Note 7).
2. Perform a DNAse I treatment.
3. Perform a Reverse transcription reaction.
4. Use the obtained cDNA as a template for PCR to evaluate transcript accumulation of candidate genes in mutant and wild type backgrounds (see Note 23).

3.4.2. Candidate Gene Evaluation

In some mutants, more than one gene can be overexpressed, and the phenotype can be the result of the overexpression of a single gene or a combination of two or more genes.

To evaluate this, use the 35S promoter to express the candidate gene(s) in the wild type background and observe whether it produces a comparable phenotype as the original mutant (recapitulation) (see Notes 24 and 25).

4. Notes

1. Finale can be purchased from Bayer (http://www.bayer.com). It contains 150 g/L glufosinate ammonium as active ingredient. Instead of Finale, other sources of active ingredient can be used adjusting to a similar final concentration. Pure phosphinotricin (PPT, Duchefa http://www.duchefa.com) can also be used at a concentration of 100 mg/L with Silwet L-77 50 μL/L (Lehle seeds http://www.arabidopsis.org), or triton 0.01%.
2. Dupont (http://www.dupont.com) owns the rights for the use of the *SU1* gene, and should be contacted for use of the gene and R7402. To make the stock R7402 solution, dissolve the powder in 10 mM KOH (or other dilute base), and then add water.
3. The DNA isolation protocol should produce good quality DNA to be used for Southern blot hybridization. As a DNA Isolation method we use the protocol published by Pereira and Aarts (38), a modified version of the method by Liu et al. (34) using either 1.5 mL tubes or 96 tube-racks for high-throughput isolation. Good quality DNA can be normally obtained from young flower buds. Green, relatively young, mature leaves in good conditions are also a good tissue to isolate good quality DNA.
4. High concentration enzyme (i.e. 50 U/μL).
5. The bar fragment can be obtained by PCR in bacteria, or plant DNA containing the binary vector or the T-DNA respectively, using the following primers:

   Bar F1: 5′-ACC ATG AGC CCA GAA CGA CGC-3′

   Bar R1: 5′-CAG GCT GAA GTC CAG CTG CCA G-3′
6. The use of a fourth primer to sequence the product obtained in the third TAIL-PCR increases the reliability of the result.
7. For mRNA transcripts, RNA can be isolated using Trizol, LiCl protocols, or commercially available RNA isolation kits. For microRNAs or small RNAs, special steps should be considered in some of these conventional protocols.
8. The WAT construct is large. For transformation, the aggressive *Agrobacterium* strain AGL0 worked well to obtain the

desired number of transformants in the Wassiliewskija ecotype, which was used successfully to build a large collection of stable Activation tag insertions. It is convenient to avoid cultivating *Agrobacterium* cultures containing the construct for many generations because the enhancer tetramer can recombine and loose copies. It is better to use direct cultures from transformed agrobacterium or from a glycerol stock (9). Transformed plants can be selected in medium (0.8% agarose, 1/2 MS medium supplemented with 50 mg/L kanamycin, or 10 mg/L PPT), or sown in soil and selected by spraying 0.7 mL/L Finale. Some transformants will not be useful due to, i.e. high T-DNA copy number or unfavorable T-DNA location, leading to high early or too low transposition frequencies. Having many initial lines allows to find suitable starter lines, and represent more chances of having starting insertions in different locations in the genome, increasing the likelihood of a better dispersion of inserts (linked transposition events will be eliminated in the double selection rounds). However, in principle, two good starting lines with unlinked original T-DNA insertions are enough to build a population.

9. Seed quantity can be calculated from the weight. Depending on the ecotype, 1 mg contains around 50 seeds. About 1,000–3,000 seeds can be sown in a tray of about 35×45×7 cm. (filled with a ~5 cm soil layer). Seeds can also be sown by resuspending in a 0.1% agarose solution, and dispersing evenly using a large (i.e. 20 mL) pipette.
10. Depending on seed age, a cold treatment helps to synchronize germination and therefore improves efficiency of the double selection assay. The treatment can be given to seeds before sowing, in a regular fridge, or to seeds sown in trays by placing the trays in a cold room.
11. An alternative to sowing in soil and spraying is to select double resistant plants in media. Plates should be supplemented with 10 mg/L PPT (Duchefa), and/or 1 μg/L R7402 (Dupont). PPT alone can be supplemented to conventional 1/2 MS, 0.8% w/v agar, 1% w/v sucrose medium. With R7402, sucrose should be replaced by 100 mg/L myoinositol.
12. Transfer plantlets together with soil using large tweezers to take out, and plant directly in the new soil. Resistant plants bear non-autonomous transposon insertions, but no T-DNA (and consequently no transposase), and therefore are considered to contain stable insertions.
13. Digest at least 4 h (overnight for best results) with 1 μL of *Eco* R I (50 U/μL) per sample. Electrophorese in a 0.8% w/v agarose gel in 1× TAE buffer (40 mM Tris-acetate and 1 mM EDTA).

Instead of using Southern blot hybridization, the independent transposition frequency can also be calculated by performing Transposon Insertion Display (30), or another strategy that allows high-throughput visualization or sequencing of transposon insertions.

14. Taking the first transformant as T1, its first progeny seed are T2, and the third generation T3.
15. Alternatively, sow T2 seed as a bulk and transfer basta resistant plantlets separately to evaluate the *SU1* phenotype. It is advisable to perform a segregation analysis to identify heterozygous plants for the T-DNA before using their T3 seed for the next stable transposition selection round, since progeny T3 seed from T2 homozygous individuals will not survive the double selection. However, when space is not limiting, F3 seed from all plants can be used directly for the large double selection assay.
16. Avoid mixing T3 or F2 seed from different parentals, progeny seed from single plants should be used to generate stable insertions. In this way, plants where an insertion has been fixed in unlinked positions to the T-DNA will not "contaminate" a pool of seed. These plants can be identified as having high stable transposition frequencies and should not be considered, or just very few double resistants should be transferred to new pots and used for the population.
17. Stable insertions can also be selected in medium as indicated in Note 11 when mutant screenings require it.
18. When using the WAT construct in *Arabidopsis*, most plants containing stable insertions had only one insertion (around 70%), and the rest had two (around 20%), or three insertions (around 10%) (13). If the identified mutant has more than one insertion, it is necessary to cross it to the wild type to segregate the different insertions to ease analysis. If this is not possible, then obtain the flanking sequences for the different insertions and assay candidate genes using recapitulation constructs.
19. If the third TAIL-PCR reaction produces a single, specific band, purify the reaction and sequence it directly. If the third reaction also produces other unspecific bands, then touch the specific band with a sterile tip or toothpick, and introduce it in a tube containing a third TAIL-PCR mix with the corresponding degenerate oligo. This should help to get a single, specific band that can be sequenced. For this re-PCR, the electrophoresis buffer should be TBE.
20. Other techniques to isolate flanking sequences can be used, for example, the walk PCR technique (39), or even high-throughput transcriptomics approaches.

21. In many cases, the activation tag does not cause constitutive expression, but increased expression of the native expression pattern. Therefore, to be able to detect overexpression, use tissues showing a mutant phenotype.
22. Consider genes located from the insertion up to 8 kb further to each side. In most cases, the responsible genes are the nearest to the insertion to either side. Different kinds of genes can be activated, including transcription factors, enzymes, and even microRNA precursors among others. In some cases, the gene identity makes it a better candidate for a specific phenotype.
23. Alternatively, northern blotting can also be used to detect overexpression.
24. There are three main strategies to recapitulate the mutant phenotype in wild type plants.
    (a) Using the 35S promoter to overexpress candidate genes.
    (b) Cloning the whole genomic region comprised from upstream the enhancer tetramer until the end of the candidate gene.
    (c) Using the candidate gene together with its native promoter, and add the 35S enhancer tetramer.
25. The use of the 35S promoter to overexpress a gene can have stronger effects than using only the enhancer tetramer in combination with native regulatory regions as in the other two options. Sometimes, a softer effect is more desirable. For example, the constitutive overexpression of certain genes will produce lethal or compromised phenotypes.

References

1. The Arabidopsis Genome Initiative. (2000) Analysis of the genome sequence of the flowering plant *Arabidopsis thaliana*. *Nature* **408**, 796–815.
2. Bouché, N., and Bouchez, D. (2001) *Arabidopsis* gene knockout: phenotypes wanted. *Current Opinion in Plant Biology* **4**, 111–17.
3. Kuromori, T., Wada, T., Kamiya, A., Yuguchi, M., Yokouchi, T., Imura, Y., Takabe, H., Sakurai, T., Akiyama, K., Hirayama, T., Okada, K., and Shinozaki, K. (2006) A trial of phenome analysis using 4000 Ds-insertional mutants in gene-coding regions of *Arabidopsis*. *Plant Journal* **47**, 640–51.
4. Walden, R., Fritze, K., Hayashi, H., Miklashevichs, E., Harling, H., Schell, J. (1994) Activation tagging: a means of isolating genes implicated as playing a role in plant growth and development. *Plant Molecular Biology* **26**, 1521–28.
5. Lewin, B. (Ed.) (2008) Genes IX, Jones and Bartlett Publishers Inc., Sudbury, MA, USA.
6. Odell, J. T., Nagy, F., and Chua, N. H. (1985) Identification of DNA sequences required for activity of the cauliflower mosaic virus 35 S promoter. *Nature* **313**, 810–12.
7. Neff, M. M., Nguyen, S. M., Malancharuvil, E. J., Fujioka, S., Noguchi, T., Seto, H., Tsubuki, M., Honda, T., Takatsuto, S., Yoshida, S., and Chory, J. (1999) BAS1: a gene regulating brassinosteroid levels and light responsiveness in *Arabidopsis*. *Proceedings of the National Academy of Sciences of the United States of America* **96**, 15316–23.
8. van der Graaff, E., Dulk-Ras, A. D., Hooykaas, P. J., and Keller, B. (2000) Activation tagging

of the LEAFY PETIOLE gene affects leaf petiole development in *Arabidopsis thaliana*. *Development* **127**, 4971–80.

9. Weigel, D., Ahn, J. H., Blazquez, M. A., Borevitz, J. O., Christensen, S. K., Fankhauser, C., Ferrandiz, C., Kardailsky, I., Malancharuvil, E. J., Neff, M. M., Nguyen, J. T., Sato, S., Wang, Z. -Y., Xia, Y., Dixon, R. A., Harrison, M. J., Lamb, C. J., Yanofsky, M. F., and Chory, J. (2000) Activation tagging in *Arabidopsis*. *Plant Physiology* **122**, 1003–14.
10. Borevitz, J. O., Xia, Y., Blount, J., Dixon, R. A., and Lamb, C. (2000) Activation tagging identifies a conserved MYB regulator of phenylpropanoid biosynthesis. *Plant Cell* **12**, 2383–94.
11. Huang, S., Cerny, R. E., Bhat, D. S., and Brown, S. M. (2001) Cloning of an *Arabidopsis* patatin-like gene, STURDY, by activation T-DNA tagging. *Plant Physiology* **125**, 573–84.
12. Kakimoto, T. (1996) CKI1, a histidine kinase homolog implicated in cytokinin signal transduction. *Science* **274**, 982–85.
13. Marsch-Martinez, N., Greco, R., Van Arkel, G., Herrera-Estrella, L., and Pereira, A. (2002) Activation tagging using the En-I maize transposon system in *Arabidopsis*. *Plant Physiology* **129**, 1544–56.
14. Schneider, A., Kirch, T., Gigolashvili, T., Mock, H. -P., Sonnewald, U., Simon, R., Flügge, U. -I., and Werr, W. (2005) A transposon-based activation-tagging population in *Arabidopsis thaliana* (TAMARA) and its application in the identification of dominant developmental and metabolic mutations. *FEBS Letters* **579**, 4622–28.
15. Wilson, K., Long, D., Swinburne, K., and Coupland, G. (1996) A dissociation insertion causes a semidominant mutation that increases expression of TINY, an *Arabidopsis* gene related to APETALA2. *Plant Cell* **8**, 659–71.
16. Ayliffe, M. A., and Pryor, A. J. (2007) Activation tagging in plants – generation of novel, gain-of-function mutations. *Australian Journal of Agricultural Research* **58**, 490–97.
17. Busov, V. B., Meilan, R., Pearce, D. W., Ma, C., Rood, S. B., and Strauss, S. H. (2003) Activation tagging of a dominant gibberellin catabolism gene (GA 2-oxidase) from poplar that regulates tree stature. *Plant Physiology* **132**, 1283–91.
18. Jeong, D. -H., An, S., Kang, H. -G., Moon, S., Han, J. -J., Park, S., Lee, H. S., An, K., and An, G. (2002) T-DNA insertional mutagenesis for activation tagging in rice. *Plant Physiology* **130**, 1636–44.
19. Mathews, H., Clendennen, S. K., Caldwell, C. G., Liu, X. L., Connors, K., Matheis, N., Schuster, D. K., Menasco, D. J., Wagoner, W., Lightner, J., and Wagner, D. R. (2003) Activation tagging in tomato identifies a transcriptional regulator of anthocyanin biosynthesis, modification, and transport. *Plant Cell* **15**, 1689–703.
20. Qu, S., Desai, A., Wing, R., and Sundaresan, V. (2008) A versatile transposon-based activation tag vector system for functional genomics in cereals and other monocot plants. *Plant Physiology* **146**, 189–99.
21. Zubko, E., Adams, C. J., Machaekova, I., Malbeck, J., Scollan, C., and Meyer, P. (2002) Activation tagging identifies a gene from *Petunia hybrida* responsible for the production of active cytokinins in plants. *Plant Journal* **29**, 797–808.
22. Ayliffe, M. A., Pallota, M., Langridge, P., and Pryor, A. J. (2007) A barley activation tagging system. *Plant Molecular Biology* **64**, 329–47.
23. Chalfun-Junior, A., Mes, J., Mlynárová, L., Aarts, M., and Angenent, G. C. (2003) Low frequency of T-DNA based activation tagging in *Arabidopsis* is correlated with methylation of CaMV 35 S enhancer sequences. *FEBS Letters* **555**, 459–63.
24. Pereira, A. (2000) A transgenic perspective on plant functional genomics. *Transgenic Research* **9**, 245–60.
25. Aarts, M. G., Corzaan, P., Stiekema, W. J., and Pereira, A. (1995) A two-element enhancer-inhibitor transposon system in *Arabidopsis thaliana*. *Molecular and General Genetics* **247**, 555–64.
26. Bancroft, I., and Dean, C. (1993) Transposition pattern of the maize element Ds in *Arabidopsis thaliana*. *Genetics* **134**, 1221–9.
27. Matsuhara, S., Jingu, F., Takahashi, T., and Komeda, Y. (2000) Heat-shock tagging: a simple method for expression and isolation of plant genome DNA flanked by T-DNA insertions. *Plant Journal* **22**, 79–86.
28. Zuo, J., Niu, Q. -W., and Chua, N. H. (2000) An estrogen receptor-based transactivator XVE mediates highly inducible gene expression in transgenic plants. *Plant Journal* **24**, 265–73.
29. Banks, J. A., Masson, P., and Fedoroff, N. (1988) Molecular mechanisms in the developmental regulation of the maize Suppressor-mutator transposable element. *Genes Dev* **2**, 1364–80.
30. Tissier, A. F., Marillonnet, S., Klimyuk, V., Patel, K., Torres, M. A., Murphy, G., and Jones, J. D. (1999) Multiple independent defective suppressor-mutator transposon insertions in *Arabidopsis*: a tool for functional genomics. *Plant Cell* **11**, 1841–52.

31. De Block, M., Botterman, J., Vanderwiele, M., Dockx, J., Thoen, C., Gossele, V., Movva, R. N., Thompson, C., Montagu, V. M., and Leemans, J. (1987) Engineering herbicide resistance in plants by expression of a detoxifying enzyme. *EMBO Journal* **6**, 2513–18.
32. Thompson, C. J., Movva, N. R., Tizard, R., Crameri, R., Davies, J. E., Lauwereys, M., and Botterman, J. (1987) Characterization of the herbicide-resistance gene bar from *Streptomyces hygroscopicus*. *EMBO Journal* **6**, 2519–23.
33. O'Keefe, D. P., Tepperman, J. M., Dean, C., Leto, K. J., Erbes, D. L., and Odell, J. T. (1994) Plant expression of a bacterial cytochrome P450 that catalyzes activation of a sulfonylurea pro-herbicide. *Plant Physiology* **105**, 473–82.
34. Liu, Y. G., Mitsukawa, N., Oosumi, T., and Whittier, R. F. (1995) Efficient isolation and mapping of *Arabidopsis thaliana* T-DNA insert junctions by thermal asymmetric interlaced PCR. *Plant Journal* **8**, 457–63.
35. Liu, Y. G., and Whittier, R. F. (1995) Thermal asymmetric interlaced PCR: automatable amplification and sequencing of insert end fragments from P1 and YAC clones for chromosome walking. *Genomics* **25**, 674–81.
36. Tsugeki, R., Kochieva, E. Z., and Fedoroff, N. V. (1996) A transposon insertion in the *Arabidopsis* SSR16 gene causes an embryo-defective lethal mutation. *Plant Journal* **10**, 479–89.
37. Altschul, S. F., Madden, T. L., Schaffer, A. A., Zhang, J., Zhang, Z., Miller, W., and Lipman, D. J. (1997) Gapped BLAST and PSI-BLAST: a new generation of protein database search programs. *Nucleic Acids Research* **25**, 3389–402.
38. Pereira, A., and Aarts, M. G. M. (1997) *in* "Arabidopsis Protocols" (Martinez-Zapater, J. M., and Salinas, J., Eds.), Vol. 82, Springer.
39. Balzergue, S., Dubreucq, B., Chauvin, S., Le-Clainche, I., Le Boulaire, F., de Rose, R., Samson, F., Biaudet, V., Lecharny, A., Cruaud, C., Weissenbach, J., Caboche, M., and Lepiniec, L. (2001) Improved PCR-walking for large-scale isolation of plant T-DNA borders. *Biotechniques* **30**, 496–8, 502, 04.

Chapter 9

Activation Tagging and Insertional Mutagenesis in Barley

Michael A. Ayliffe and Anthony J. Pryor

Abstract

The process of activation tagging in plants involves the random distribution of plant regulatory sequences throughout the genome. The insertion of a regulatory sequence in the vicinity of an endogenous gene can alter the transcriptional pattern of this gene resulting in a mutant phenotype that arises from excess functional gene product. Activation tagging has been undertaken extensively in a number of dicot plants and also in rice. This has been achieved primarily by high-throughput plant transformation using T-DNA sequences that encode regulatory elements. Apart from rice, most cereals do not have a suitably efficient transformation system for high-throughput transformation. In this article, we detail an activation tagging system in barley that exploits the mobility of the maize *Ac/Ds* transposable element system to distribute a highly expressed promoter throughout the barley genome. The advantage of this approach in this species is that a relatively small number of primary transgenics are required to generate an activation tagging population. Insertion of this transposable element into genes can also generate insertional inactivation mutants enabling both gene overexpression and gene knockout mutants to be identified in the same population.

Key words: Activation, Tagging, Barley, Transposon, Ac, Ds

1. Introduction

In spite of the ever increasing number of sequenced plant genomes, the functions of genes identified in these sequencing efforts are generally unknown or inferred at best. A large number of genes have no assigned function while many genes are assigned function based solely upon sequence homology. This situation is exemplified in rice were the function of only a few genes have been biologically confirmed (1). Homology-based descriptions are also often of limited use. For example, the sequence identification of a gene encoding a transcription factor may give no indication of the processes this gene product contributes to, while the

Andy Pereira (ed.), *Plant Reverse Genetics: Methods and Protocols*, Methods in Molecular Biology, vol. 678, DOI 10.1007/978-1-60761-682-5_9, © Springer Science+Business Media, LLC 2011

identification of a conserved domain in an enzyme does not indicate the particular substrate of that enzyme.

Mutation remains one of the primary methods in elucidating gene function. Mutagenesis using chemical and radiation mutagens has been very successfully employed for many years; however, the isolation of genes mutated by these processes remains laborious in forward genetic screens. In contrast, the development of TILLING technologies has revolutionised the use of these mutagens in reverse genetics by enabling the rapid, targeted identification of mutant alleles in mutagenised populations (2). To improve forward genetic screens, increasingly sophisticated insertional mutation systems have been developed by using DNA elements, such as T-DNA sequences (3–8) or transposable elements (9–13), to inactivate genes. The resultant mutations are readily cloned by using the insertion sequence as a molecular tag. Insertional mutagenesis has subsequently advanced to include gene traps, promoter traps, trans-activation systems and activation tagging systems. These systems can also be utilised in reverse genetics by isolating genomic sequences that flank new DNA insertions and constructing flanking sequence databases. Plant lines containing an insertion sequence in a gene of interest can then be identified by BLAST searches (reviewed in 14).

Activation tagging systems can be used in both forward and reverse genetic screens. In forward screens, these systems differ from insertional mutagenesis in that they rely on gene overexpression as opposed to gene inactivation (reviewed in 15). Gene overexpression is achieved by incorporating promoter or enhancer sequences into DNA elements that are then distributed throughout the genome either by exploiting transposon mobility or using high-throughput T-DNA transformation. The insertion of a DNA element containing a regulatory sequence immediately 5′ of a gene can lead to gene overexpression which in turn can generate a mutant phenotype by the production of excess functional gene product. Alternatively, insertion of these DNA sequences into a gene ORF can generate mutations through insertional inactivation. As for other DNA insertion elements, large databases of flanking sequences can be constructed to enable the identification of a plant line containing an insertion in a gene of interest. Alternatively, these flanking sequence databases in, combination with a genome sequence, enable the identification of genes in the close vicinity of an activation element. Potentially, these genes may be overexpressed by the localised insertion of this activation element enabling reverse genetics activation tagging (16).

We have developed an activation tagging system in barley that utilises a maize *Ds* transposable element that contains two highly expressed cereal promoter sequences (17). Insertion of this element into the genome leads to high levels of adjacent sequence transcription at 75% of insertion sites. Insertion of this element into genes leads to gene inactivation (17) while insertion 3′ of a gene can generate silencing mutations through the production of anti-sense transcripts (18). In all cases, the insertion site is molecularly tagged making subsequent cloning a relatively simple process. For reverse genetic screens, in addition to generating flanking sequence databases to identify mutations in or near known genes, we propose a novel reverse genetic strategy for identifying overexpression mutants of known genes.

2. Materials

1. Glasshouse facilities for growing transgenic barley plants – 16 h light, 21°C/8 h dark, 16°C, 15 cm pots of soil:compost mix (1:1), aquasol liquid fertiliser applied fortnightly.
2. Tissue culture facilities for generating transgenic plants – laminar flow cabinets, tissue culture growth rooms at 24°C and a 12 h light/dark cycle.
3. Barley cultivar Golden Promise plants as a source of scutellum tissue for transformation. Plants are grown in growth cabinets at a constant 13°C 16 h light/8 h dark light regime.
4. Standard molecular biological reagents for PCR, sequencing, cloning, DNA extraction and DNA blotting.
5. For mutant screening – 10×15 in. paper sheets (Hoffman Manufacturing Inc), metal trays 80 cm×60 cm containing a 50:50 mix of soil and perlite, 0.2% BASTA solution in water containing a suitable wetting agent (Hoechst Ltd.).
6. For agarose gel electrophoresis – agarose, 1XTAE buffer (0.04 M Tris-acetate, 0.001 M EDTA), power pack (300 V, 400 mA) and electrophoresis apparatus, 10 mg/ml ethidium bromide DNA stain, UV transilluminator.
7. For DNA blot analysis – Amersham Hybond N+membrane, 0.4 M NaOH transfer solvent, 2× SSC (300 mM NaCl, 30 mM sodium citrate pH 7.0) buffer for membrane rinsing, 3 mm filter paper for solvent wick, paper towel for capillary action, ultraviolet crosslinker (e.g. UVP, USA).
8. For DNA hybridisation – hybridisation buffer (7% SDS, 1% BSA, 0.5 M $NaHPO_4$ pH 7.2, 1 mM EDTA), washing buffers (2× SCC/0.1% SDS, 1× SSC/0.1% SDS, 0.5× SSC/0.1%SDS), 65°C hybridisation oven capable of sample rotation (e.g. ThermoHybaid), bottles for membrane incubation (e.g. Hybaid HB-OV-BM), shaking 65°C incubator for high stringency washing, X-ray film and cassettes (e.g. Kodak Biomax) and X-ray film developing facilities, membrane stripping solution (5 mM Tris-HCl pH 8/0.1% SDS).

9. For labelling of DNA probes – ^{32}P-dCTP, Megaprime™ DNA labelling system (Amersham), sepharose G50 beads for the separation of unincorporated nucleotides, appropriate facilities for handling and disposing of radioactive compounds.
10. For GUS histochemical staining – staining buffer (0.1 M NaPO4 pH 7.0, 10 mM EDTA, 10% methanol, 200 μM potassium ferricyanide, 200 μM potassium ferrocyanide, 0.75 mg/ml of X-gluc substrate, 0.01% silwet), vacuum pump and bell jar for stain infiltration, 100% ethanol for tissue decolourisation following staining.
11. For RNA reverse transcription – reverse transcriptase and 10× buffer supplied by the manufacturer, 0.5 mM deoxyribonucleotide triphosphates (dATP, dCTP, dTTP, dGTP), appropriate DNA primer, 37°C incubator.
12. For PCR reactions –thermostable DNA polymerase and 10× buffer supplied by the manufacturer, 0.5 mM deoxyribonucleotide triphosphates (dATP, dCTP, dTTP, dGTP), appropriate DNA primers, PCR thermocycler and appropriate tubes. PCR purification columns (Qiagen, Germany).

3. Methods

3.1. Construct Design

Figure 1 shows a schematic representation of the modified *Ds* element (*UbiDs*) and transposase source (*Ubi-transposase*) we have used in barley and other grasses (i.e. wheat and maize). The general features of these constructs are applicable to the design of other *Ds*-based activation tagging systems. These features are as follows:

1. The *UbiDs* element was constructed by genetic modification of the *Ac* element present in the maize waxy gene (*Ac wx-m9*) (Genbank K01964) (19).
2. It is essential to maintain a minimum size of *Ac* terminal inverted repeats for efficient levels of transposition (20). The *UbiDs* construct employs *Ac* 5′ and 3′ termini of 270 bp and 222 bp, respectively.
3. *UbiDs* sequences derived from *Ac wx-m9* are as follows. Nucleotide positions are provided for the reverse complement of this sequence (Genbank K01964). 270 bp 5′ *Ac wx-m9* terminus (nucleotides 103–373), the 909 bp internal *Ac EcoRI/HindIII* fragment (nucleotides 2610–3519) and a 222 bp 3′*Ac wx-m9* terminus (nucleotides 4469–4691). The 909 bp *AcXE* fragment was placed in juxtaposition to the *BAR* gene to increase the distance between the inverted repeat created by the two polyubiquitin promoter sequences.

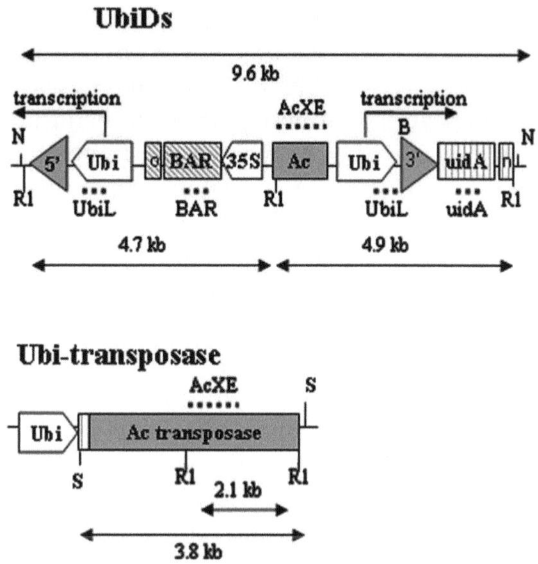

Fig. 1. *UbiDsGUS* and *Ubi-transposase* constructs. The *UbiDs* construct includes 270 bp and 222 bp of the 5′ and 3′ terminus of the maize *Ac* element found in the waxy gene (*Ac wx-m9*). These sequences correspond to nucleotides 103–373 and 4469–4691 of the published *Ac wx-m9* sequence (GenBank K01964 reverse complement) and are indicated by *grey triangles*. Also included in the *UbiDs* construct is a 909 bp *Eco*RI/*Hin*dIII fragment from the *Ac wx-m9* element (nucleotides 2610–3519, reverse complement). In juxtaposition to each *Ac* terminus is a 1.8 kb fragment derived from the maize polyubiquitin gene (labelled Ubi) that includes the promoter, 5′ untranslated leader sequence and first intron of this gene (GenBank J944464). The direction of transcription of these promoter elements is indicated. Also contained in the *UbiDs* element is a *BAR* gene and *OCS* terminator sequence (obtusely *hatched boxes*) under the regulatory control of a CaMV 35S promoter (labelled 35S). Immediately, adjacent to the *UbiDs* element is the ORF of a *uidA* reporter gene and *NOS* transcriptional termination sequence (*vertically hatched boxes*). The *Ubi-transposase* gene (*22*) consists of the same maize polyubiquitin promoter (labelled Ubi) fused to the *Ac* transposase coding sequence and extending to the 3′ terminus of the transposon (nucleotides 1093–489 of *Ac wx-m9* (reverse complement) with an *E. coli omega* transcriptional enhancer sequence inserted between the polyubiquitin promoter and first exon of the transposase coding sequence (*small white box*). For both constructs, *Eco*RI (RI), *Not*I (N) and *Sac*I (S) restriction sites are indicated as is the size of relevant restriction sites. The binding sites of AcXE, UbiL, uidA and BAR probes are shown as *dotted lines*.

4. The maize polyubiquitin promoter (21) (GenBank J944464) was placed at each end of *UbiDs* to enable the adjacent transcription from both ends of the element. These 2 kb sequences includes the ubiquitin promoter, the 5′ untranslated leader sequence (5′ UTR) and the first intron of the gene. The leader sequence encoded on this fragment terminates at the last base 5′ of the translation initiation codon of the polyubiquitin gene. Transcriptional read-through by this promoter generates transcripts that encode the adjacent flanking sequence in addition to the *Ac* terminus and polyubiquitin 5′ untranslated leader sequence (17) (see Note 1).

5. The inclusion of a *35SBAR* gene in the *UbiDs* element enables the selection of *Ds* containing plants by herbicide application. *UbiDs* plants can be selected in large populations by leaf painting with the herbicide BASTA. The adaxial surface of barley leaves are thoroughly wetted by painting with a 0.2% BASTA solution in water containing a suitable wetting agent (Hoechst Ltd.).
6. A promoterless *uidA* gene (the uidA protein product converts a chemical substrate, 5-bromo-4-chloro-3-indoxyl-beta-D-glucuronide, into a blue GUS stain) derived from plasmid pBI101.2 (CloneTech) was cloned as a *Sma/Eco*RI fragment. The promoterless reporter gene in this position is used to confirm the ability of the polyubiquitin regulatory sequence in *UbiDs* to transcriptionally activate an adjacent gene by GUS staining. In addition, plants containing the *UbiDsGUS* element can be rapidly identified by GUS staining while evidence of DNA transposition can be observed in somatic tissue sectors by an absence of GUS staining.
7. The *Ubi-transposase* gene (Fig. 1) has been previously described (22) and encodes an *Ac transposase* protein. It consists of the maize *Ac* element with the first 965 bp of this element replaced with a maize polyubiquitin promoter, including 5′ leader and first intron, in addition to an *Escherichia coli* omega transcriptional enhancer sequence (see Note 2). The polyubiquitin promoter transcribes the transposase coding sequence encoded within the *Ac* element.

3.2. Production of Barley Transgenics

Barley transgenics were made essentially as described by Tingay et al. (23) using *Agrobacterium*-mediated transformation of scutellum tissue except transgenics were selected based upon the resistance to the antibiotic hygromycin. The barley cultivar Golden Promise was used for transformation.

3.3. Molecular Analysis of T0 Barley Plants: The Identification of Plants with Full Length Transgenes and Copy Number Estimates

The restriction endonucleases used in the following protocols are diagnostic for the *UbiDsGUS* and *Ubi-transposase* transgenes. The basic principles of this analysis are applicable to any two component *Ds/transposase* system, however, diagnostic enzymes will differ for other constructs.

3.3.1. Identifying T0 Plants with Full Length UbiDsGUS Transgenes

1. Digest T0 barley DNAs and a control Golden Promise DNA with *Not*I.
2. Run DNAs on 0.8% agarose gels using 1× TAE buffer for 16 h at 40 V.
3. Transfer digested DNAs to a Hybond N+ membrane (Amersham Life Science, UK) by alkaline transfer as described by the manufacturer. Transfer proceeds for 4 h.

Table 1
Probes used for hybridisation

Probe name	Sequence description	Primers used for PCR amplification	Target sequence(s)
AcXE	909 bp *Hin*dIII/*Eco*RI fragment corresponding to reverse complement nucleotides 2610–3519 of *Ac wx-m9*	M13 forward and reverse primers used to amplify a cloned fragment	*UbiDs* element and *Ubi-transposase* gene
BAR	Nucleotides 91–534 of the *Streptomyces hygroscopicus BAR* gene sequence (GenBank X17220)	BAR1 and BAR2	*UbiDs* element
uidA	Nucleotides 2498–3095 of the *uidA* gene sequence from *pBI101.2* (GenBank U12668)	uidA1 and M13 reverse. A *uidA* gene cloned into BlueScript was used as a template	Promoterless *uidA* gene flanking the *UbiDs* element
UbiL	Nucleotides 918–1973 of the maize polyubiquitin gene which encode the leader sequence and first intron (GenBank S94464)	UbiL1 and UbiL2	Polyubiquitin promoters in *UbiDs* and *Ubi-transposase*
hyg	Nucleotides 3099–4017 of the hygromycin phosphotransferase gene ORF that in addition contains a catalase intron (GenBank AY225224.1)	Hyg1 and Hyg2	Hygromycin selectable marker gene used for barley transformation

4. Prepare a radioactively labelled probe consisting of the 909 bp AcXE fragment (Fig. 1) using a Megaprime DNA labelling kit. The probe template is prepared by PCR amplification of the AcXE fragment cloned in BlueScript using M13 forward and reverse primers (Tables 1 and 2). This PCR product (100 μl) is purified using Qiagen PCR purification columns as recommend by the manufacturer (Qiagen, Germany). 200 ng of purified PCR product is labelled by random hexamer priming using P^{32}-dCTP and the klenow fragment of DNA polymerase I. Unincorporated nucleotides are removed using sepharose G50 columns.
5. Transferred DNA and membranes are crosslinked using an ultraviolet crosslinker, and then rinsed three times for 10 min in a 2× SSC solution.
6. Membranes are pre-hybridised at 65°C for 4 h.
7. The DNA probe is boiled for 2 min before adding to the pre-hybridisation solution.

Table 2
Primer sequences

Primer	Sequence
M13 forward	CGCCAGGGTTTTCCCAGTCACGA
M13 reverse	AGCGGATAACAATTTCACACAGGA
BAR1	GTCTGCACCATCGTCAACC
BAR2	GAAGTCCAGCTGCCAGAAAC
uidA1	GTTCGGCGTGGTGTAGAG
UbiL1	GTTCGGAGCGCACACACA
UbiL2	AACAGGGTGAGCATCGAC
Hyg1	GCGCGTCTGCTGCTCCATACA
Hyg2	GAACTCACCGCGACGTCTGTC

8. After hybridisation overnight at 65°C, DNA membranes are washed for 15 min at 65°C in 2× SCC/0.1% SDS, followed by 1× SSC/0.1% SDS and finally 0.5× SSC/0.1%SDS.
9. Membranes are air dried, wrapped in plastic and exposed to X-ray film (Biomax, Kodak).

An intact *UbiDsGUS* transgene is identified as a 9.6 kb fragment (see Figs. 1 and 2a). Additional hybridising bands not of this size indicate partial transgene integrants (Fig. 2a).

An alternative to *Not*I digestion is as follows;

1. T0 DNAs are restricted with *Eco*RI, separated by agarose gel electrophoresis and hybridised as described above with an AcXE probe (Table 1).
2. After autoradiography membranes are stripped of the AcXE probe by two applications of a 500 ml boiling solution of membrane stripping buffer.
3. The stripped membrane is rinsed in 2× SSC, and then re-hybridised as described above with a BAR specific probe (Table 1).

For transgenic plants containing a complete *UbiDsGUS* transgene, the AcXE probe detects a 4.9 kb *Eco*RI restriction fragment (described later in Fig. 5), while the BAR probe detects a 4.7 kb restriction fragment. The presence of both fragments *Eco*RI in the same DNA sample is consistent with the presence of a full length transgene. Additional hybridising fragments indicate partial transgene insertions in the genome.

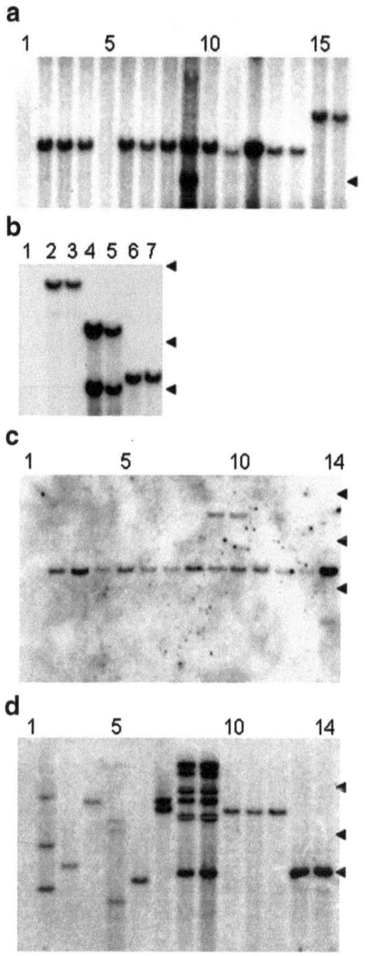

Fig. 2. Screening *UbiDsGUS* and *Ubi-transposase* barley trangenics for intact transgenes and copy number. (**a**) Barley T0 transgenic plant DNAs were restricted with *Not*I and hybridised with the AcXE probe. A common fragment of 9.6 kb indicates an intact transgene. DNA in lane 9 indicates that this plant contains at least one full length transgene and one truncated transgene. Lanes 15 and 16 contain DNAs from clonal plants (i.e. two re-generants from the same callus) that contain a single truncated *UbiDsGUS* transgene. The *arrow head* indicates a molecular weight mobility of 8.5 kb. (**b**) Estimation of *UbiDsGUS* transgene copy number. DNAs were digested with *Bam*HI and hybridised with a *uidA* probe. DNAs in lanes 2+3, 4+5 and 6+7 were from clonal plants. DNAs in lanes 2+3 and 6+7 contain a single *UbiDsGUS* transgene insertion, while DNAs in lanes 4+5 contain two copies of the *UbiDsGUS* transgene. This blot does not confirm that the transgenes in these plants are full length; it only indicates the number of transgene integrations into the genome. (**c**) Identification of potentially full length *Ubi-transposase* transgenes in T0 barley lines. DNAs were restricted with *Sac*I and membranes hybridised with the AcXE probe. A conserved 3.8 kb fragment indicates an intact *Ac* coding sequence. DNAs in lanes 2+3, 6+7+8 and 9+10 were isolated from clonal plants. (**d**) Identification of probable transgene copy numbers in *Ubi-transposase* T0 barley lines. DNAs were restricted with *Sac*I and hybidised with a probe specific for the hygromycin selectable marker gene used for transformation. DNAs in lane 8+9, 10+11+12 and 13+14 were isolated from clonal plants. *Arrow heads* in *panels B–D* indicate molecular weight mobilities of 8.5, 4.8 and 3.6 kb, respectively. Lane 1 in each panel contains wild type Golden Promise DNA.

3.3.2. Estimating UbiDsGUS Transgene Copy Number

1. Restrict T0 DNAs with *Bam*HI and run on 0.8% agarose gel.
2. Transfer DNA to a nylon membrane and hybridise with a *uidA*-specific probe.

Each band indicates a separate *UbiDsGUS* transgene introduced during transformation (Fig. 2b). The presence of unique restriction fragments confirms the independence of transgenic lines.

3.3.3. Identifying T0 Plants with Potentially Full Length Ubi-transposase Transgenes and Estimating Copy Number

1. Restrict T0 DNAs with *Sac*I and run on 0.8% agarose gel.
2. Transfer DNA to a nylon membrane and hybridise with AcXE probe. A conserved 3.8 kb fragment indicates a full length *Ac transposase* ORF (see Fig. 2c). However, it does not indicate if the promoter sequence is also present.
3. Strip the membrane as described in Subheading 3.3.1 and re-hybridise with a probe specific for the hygromycin gene (Table 1). This gene lies immediately adjacent to *Ubi-transposase* on the binary vector and is used as a selectable marker for barley transformation. The number of bands homologous to the hygromycin probe indicates the copy number of the selectable marker gene and likely copy number of the *Ubi-transposase* transgene (see Fig. 2d).

3.4. Generating Hybrid Plants Containing UbiDsGUS and Ubi-transposase Transgenes

1. Cross barley lines containing full-length *UbiDsGUS* elements and *Ubi-transposase* transgenes to generate hybrid progeny.
2. Identify progeny containing both transgenes by DNA blot analysis of *Eco*RI restricted DNAs hybridised with the AcXE probe.

Hybrid plants containing a 2.1 kb fragment (*Ubi-transposase*) and 4.9 kb fragment (*UbiDsGUS*) homologous to this probe were identified as containing both transgenes and kept (Fig. 3a) (see Note 3). An alternatively approach is to generate homozygous *UbiDsGUS* and *Ubi-transposase* lines prior to making hybrid plants, thereby removing the necessity for the above molecular analysis of hybrid plants.

3.5. Detecting Somatic Transposition Events

Evidence of *UbiDs* transposition occurring in somatic tissues of F1 plants can be detected by DNA blot analysis and GUS staining.

3.5.1. DNA Blot Analysis

1. Restrict DNAs from hybrid plants with *Eco*RI.
2. Transfer DNAs to membranes and hybridise with a *uidA*-specific probe.

DNA transposition can be detected by the presence of a 1.8 kb *uidA* homologous fragment in hybrid plants that show

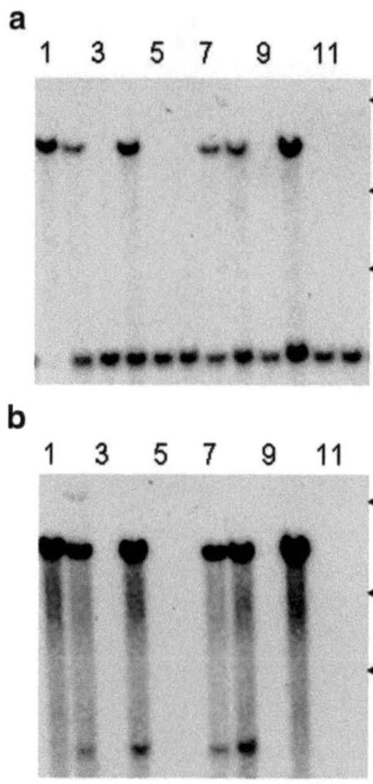

Fig. 3. *Evidence of UbiDs transposition in somatic cells of hybrid barley plants.* (**a**) DNAs (lanes 2–12) were isolated from barley seedlings derived from a cross between a single *UbiDsGUS* T0 parent (lane 1) and a plant homozygous for the *Ubi-transposase* transgene. DNAs were restricted with *Eco*RI and hybridised with the AcXE probe. All plants inherited the transposase gene (*lower band*) while only five plants inherited the *UbiDsGUS* transgene (*upper band*). (**b**) Hybridisation of the same membrane shown in **a** with a *uidA* probe identifies the same *UbiDsGUS* transposon fragment in these plants (*upper band*). An additional smaller band is apparent in four out of five of these hybrid plants that arises from transposition of the *UbiDs* element to leave behind a 1.8 kb fragment (*see* Fig. 1) that encodes the *uidA* ORF. This fragment is absent in the parent DNA shown in lane 1. The reduced stoichiometry of this 1.8 kb excision fragment is due to only some cells in each tissue containing a *Ds* excision event. *Arrow heads* in each panel indicate molecular weight mobilities of 8.5, 4.8 and 3.6 kb, respectively.

evidence of transposition (Fig. 3b). This DNA fragment (see Fig. 1) is apparent in somatic sectors of tissue in which the *UbiDs* element has excised. In tissues on the same plant in which *UbiDs* has not moved a 4.9 kb *Eco*RI fragment with homology to this probe remains (Fig. 3b) (see Note 4).

3.5.2. GUS Staining

Evidence of transposition activity in hybrid plants containing both transgenes can be detected by GUS staining due to the juxtaposition of the *UbiDs* element and the *uidA* ORF. Excision of *UbiDs* generates somatic sectors of tissue that no longer express *uidA* (Fig. 4a, b).

1. Small segments of leaf tissue (1–2 cm) are harvested in eppendorf tubes.
2. 1 ml of GUS staining solution is added and samples are placed under vacuum for 2 min before releasing the vacuum.

Fig. 4. GUS staining and mutant identification in *UbiDs* barley plants. (**a**) A GUS stained leaf segment from a *UbiDsGUS* T0 transgenic showing constitutive GUS expression (*dark grey tissue*). (Reprinted from ref. (17)). (**b**) Leaf segments from two hybrid barley plants containing the *UbiDsGUS* transgene and *Ubi-transposase* gene. Distinct sectors of staining (*dark grey*) and non-staining (*light coloured*) tissue are evident due to the excision of *UbiDs* in some tissue, thereby leaving behind a promoterless *uidA* ORF. The stripes associated with this staining are due to linear files of cell division during leaf development. (Reprinted from ref. (17)). (**c**) Paper roll method used for barley mutant identification. Seed were rolled up in paper that was then placed seed end upper most in a half filled beaker of water. Seed were germinated and grown for 10 days prior to photography. Unrolling the paper enables morphological inspection of both shoots and roots. (**d**) Trays of barley seedlings from the *UbiDs* activation tagging population for screening. (**e**) Identification of a barley dwarf mutant potentially tagged due to the presence of a new *UbiDs* insertion. On the right hand side of the figure is a DNA blot containing parental DNA and mutant DNA restricted with *Eco*RI and hybridised with the UbiL probe. A new *UbiDs* insertion in the mutant that is not apparent in the parent is highlighted with an *arrow*.

This process is repeated five times. Release of the vacuum forces staining solution into the leaf.

3. Samples are incubated overnight at 37°C.
4. Following incubation, the staining solution is removed followed by the addition of several changes of 100% ethanol to remove plant pigments. Tissue showing *uidA* activity can be then seen as blue GUS stained regions.
5. Samples can be stored for months to years in ethanol.

From the above analyses, it is possible to determine which transposon and transposase lines are likely to show the highest levels of activity in hybrid plants. However, confirmation in progeny of these hybrid plants is sought.

3.6. Identification of Newly Transposed UbiDs Elements in Progeny of Hybrid Plants

1. DNAs are isolated from progeny of hybrid plants containing both transposon and transposase transgenes, the original T0 *UbiDsGUS* and *Ubi-transposase* lines and Golden Promise control plants.
2. DNAs are restricted with *Eco*RI, separated by agarose gel electrophoresis and transferred to nylon membranes as described above.
3. Membranes are hybridised with a probe specific for the maize polyubiquitin leader sequence (Table 1) (see Note 5) or alternatively the AcXE or BAR probe.

As shown in Fig. 5, newly transposed *UbiDs* insertions are detected by the presence of new restriction fragments with homology to these probes, which are not present in original hybrid parent DNAs. A number of considerations are required when interpreting this data (see Note 6).

3.7. Mutant Identification: Forward Genetic Screening

Two methods have been undertaken for the phenotypic identification of barley mutants from this activation tagging population in forward genetic screens.

3.7.1. Paper Roll Method

1. Place 20 seed 3 cm from the top of a sheet of paper. Distribute seeds uniformly across the top of the sheet.
2. Roll the paper into a cylinder and place the paper cylinder in a beaker of water with the seed end upper most.
3. Allow seed to germinate for 10–14 days.
4. Remove paper cylinders from the beaker and unroll for morphological assessment of seedling shoots and roots (Fig. 4c).

Although rolling seed in paper cylinders is laborious, the advantage of this approach is the potential to identify morphological mutants in root development.

3.7.2. Seedling Tray Method

1. Sow approximately 800 seed per 80 cm × 60 cm tray and water.
2. Allow seed to germinate for 10–14 days (Fig. 4d).
3. Examine seedlings for morphological phenotypes. Remove wild type seedlings as they are screened to avoid confusion.

Families derived from potential mutant plants can also be screened in these trays with ten rows of approximately 25 seedlings per tray (see Notes 7 and 8).

3.8. Co-segregation of a Newly Transposed Ds and the Mutant Phenotype

Upon mutant identification, it is necessary to determine firstly if the mutant has a new, unique *Ds* insertion compared with the parent line (Fig. 4e).

1. Restrict putative mutant and parent DNAs with *Eco*RI and hybridise with either UbiL, AcXE or BAR probes to identify

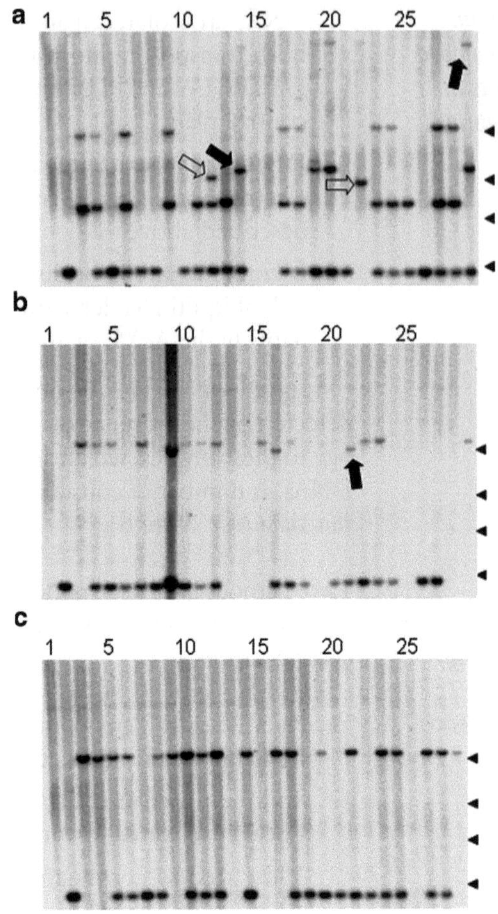

Fig. 5. *UbiDs* transposition patterns in progeny derived from plants that are hemizygous for *UbiDsGUS* and *Ubi-transposase*. Each panel contains DNA restricted with *Eco*RI and hybridised with the AcXE probe. DNAs were isolated from (1) Golden Promise, (2) a *Ubi-transposase* line, (3) a *UbiDsGUS* line while the remaining lanes contain DNAs from progeny derived from a *UbiDsGUS*/*Ubi-transposase* hybrid plant produced by crossing the plants whose DNAs are shown in lanes 2 and 3. *Panel a* shows a highly active *UbiDs* family, *panel b* shows a family containing a new *UbiDs* insertion inherited from the hybrid parent while *c* shows a family in which no activity is apparent, even though the same *Ubi-transposase* parent was used in each cross. *Ds* transpositions inherited from hybrid parents that are present in more than one sib are indicated with *black arrows*. Unique *Ds* transpositions present in a single plant are indicated with no-fill arrows. These latter insertions may have arisen in parental germline cells or early in plant development in somatic cells. *Arrow heads* indicate molecular weight sizes of 4.9, 3.6, 2.8 and 1.9 kb in each panel.

a newly transposed *Ds* element. Additional enzymes can be used to maximise the likelihood of detecting a new insertion event.

2. Cross the putative mutant (m) to wild type Golden Promise (+) (see Note 9).
3. Generate a selfed family from the hybrid plant (m/+) and score progeny for the mutant phenotype.

4. Extract DNAs from progeny, restrict with *Eco*RI and hybridise with the probe that detects the new *UbiDs* insertion.
5. Determine if the new *Ds* element co-segregates with the mutant phenotype in this family (see Notes 10 and 11).

3.9. Isolation of the UbiDs Insertion Site

Having shown co-segregation between the *UbiDs* insertion and the mutant phenotype, genomic sequence flanking the insertion is then sought.

1. Isolate sequences flanking the *UbiDs* insertion site by TAIL PCR as described by Liu et al. (24).
2. Use either the 5′ or 3′ flanking sequence as a probe on DNA blots (described in Subheading 3.8) that detect the new *UbiDs* insertion with one of the *UbiDs* specific probes (e.g. AcXE, UbiL and BAR).
3. Confirm that the flanking sequence probe detects that same restriction fragment as that detected by the above *Ds* specific probe. As the flanking sequence probe detects both an endogenous sequence and the new *UbiDs* insertion allele, it will indicate if the mutant is homozygous or heterozygous for the *Ds* insertion event (see Note 12).
4. Ensure that the fragment detected by the flanking sequence probe co-segregates with the mutant phenotype by re-hybridising the membranes described in Subheading 3.8.
5. Having confirmed the flanking sequence as *bona fide*, use it as a probe to screen a genomic library (e.g. Morex BAC library or lambda genomic library).
6. Sequence the region surrounding the *Ds* insertion site present in the library clone.
7. Analyse sequence to determine if the *UbiDs* insertion is located upstream, downstream or within a gene ORF (see Note 13).

3.10. Confirmation that a UbiDs Insertion is Responsible for a Mutant Phenotype

Additional analyses are required for proof that a new *Ds* insertion that co-segregates with a mutant phenotype is responsible for the mutation (see Note 14). The analysis that is most relevant will depend upon whether the mutation is likely to be caused by insertional gene inactivation, gene overexpression or gene silencing.

3.10.1. Identifying Reversion Events

Reversion events are most readily detected in progeny from selfed mutants if they are recessive mutations. Revertants of dominant mutations are more readily identified by testcrossing homozygous mutant plants.

1. Plant out a large number of seeds derived from the mutant plant, or mutant offspring, *that still contains the Ubi-transposase gene* (i.e. has not been lost by segregation).

2. Examine the mutant family for the presence of wild type revertant plants.
3. Extract DNA and undertake DNA blot analysis on mutant and wild type revertant sibs using the flanking sequence probe and enzyme combinations describe in Subheading 3.9.
4. Confirm the presence of at least one "wild type" allele in this revertant plant (a *Ds* excision footprint that does not disrupt gene function may be present in this allele so it may not be identical in sequence to the endogenous allele) (see Note 15).

3.10.2. Complementation

Complementation is only applicable for mutations arising from insertional gene inactivation.

1. Isolate the wild type sequence surrounding the *UbiDs* insertion site as described in Subheading 3.9, item 5.
2. Clone into a suitable barley transformation binary vector.
3. Transform the wild type sequence into the mutant barley plant.
4. Look for complementation of a wild type phenotype in subsequent transgenics.

3.10.3. Gene Overexpression

Gene overexpression is used to confirm mutants arising from gene activation.

1. Having identified a gene ORF adjacent to *UbiDs*, genetically engineer the wild type barley gene ORF to be under the regulatory control of the maize polyubiquitin gene promoter.
2. Insert this overexpression construct into a barley transformation binary.
3. Generate transgenic barley plants containing this construct.
4. Look for recapitulation of the mutant phenotype in those transgenic plants overexpressing the gene.
5. Confirm gene overexpression in these lines by RNA blot analysis.

3.10.4. Gene Silencing

Gene silencing is a strategy used to confirm mutants arising from the production of antisense transcripts by the *UbiDs* element.

1. Generate a hairpin construct (25) of the gene ORF that is in juxtaposition to the *UbiDs* insertion site.
2. Engineer the hairpin construct to be under the regulatory control of the maize polyubiquitin promoter.
3. Insert this RNAi construct into a barley transformation binary.
4. Generate transgenic barley plants containing the RNAi construct.

5. Examine transgenics for a re-capitulation of the mutant phenotype.
6. Confirm reduced levels of endogenous target gene expression by RNA blot analysis (see Note 16).

3.11. Mutant Identification: Reverse Genetics

In practice, we have not pursued reverse genetic strategies much in this barley population; however, it is certainly feasible to do so. Two methods are suggested for reverse genetic studies.

3.11.1. Isolation of Large Numbers of Barley Plants that Contain New UbiDs Insertions Sites

Large collections of stable trap lines (new *UbiDs*, no transposase) can be produced and flanking sequences rescued from these plants and sequenced. This produces a library of insertion mutants with known flanking sequences. This library can then be screened by BLAST searching the gene of interest sequence against the flanking sequence library to identify a plant line containing a *UbiDs* insertion in or near the gene.

1. Grow trays of 800 barley seedlings as described in Subheading 3.7.
2. Spray seedlings with BASTA herbicide to select only those seedlings containing a *UbiDs* element.
3. Stain a segment of surviving seedlings for GUS activity. Keep only those plants that are GUS negative.
4. Undertake PCR analyses on GUS negative seedlings with a primer pair specific for the *Ubi-transposase* gene. Include primers BAR1 and BAR2 as a positive PCR control. Keep those plants that are BAR positive and transposase negative. These plants should contain a newly transposed *UbiDs* element that has segregated away from the *Ubi-transposase* gene and is now stable.
5. Harvest and store seed individually from each plant.
6. Isolate flanking sequences from stable trap lines as described by Liu et al., (24), clone and sequence.

3.11.2. Identification of Overexpression Mutants of a Gene of Interest Among Seedling Pools

Transcripts initiated by the *UbiDs* element contain the 5′ polyubiquitin leader sequence at their 5′ end (17, 18), thereby making these transcripts unique in the plant cell (see Note 17). These unique transcripts can therefore be detected among large seedling populations using RNA pools and RT-PCR.

1. Grow 800 seedlings in large trays and spray with BASTA at the second leaf stage.
2. Harvest the top of all surviving seedling leaves and make a bulk RNA extraction (see Note 18) for each tray.
3. Undertake RT-PCR analysis on the RNA pools using a polyubiquitin leader sequence primer (Table 2, primer UbiL1) and a primer specific for the gene of interest.

4. Run PCR reactions on an agarose gel to identify an RNA sample derived from a tray in which a seedling is overexpressing the gene of interest, i.e. a PCR product is apparent.
5. GUS stain surviving seedlings in the tray that contains the seedling overexpressing the gene of interest.
6. Isolate RNA from individual seedlings that are GUS negative and screen by RT-PCR for the plant overexpressing the gene of interest (see Note 19).
7. Recover seed from the identified plant.

We have not undertaken this screening procedure ourselves; however, we have shown by RT-PCR that we can detect transcripts of specific flanking sequences that are expressed by an adjacent *UbiDs* element for a number of independent insertions (17). These novel transcripts can readily be detected by RT-PCR when this RNA is diluted 1,000-fold and in some cases 100,000-fold, suggesting the above technique is feasible for detecting an overexpressing plant among thousands of sibs.

4. Notes

1. An alternative strategy to using promoter sequences are enhancer sequences. In rice, multimers of 35S enhancer sequences have been successfully used to produce overexpression mutants (26, 27). Unlike entire promoter sequences, enhancers have been suggested to alter transcript levels rather than transcription patterns while constitutive promoters impose ectopic transcription patterns (28).
2. Rather than using an active *Ac* transposon as a source of transposase, this engineered transposase source is used as its inability to move, due to the absence of the *Ac* 5′ terminus, greatly simplifies subsequent genetic analyses and maintains the integrity of transposase genetic stocks.
3. At this stage of analysis, it is necessary to identify *Ubi-transposase* lines that promote the highest levels of *Ds* transposition and *UbiDsGUS* lines that are capable of transposition. Not all transgenic plants will contain transgenes capable of function, in spite of these genes being intact (Fig. 5a–c). This point is illustrated in Fig. 5c and presumably reflects T-DNA integration sites that are non-condusive for gene expression and transposition. A *Ubi-transposase* line that promotes the highest levels of *Ds* transposition is essential as only one or two of these lines are used as genetic stocks. In contrast, it is advantageous to have as many independent transgenics that contain *UbiDs* elements that are capable of transposition as *Ds* elements tend to jump to linked locations (29).

4. The stoichiometry of the 1.8 kb excision band (Fig. 3b) is reduced due to the somatic nature of this event. The advantage of targeting an excision band is that every *Ds* transposition will produce the same restriction fragment. Even minor somatic sectors will contribute to this band intensity.
5. *Eco*RI in combination with the UbiL probe is advantageous because the enzyme cuts the *UbiDs* element in half and each end of the *Ds* element is recognised by the probe (see Fig. 1). Each new *Ds* transposition event therefore generates two unique restriction fragments detected by this probe, thereby maximising the likelihood of detecting a new restriction fragment. Other probes that can be used in conjunction with *Eco*RI digests (AcXE and BAR probes) detect only a single restriction fragment from each *Ds* insertion event.
6. DNA isolated from progeny plants can exhibit a range of transposition phenotypes depending on the timing of transposition during development. Transposition events that occur in the zygote, in gametes or prior to gamete formation in the parent plant will exhibit the newly transposed *UbiDs* fragment in all tissues of the progeny plant. Transpositions that occur prior to gamete differentiation can produce multiple plants all with the same transposition (Fig. 5b). In those progeny that also inherit the transposase gene, there is the possibility of transpositions from both the original *UbiDs* location and secondarily from a new location leading to progeny plants containing somatic sectors of larger or smaller size again depending on timing. Such events could explain the pattern in Fig. 5a, where the primary transposition gave rise to a new 3.7 kb *UbiDs* band in several sibs but in one plant (lane 19) secondary transposition could account for the lower stoichiometry of this band relative to the 1.8 kb transposase band.
7. It is possible to spray these trays with BASTA herbicide to select only those plants that contain a *Ds* element among these segregating populations. However, we have found little advantage in this approach when screening for mutants as the bulk of the labour is in the initial seed planting and germination. However, this strategy is likely to be very useful when selecting for large numbers of wild type plants that contain new *Ds* insertions.
8. Both the paper roll and seedling tray screening methods are limited to early vegetative tissue. No fertilisers are required as seedlings are screened around the 2–3 leaf stage. Mutant screening of mature plants requires far more glasshouse space which is a difficulty if strict regulatory requirements for growing transgenic plants are in place as in Australia.
9. We routinely backcross any potential mutants to wild type barley in case the mutant plant is homozygous for the insertion element.

10. A family of 80–100 plants are analysed for co-segregation. Co-segregation of *UbiDs* and a mutant phenotype is encouraging, *but does not constitute definitive proof that the Ds insertion is responsible for the phenotype.*
11. The inheritance of the phenotype may indicate the type of mutation. Mutations arising from gene overexpression or gene silencing are likely to be dominant while insertional inactivation mutants are usually recessive. However, this is not always the case as we have demonstrated recessive inheritance of an activation tagging gene silencing mutant (18).
12. This assay depends on using a 3′ flanking sequence probe on a DNA blot that previously detected a new *UbiDs* insertion with a restriction enzyme/*Ds* probe combination that detected the 3′ end of the *Ds* element and vice versa when using a 5′ flanking sequence probe.
13. The location of the *UbiDs* insertion relative to a gene ORF will also give insight as to whether the mutant phenotype is due to gene overexpression, gene silencing or insertional gene inactivation. Potential expression based mutations can be further examined using RNA blots. Gene overexpression will result in increased transcript levels of the endogeneous gene, gene silencing will cause reduced levels of transcript accumulation.
14. We have detected close linkage between a mutation and a new *Ds* insertion which highlights the potential hazard of relying on co-segregation data in small families as causal proof.
15. Phenotypes caused by insertional inactivation of a gene are generally recessive. Restoration of one mutant allele by re-excision of the *Ds* element can restore gene function provided a DNA footprint is not left by the excising element that maintains gene inactivity. For revertants to be identified the initial mutant parent plant must maintain the transposase source in addition to the new *Ds* element. An absence of revertant plants among mutant progeny does not disprove that the *Ds* insertion is responsible for the mutant phenotype. *Ds* elements frequently insert into the genome with the loss of terminal *Ds* sequences, thereby rendering them incapable of further transposition.
16. Virus-induced gene silencing is an alternative strategy (30).
17. Only promoter based activation tagging elements generate these unique transcripts, enhancer-based activation tagging systems generate transcripts that are identical to endogenous transcripts.
18. Seedlings will regrow and can be rescued provided they are not cut below the apical meristem.

19. Rejection of GUS staining seedlings will lead to the loss of those plants that over express the gene of interest due to a new *Ds* insertion, but also contain an unjumped *UbiDs* element and hence are GUS positive. An alternative strategy would be to screen all BAR positive seedlings in the PCR positive tray by RT-PCR for gene overexpression.

References

1. Jung, K.-H., An, G., Ronald, P.C. (2008) Towards a better bowl of rice: assigning function to tens of thousands of rice genes. *Nat. Rev. Genet.* **9**, 91–101.
2. Gilchrist, E.J., Haughn, G.W. (2005) TILLING without a plough: a new method with applications for reverse genetics. *Curr. Opin. Plant Biol.* **8**, 211–215.
3. Koncz, C., Martini, N., Mayerhofer, R., Konzc-Kalman, Z., Korber, H., Redei, G.P., Schell, J. (1989) High frequency T-DNA-mediated gene tagging in plants. *Proc. Nat. Acad. Sci. U. S. A.* **86**, 8467–8471.
4. Feldmann, K.A. (1991) T-DNA insertion mutagenesis in *Arabidopsis*: mutational spectrum. *Plant J.* **1**, 71–82.
5. Krysan, P.J., Young, J.C., Sussman, M.R. (1999) T-DNA as an insertional mutagen in *Arabidopsis*. *Plant Cell* **11**, 2283–2290.
6. Jeon, J.-S., Lee, S., Jung, K.-H., Jun, S.-H., Jeong D.-H., Lee, J., Kim, C., Jang, S., Lee, S., Yang, K., Nam, J., An, K., Han, M.-J., Sung, R.-Y., Choi, H.-S., Yu, J.-H., Choi, J.-H., Cho, S.-Y., Cha, S.-S., Kim, S.-I., An, G. (2000) T-DNA insertional mutagenesis for functional genomics in rice. *Plant J.* **22**, 561–570.
7. Walden, R. (2002) T-DNA tagging in a genomics era. *Crit. Rev. Plant Sci.* **21**, 143–165.
8. An, G., Lee, S., Kim, S.-H., Kim, S.-R. (2005) Molecular genetics using T-DNA in rice. *Plant Cell Physiol.* **46**, 14–22.
9. Long, D., Martin, M., Sundberg, E., Swinburne, J., Puangsomlee, P., Coupland, G. (1993) The maize transposable element system *Ac/Ds* as a mutagen in *Arabidopsis*: identification of an albino mutation induced by *Ds* insertion. *Proc. Nat. Acad. Sci. U. S. A.* **90**, 10370–10374.
10. Izawa, T., Ohnishi, T., Nakano, T., Ishida, N., Enoki, H., Hashimoto, H., Itoh, K., Terada R., Wu, C., Miyazaki, C., Endo, T., Iida, S., Shimamoto, K. (1997) Transposon tagging in rice. *Plant Mol. Biol.* **35**, 219–229.
11. Upadhyaya, N.M., Zhou, X.-R., Zhu, Q.-H., Ramm, K., Wu, L., Eamens, A., Sivakumar, R., Kato, T., Yun, D.-W., Santhoshkumar, C., Narayanan, K.K., Peacock, J.W., Dennis, E.S. (2002) An *iAc/Ds* gene and enhancer trapping system for insertional mutagenesis in rice. *Funct. Plant Biol.* 29, 547–559.
12. Kim, C.M., Piao, H.L., Park, S.J., Chon, N.S., Je, B.I., Sun, B., Park, S.H., Park, J.Y., Lee, E.J., Kim, M.J., Chung, W.S., Lee, K.H., Lee, Y.S., Lee, J.J., Won, Y.J., Yi, G.H., Nam, M.H., Cha, Y.S., Yun, D.W., Eun, M.Y., Han, C.-D. (2004) Rapid, largescale generation of *Ds* transposant lines and analysis of the *Ds* insertion sites in rice. *Plant J.* **39**, 252–263.
13. Kolesnik, T., Szeverenyi, I., Bachmann, D., Kumar, C.S., Jiang, S., Ramamoorthy, R., Cai, M., Ma, Z.G., Sundaresan, V., Ramachandran, S. (2004) Establishing an efficient *Ac/Ds* tagging system in rice: large-scale analysis of *Ds* flanking sequences. *Plant J.* **37**, 301–314.
14. An, G., Jeong, D.-H., Jung, K.-H., Lee, S. (2005) Reverse genetic approaches for functional genomics of rice. *Plant Mol. Biol.* **59**, 111–123.
15. Ayliffe, M.A., Pryor, A.J. (2007) Activation tagging in plants – generation of novel, gain-of-function mutations. *Aust. J. Agric. Res.* **58**, 490–597.
16. Jeong, D.H., An, S., Park, S., Kang, H.G., Park, G.G., Kim, S.R., Sim, J., Kim, Y.O., Kim, M.K., Kim, S.R., Kim, J., Shin, M., Jung, M., An, G. (2006) Generation of a flanking sequence database for activation-tagging lines in japonica rice. *Plant J.* **45**, 123–132.
17. Ayliffe, M.A., Pallotta, M., Langridge, P., Pryor, A.J. (2007) A barley activation tagging system. *Plant Mol. Biol.* **64**, 329–347.
18. Ayliffe, M.A., Agostino, A., Clarke, B.C., Furbank, R., von Caemmerer, S., Pryor, A.J. (2009) Suppression of the barley *uroporphyrinogen III synthase* gene by a *Ds* activation tagging element generates developmental photosensitivity. *Plant Cell* **21**, 814–831.

19. Polhman, R.F., Fedoroff, N.V., Messing, J. (1984) The nucleotide sequence of the maize controlling element Activator. *Cell* **37**, 635–643.
20. Coupland, G., Plum, C., Chatterjee, S., Post, A., Starlinger, P. (1989) Sequences near the termini are required for transposition of the maize transposon *Ac* in transgenic tobacco plants. *Proc. Natl. Acad. Sci U. S. A.* **86**, 9385–9388.
21. Christensen, A.H., Sharrock, R.A., Quail, P.H. (1992) Maize polyubiquitin genes: structure, thermal perturbation of expression and transcript splicing, and promoter activity following transfer to protoplasts by elctroporation. *Plant Mol. Biol.* **18**, 675–689.
22. Kumar, S.C., Narayanan, K.K. (1997) Gene and enhancer trap constructs for isolating genetic regions from rice. *Rice Biotechnol. Q.* **31**, 17–18.
23. Tingay, S., McElroy, D., Kalla, R., Fieg, S., Wang, M., Thorton, S., Brettell, R. (1997) *Agrobacterium tumefacians*-mediated barley transformation. *Plant J.* **11**, 1369–1376.
24. Liu, Y.-G., Mitsukawa, N., Oosumi, T., Whittier, R.F. (1995) Efficient isolation and mapping of *Arabidopsis thaliana* T-DNA insert junctions by thermal asymmetric interlaced PCR. *Plant J.* **8**, 457–463.
25. Helliwell, C.A., Waterhouse, P.M. (2005) Constructs and methods for hairpin RNA-mediated gene silencing in plants. *Meth. Enzymol.* **392**, 24–35.
26. Mori, M., Tomita, C., Sugimoto, K., Hasegawa, M., Hayashi, N., Dubouzet, J.G., Ochiai, H., Sekimoto, H., Hirochika, H., Kikuchi, S. (2007) Isolation and molecular characterisation of a *Spotted leaf 18* mutant by modified activation tagging in rice. *Plant Mol. Biol.* **63**, 847–860.
27. Hsing, Y., Chern, C.G., Fan, M.-J., Lu, P.-C., Chen, K.-T., Lo, S.-F., Sun, P.-K., Ho, Sh.-L., Lee, K.-W., Wang, Y.-C., Huang, W.-L., Ko, S.-S., Chen, S., Chen, J.-L., Chung, C.-I., Lin, Y.-C., Hour, A.-L., Wang, Y.-W., Chang, Y.-C., Tsai, M.-W., Lin, Y.-S., Chen, Y.-C., Yen, H.-M., Li, C.-P., Wey, C.-K., Tseng, C.-S., Lai, M.-H., Huang, S.-C., Chen, L.-J., Yu, S.-M. (2007) A rice activation/knockout mutant resource for high throughput functional genomics. *Plant Mol. Biol.* **63**, 351–364.
28. Weigel, D., Ahn, J.H., Blazquez, M.A., Borevitz, J.O., Christensen, S.K., Fankhauser, C., Ferrandiz, C., Kardailsky, I., Malancharuvil, E.J., Neff, M.M., Nguyen, J.T., Sato, S., Wang, Z.-Y., Xia, Y., Dixon, R.A., Harrison, M.J., Lamb, C.J., Yanofsky, M.F., Chort, J. (2000) Activation tagging in *Arabidopsis*. *Plant Physiol.* **122**, 1003–1013
29. Dooner, H., Belachew, A. (1989) Transposition patterns of the maize *Ac* form by bz-M2 (*Ac*) allele in maize. *Genetics* **122**, 447–457.
30. Holzberg S., Brosio P., Gross C., Pogue, G.P. (2002) Barley stripe mosaic virus-induced gene silencing in a monocot plant. *Plant J.* **30**, 315–327.

Chapter 10

Methods for Rice Phenomics Studies

Chyr-Guan Chern, Ming-Jen Fan, Sheng-Chung Huang, Su-May Yu, Fu-Jin Wei, Cheng-Chieh Wu, Arunee Trisiriroj, Ming-Hsing Lai, Shu Chen, and Yue-Ie C. Hsing

Abstract

With the completion of the rice genome sequencing project, the next major challenge is the large-scale determination of gene function. A systematic phenotypic profiling of mutant collections will provide major insights into gene functions important for crop growth or production. Thus, detailed phenomics analysis is the key to functional genomics. Currently, the two major types of rice mutant collections are insertional mutants and chemical or irradiation-induced mutants. Here we describe how to manipulate a rice mutant population, including conducting phenomics studies and the subsequent propagation and seed storage. We list the phenotypes screened and also describe how to collect data systematically for a database of the qualitative and quantitative phenotypic traits. Thus, data on mutant lines, phenotypes, and segregation rate for all kinds of mutant populations, as well as integration sites for insertional mutant populations, would be searchable, and the collection would be a good resource for rice functional genomics study.

Key words: Chemical or irradiation induced mutants, Insertional mutants, Phenotype, Rice, Seed handling, Seed storage

1. Introduction

Classical genetics is usually based on screening of collections of mutant plants and isolating the mutated gene. However, in the postgenomics era, nonbiased large-scale phenotype monitoring of mutant collections is an efficient approach. For the rice studies, several vectors, including T-DNA (1, 2), *Tos17* (3), *Ac/Ds* (4, 5), and *En/Spm* (6), have been used to generate insertional mutants. For chemically or physically induced mutants, researchers have

Andy Pereira (ed.), *Plant Reverse Genetics: Methods and Protocols*, Methods in Molecular Biology, vol. 678, DOI 10.1007/978-1-60761-682-5_10,

used fast neutron (7), γ-ray (7), ethyl methanesuphonate (EMS, (7)), *N*-methyl-*N*-nitrosourea (MNU, (8)), and sodium azide (9) to induce mutations. Any one of the resulting mutant population can be used for a detailed phenomics study.

Here we describe methods for rice mutant phenotype studies, including the field preparation and management for growth of mutant lines. A field sampling sheet is provided to code for more than 60 traits, including overall growth condition, leaf color, leaf morphology, plant morphology, mimic response, tiller, heading date, flower, panicle, seed fertility, and seed morphology. Handling and storage of the collected seeds are also described. All the data collected can be used to create a user-friendly database for detailed phenomics study.

2. Materials

2.1. Field Preparation

1. For chemically or physically induced mutants and the *Tos17* insertional mutants, a regular field is used for growth and propagation. For other insertional mutant populations, an isolated field specific for genetically modified (GM) crops is used.
2. The GM field is surrounded by two layers of net: a 32-mesh net to 2 m from the ground, and a 24-mesh net to 5 m to reduce pollen spread from the field. A bird net with a mesh of 2 × 2 cm at the top covers the whole area. The entrance gate is lockable to fulfill the requirements of the GM field (see Note 1).
3. For both GM and non-GM fields, the field is divided into several regions. Each region is divided by ribs to allow for walking between the regions, and the wild-type rice variety is planted as border lines to serve as the control plants for measuring qualitative and quantitative traits. The main rib is broad to allow for the mechanical tractor working in the field. Field management, including the application of fertilizers and pesticides, is the same as that for a regular paddy rice field (see Note 2).

2.2. Rice Mutant Lines

1. Both *japonica* and *indica* varieties can be used (see Notes 3–5).
2. Because of the concern of heterogeneity, the seeds used are originally derived from one rice plant. From the seeds of that single plant, a permanent seed stock is generated by growing several generations. These seeds are used for transformation or mutagen treatment, as well as for controlled plants for phenomics study.

3. Methods

3.1. Field Management

1. The planting density is 25 × 25 cm, and all plants are derived by single-seed descent. For insertional mutant population, a 1-ha paddy field is divided into two areas; each of the two regions can hold approximately 4,000 M_0 plants and 2,500 M_1 lines (12 plants per line). Therefore, approximately one-eighth of the area is used for M_0 plants and the rest for M_1 plants. M_0 plants are often sent from the tissue culture laboratory and can be planted in the region closest to the entrance. The M_1 lines are planted in the rest of the area.
2. The M_0 seedlings with four to five leaves are transplanted to the field. For insertional mutants, leaf samples are collected at this stage for subsequent flanking sequence analysis. In addition, all seedlings should be tagged with a barcode before transplanting.
3. A total of 30 seeds per M_1 line are used for germination. The seedling phenotypes can be recorded at three-leaf stage, and 12 seedlings are then transplanted in the field in blocks of 3 × 4 plants. Do not perform selection during transplanting; transplant some healthy seedlings and some weaker ones, according to the ratio of these plants.
4. Prepare at least 1,000 purple rice seedlings during the same period. These are mutant lines such as IRGC accession number 66712, 62133, or the equivalent, which have purple leaf blade and sheath, but with similar plant growth and yield as the wild type. If some of the M_2 seedlings do not grow, transplant the purple rice in that position (see Note 6).
5. All the M_0 and M_1 plants are planted in a 25 × 25 cm array so that plants growing at unexpected locations can be recognized and discarded. Contaminated plants are removed at least once a week before the heavy tiller stage (i.e., the close of canopy). The empty position can be replaced by the purple rice plant.

3.2. Phenotype Scoring

1. A 1-ha paddy rice field requires two senior breeders and another four to five breeders to take care of the daily screening, recording, and field management.
2. We recommend a numerical code system for phenotype scoring. The phenotypes are divided into 11 categories of 61 subcategories (10).
3. For the M_0 plants, the phenotypes may be recorded and used as a reference. Some growth defects may be caused by tissue culture or mutagen treatment and thus are not heritable.
4. For the M_1 plants, the seedlings are scored for phenotype (Subheading 3.3, item 6) before transplanting.

5. Breeders can start the M_1 plants' phenotype scoring about 1 month after transplanting, at the early tillering stage. We recommend the use of a recording sheet for each mutant line (Table 1). Information about cropping season, mutant ID, quantitative trait loci, and the 61 subcategories are listed in Table 1. The breeders should check the mutant lines once a week or every 2 weeks, record the phenotypes according to the subcategory code numbers, and write notes if necessary. Examples of some phenotype traits are shown in Figs. 1 and 2. Many other examples were presented in our previous paper (10) (see Note 7).
6. The mutant traits segregate in the M_1 population. Thus, the information for the 3×4 plants for each line should be recorded separately.
7. The 12 M_1 plants may have several mutant subgroups. The sampling sheet allows for four subgroups (i.e., wild type and subgroups B–D). Once the subgroups are well classified, their position in the 3×4 array is indicated on the datasheet (see Note 8).
8. About 1 week before harvesting, three important agronomic quantitative traits – heading date, plant height, and panicle number – of each subgroup in mutant lines are recorded. The data for the wild-type plants grown in border lines are also recorded. We usually record the data for four plants in each subgroup and four wild-type plants in each block (see Note 8).
9. All the data are stored in a database. A website may be constructed with all the data collected so that the line number, phenotype traits (quality and quantity traits), flanking sequence, and segregation ratio can be used as parameters in the search engine (see Note 9).

3.3. Seed Handling

1. Seeds from each M_0 plants are individually harvested. For the M_1 population, the seeds from the plants with the same phenotype subgroup should be harvested together.
2. The total seed weight before and after wind selection is recorded. The yield of each mutant line can be estimated by total seed weight/plant number.
3. The harvested seeds are transferred to a quarantined head house for cleaning and drying. Seeds are cleaned and selected by hand to eliminate unfilled and bad seeds. After cleaning the seeds, seed lots are transferred to a seed drying room under 20 ± 2.5°C and 8–10% relatively humidity (RH) to reduce the seed moisture content (Fig. 3). When the seed water content drops to 7%, the seeds are immediately transferred to a seed packing room under 20 ± 2.5°C and 50 ± 3% RH.

Table 1
Phenotype sampling sheet

Cropping season	○	○	○	○	Notes
ID	○	○	○	○	
Field ID	○	○	○	○	
	WT	*Type B*	*Type C*	*Type D*	
Panicles (#)					
Heading days					
Height (cm)					
Phenotype					
Development					(1) Germination rate, (2) Lethal, (3) Abnormal plants, (4) Weak
Leaf color					(11) Albino, (12) Yellow leaf, (13) Dark green leaf, (14) Pale green leaf, (15) Bluish green leaf, (16) Stripe, (17) Zebra, (18) Others
Leaf morphology					(21) Wide leaf, (22) Narrow leaf, (23) Long leaf, (24) Short leaf, (25) Drooping leaf, (26) Rolled leaf, (27) Spiral leaf, (28) Brittle leaf/culm, (29) Thin lamina joint, (30) Withering, (31) Others
Plant type					(41) Semidwarf, (42) Dwarf, (43) Extremely dwarf, (44) Long culm, (45) Erect, (46) Spread-out, (47) Thin culm, (48) Thick culm, (49) Lazy
Lesion					(51) Lesion mimic
Tiller					(61) High tiller position, (62) Low tiller position, (63) Monoculm, (64) Few panicle, (65) Many panicle
Heading date					(71) Early heading, (72) Late heading, (73) No heading
Glume					(75) Abnormal hull, (76) Abnormal floral organ, (77) With awn, (78) Abnormal hull, (79) Abnormal hull color
Panicle					(81) Long panicle, (82) Short panicle, (83) Sparse panicle, (84) Dense panicle , (85) Vivipary, (86) Shattering, (87) Neck leaf, (88) Abnormal panicle shape , (89) Others
Fertility					(91) Sterile, (92) Low fertility
Grain					(101) Large grain, (102) Small grain, (103) Slender grain, (104) Others

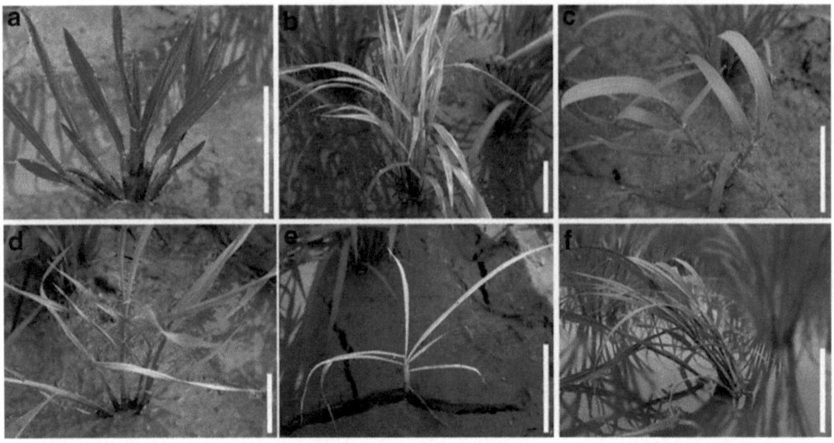

Fig. 1. Examples of variation in young plant morphology. (**a**) Wide leaf, short leaf; (**b**) pale green leaf, lesion mimic; (**c**) drooping leaf; (**d**) spiral leaf; (**e**) albino; (**f**) abnormal plant growth. Bar = 10 cm in each panel.

Fig. 2. Examples of variation in panicle morphology. (**a**) Wild type; (**b**) small grain, dense panicle, short panicle; (**c**) sparse panicle; (**d**) abnormal panicle; (**e**) small grain, dense panicle. Bar = 1 cm in each panel.

4. All the materials are registered and labeled with a barcode.
5. The 10-seed weight of each subgroup is measured, with three duplicates. Total seed numbers can be estimated by using the formula "total seed weight/average 10-seed weight × 10".
6. The seed length, width, thickness, and kernel color of each line are recorded, with three duplicates and ten seeds for each duplicate. For germination and seedling test, three duplicates, with ten seeds for each duplicate, are kept in a growth chamber (day time temperature 30°C, night time temperature 20°C) and scored 14 days later. Germination rate, seedling lethal rate, seedling height, root length, and special characters are recorded. Photographs of seeds and seedlings with specific morphology are taken. Examples of some seed traits were shown previously (10).

Fig. 3. Stacks of seeds in the drying room.

7. All the information are stored in a database so that information about seed length, width, and height; ratio of seed length to width; germination rate; and the average weight of 10 seeds, for example, are searchable.

3.4. Seed Storage

1. The M_1 seeds are packed into aluminum cans, which are labeled with a barcode, for storage in a long-term storage room under −12 ± 2°C and 30 ± 3% RH.
2. For M_2 seeds, 30 seeds are packed into aluminum foil bags. Bags are packed into an aluminum can for storage in a medium-term storage room under 1 ± 2°C and 40 ± 3 RH, ready for distribution. The remaining seeds are then packed in several bags and stored in a long-term storage room.

4. Notes

1. The regulation of GM plants differs by country. Thus, the GM field practice should be adapted for each country.
2. Pay attention to field management. For instance, the fertilizer should be evenly distributed so that the differences in plant growth and yield between wild-type and mutant plants can be interpreted correctly.
3. The genome sequence information for one *japonica* variety, Nipponbare, and one *indica* variety, 93-11, are available (11, 12),

and the SNP rates of several varieties versus Nipponbare are relatively low (13, 14). Thus, the integration site for each insertional mutant line can be allocated to the rice chromosome for most of the *japonica* and *indica* varieties.

4. Rice is a short-day plant, with a critical day length of approximately 15 h. In addition, the critical day length differs for different varieties. Nipponbare, the variety used for the international genome sequencing work, is sensitive to both temperature and day length. To obtain enough seeds in a reasonable period, the growth condition must be carefully controlled. Alternatively, varieties not sensitive to environment can be used.
5. We use an elite local *japonica* rice variety, Tainung 67 or TNG67, to generate the T-DNA tagged population in Taiwan (1). This variety is insensitive to both temperature and photoperiod, and sets seed in a reasonable time (4 months), so it can grow for two cropping seasons each year. Thus, use of a rice variety insensitive to photoperiod and temperature, such as TNG67, doubles the efficiency of field utilization, and does not require additional artificial light for the promotion or prevention of heading.
6. Purple rice plants should have growth rates similar to that of the wild type. The use of these plants allows for (1) determining each mutant in the block easily and (2) eliminating the position effect caused by larger growth spaces.
7. For the T-DNA-tagging population we work with, about 18% of the T_1 lines show at least one clearly visible mutant phenotype under normal condition. Each line with obvious mutated phenotypes contains a mean of three mutated phenotypes (range 1–12) (10). Thus, the detailed phenotype scoring is very important.
8. For a T-DNA-tagged population, the insertion copies are 1–4 for T-DNA and 0–3 for *Tos17* (1, 15). For the *Tos17*-tagged population, the mean insertion sites are ten (3). In addition, the insertional mutants contain many somaclonal variations (16). Mutagen-induced mutant populations contain even more mutation sites (17). Thus, each line of the M_1 population will have several mutant subgroups.
9. Rice mutant phenotype databases are available for the mutant populations *Tos17*-tagged Nipponbare (3), T-DNA-tagged Nipponbare (18), TNG67 (10), or Zhonghua 11 (19) and for chemically and irradiation-induced IR64 mutant population (4). Each group uses different descriptions for mutant traits. A unified vocabulary for plant structure ontology was recently suggested (10, 20, 21). The comparison among these groups is available at http://ipmb.sinica.edu.tw/soja/rice/phenomics_comparison/. Development of a cross-talk or even a unified vocabulary should be accelerated so that the mutant traits from different groups may be compared.

Acknowledgments

The authors acknowledge the contributions from Drs. Richard Bruskiewich, International Rice Research Institute, and Chih-Wei Tung, Cornell University, about the phenotype terms of IRRI, PO, PATO, and TO shown in the supplementary table at http://ipmb.sinica.edu.tw/soja/rice/phenomics_comparison/. We also acknowledge Ms. Laura Heraty for critical review of this manuscript. This work was supported by grants from Academia Sinica and the Taiwan National Science Council to CGC, MJF, SCH, SMY, and YICH.

References

1. Hsing, Y. I., Chern, C. G., Fan, M. J., Lu, P. C., Chen, K. T., Lo, S. F., et al. (2007) A rice gene activation/knockout mutant resource for high throughput functional genomics. *Plant Mol. Biol.* **63**, 351–364.
2. Jeong, D. H., An, S., Kang, H. G., Moon, S., Han, J. J., Park, S., et al. (2002) T-DNA insertional mutagenesis for activation tagging in rice. *Plant Physiol.* **130**, 1636–1644.
3. Miyao, A., Tanaka, K., Murata, K., Sawaki, H., Takeda, S., Abe, K., et al. (2003) Target site specificity of the *Tos17* retrotransposon shows a preference for insertion within genes and against insertion in retrotransposon-rich regions of the genome. *Plant Cell* **15**, 1771–1780.
4. He, C., Day, M., Lin, Z., Duan, F., Li, F., and Wu, R. (2007) An efficient method for producing an indexed, insertional mutant library in rice. *Genomics* **89**, 532–540.
5. Upadhyaya, N. M., Zhu, Q. H., Zhou, X. R., Eamens, A. L., Hoque, M. S., Ramm, K., et al. (2006) *Dissociation* (*Ds*) constructs, mapped *Ds* launch pads and a transiently-expressed transposase system suitable for localized insertional mutagenesis in rice. *Theor. Appl. Genet.* **112**, 1326–1341.
6. Kumar, C. S., Wing, R. A., and Sundaresan, V. (2008) Efficient insertional mutagenesis in rice using the maize *En/Spm* elements. *Plant J.* **44**, 879–892.
7. Wu, J. L., Wu, C., Lei, C., Baraoidan, M., Bordeos, A., Madamba, M. R., et al. (2005) Chemical- and irradiation-induced mutants of *indica* rice IR64 for forward and reverse genetics. *Plant Mol. Biol.* **59**, 85–97.
8. Kurata, N., Miyoshi, K., Nonomura, K. I., Yamazaki, Y., and Ito, Y. (2005) Rice mutants and genes related to organ development, morphogenesis and physiological traits. *Plant Cell Physiol.* **46**, 48–62.
9. Ma, J. F., Tamai, K., Ichii, M., and Wu, G. F. (2002) A rice mutant defective in Si uptake. *Plant Physiol.* **130**, 2111–2117.
10. Chern, C. G., Fan, M. J., Yu, S. M., Hour, A. L., Lu, P. C., Lin, Y. C., et al. (2007) A rice phenomics study – phenotype scoring and seed propagation of a T-DNA insertion-induced rice mutant population. *Plant Mol. Biol.* **65**, 427–438.
11. IRGSP (2005) The map-based sequence of the rice genome. *Nature* **436**, 793–800.
12. Yu, J., Hu, S., Wang, J., Wong, G. K., Li, S., Liu, B., et al. (2002) A draft sequence of the rice genome (*Oryza sativa* L. ssp. *indica*). *Science* **296**, 79–92.
13. Feltus, F. A., Wan, J., Schulze, S. R., Estill, J. C., Jiang, N., and Paterson, A. H. (2004) An SNP resource for rice genetics and breeding based on subspecies *indica* and *japonica* genome alignments. *Genome Res.* **14**, 1812–1819.
14. Hour, A. L., Lin, Y. C., Li, P. F., Chow, T. Y., Lu, W. F., Wei, F. J., et al. (2007) Detection of SNPs between Tainung 67 and Nipponbare rice cultivars. *Bot. Stud.* **48**, 243–253.
15. Hirochika, H., Guiderdoni, E., An, G., Hsing, Y. I., Eun, M. Y., Upadhyaya, N., et al. (2004) Rice mutant resources for gene discovery. *Plant Mol. Biol.* **54**, 325–334.
16. An, G., Lee, S., Kim, S. H., and Kim, S. R. (2005). Molecular genetics using T-DNA in rice. *Plant Cell Physiol.* **46**, 14–22.
17. Suzuki, T., Eiguchi, M., Kumamaru, T., Satoh, H., Matsusaka, H., Moriguchi, K., et al. (2008) MNU-induced mutant pools and high

performance TILLING enable finding of any gene mutation in rice. *Mol. Genet. Genomics* **279**, 213–223.

18. Larmande, P., Gay, C., Lorieux, M., Perin, C., Bouniol, M., Droc, G., et al. (2008) *Oryza* Tag Line, a phenotypic mutant database for the Genoplante rice insertion line library. *Nucleic Acids Res.* **36**, D102–1027.
19. Zhang, J., Li, C., Wu, C., Xiong, L., Chen, G., Zhang, Q., et al. (2006) RMD: a rice mutant database for functional analysis of the rice genome. *Nucleic Acids Res.* **34**, D745–748.
20. Ilic, K., Kellogg, E. A., Jaiswal, P., Zapata, F., Stevens, P. F., Vincent L. P., et al. (2007) The plant structure ontology, a unified vocabulary of anatomy and morphology of a flowering plant. *Plant Physiol.* **143**, 587–599.
21. Yamazaki, Y., and Jaiswal, P. (2005) Biological ontologies in rice databases. An introduction to the activities in Gramene and Oryzabase. *Plant Cell Physiol.* **46**, 63–68.

Chapter 11

Development of an Efficient Inverse PCR Method for Isolating Gene Tags from T-DNA Insertional Mutants in Rice

Sung-Ryul Kim, Jong-Seong Jeon, and Gynheung An

Abstract

The central goal of current genomics research in plants, as in other organisms, is to elucidate the functions of every gene. Insertional mutagenesis using known DNA sequences such as T-DNA is a powerful tool in functional genomics. Development of efficient methods for isolating the genomic sequences flanking insertion elements accelerates the systematic cataloging of insertional mutants, and thus allows functions to be assigned to uncharacterized genes via reverse genetic approaches. In our current study, we report a rapid and efficient inverse PCR (iPCR) method for the isolation of gene tags in T-DNA mutant lines of rice (*Oryza sativa*), a model monocot plant.

Key words: Functional genomics, Gene tag, Inverse PCR, Rice

1. Introduction

During the last decade, whole-genome sequencing projects have increased the amount of available molecular information on plant genomes including rice (1–3). Recent studies have used this sequence information to ultimately elucidate the functions of all genes. One approach to rapidly and efficiently obtain information on gene function is to generate mutations, and then study the resulting phenotypes. Random large-scale mutagenesis using known DNA sequences as insertion elements is one of the most effective strategies in this regard. The gene knockout and activation of mutant lines generated by this approach are valuable resources that are widely used for assigning functions to a large number of genes (4–8).

The development of efficient PCR-based methods, such as inverse PCR (iPCR) (9, 10), thermal asymmetric interlaced (TAIL) PCR (11, 12), or adaptor-ligated PCR (13, 14), has accelerated the

Andy Pereira (ed.), *Plant Reverse Genetics: Methods and Protocols*, Methods in Molecular Biology, vol. 678,
DOI 10.1007/978-1-60761-682-5_11, © Springer Science+Business Media, LLC 2011

large-scale isolation of sequences flanking gene tags, thereby facilitating the systematic cataloging of insertional mutants. Hence, in-silico mutant screenings are currently possible by performing BLAST searches of these flanking sequences in thousands of mutant lines (15–27).

Rice is a model organism for functional genomics studies of monocot plants including cereals because of its many advantages as a system for genetic analysis, as well as the worldwide development of resources. We have generated approximately 100,000 T-DNA insertional mutant lines of rice in our laboratory to date (6, 17, 22, 23, 28, 29). To isolate the gene tags on a large-scale from these mutant lines, we have also established a rapid and efficient iPCR method (17, 23). In this technique, we first digest genomic DNA from each mutant line using appropriate restriction enzymes to obtain an intact fragment containing the known insertion sequence and its flanking region. We then circularize the resulting, several thousand restriction fragments by self-ligation for use as PCR templates. Finally, we amplify the unknown flanking DNA segments with primers specific for the ends of the known sequences. In our present report, we summarize the details of this iPCR procedure that is now being effectively used in our laboratory to isolate gene tags from T-DNA mutant lines of rice.

2. Materials

2.1. Preparation of Genomic DNA for Inverse PCR

1. DNA extraction buffer: 2% (w/v) Cetyl trimethyl ammonium bromide (CTAB; Amresco, Solon, OH), 1.42 M NaCl, 20 mM ethylenediaminetetraacetic acid (EDTA), 100 mM Tris–HCl (pH 8.0), 2% (w/v) polyvinylpolypyrrolidone (PVPP; Sigma, St. Louis, MO), and 5 mM ascorbic acid. Store the reagents at room temperature. Because of the sedimentation of water insoluble PVPP, resuspend this buffer before use.
2. Ribonuclease A from bovine pancreas (RNase A, Sigma): Dissolve RNase A at a concentration of 10 mg/ml in 10 mM sodium acetate (pH 5.2). Boil for 15 min and then cool slowly to room temperature. Add 0.1 volume of 1 M Tris–HCl (pH 7.4). Store the enzyme in aliquots at –20°C.
3. Tris EDTA (TE) buffer (1×): 10 mM Tris–HCl (pH 8.0), and 1 mM EDTA.

2.2. iPCR for Isolation of the T-DNA Flanking Sequences

1. Restriction endonucleases (Enzynomics, Daejeon, Korea).
2. T4 DNA ligase and supplied 10× ligation buffer (Takara Bio, Shiga, Japan).
3. DNA polymerase and supplied PCR reagents (SolGent, Daejeon, Korea): EF-Taq DNA polymerase, 10 mM dNTP mix, 10× EF-Taq buffer, and Band Doctor™.

4. Tris–borate–EDTA (TBE) electrophoresis buffer (5×): Dissolve 54 g of Tris base and 27.5 g of boric acid, and add 20 mL of 0.5 M EDTA (pH 8.0). Adjust to 1 L with water. Store the buffer at room temperature.
5. DNA purification from an agarose gel band using Wizard® SV Gel and PCR Clean-Up System (Promega, Wisconsin, USA).

3. Methods

3.1. Preparation of Genomic DNA for iPCR

1. Prepare 20 seeds from the primary transgenic plants in a small net. To sterilize the seeds, soak the net in 0.025% (v/v) prochloraz (Aventis Crop Science, Yongin, Korea) diluted with water for 24 h at room temperature. Next, wash the seeds with tap water for one hour and imbibe in water for 24 h.
2. Prepare a rice nursery in a greenhouse. Put a 50-well plate (5 × 10 wells, each well is 50 mm in diameter and 50 mm in height, and has a small hole for water uptake) on a vessel filled with water to a 20 mm height, and then fill in the wells with soil. Sow 20 seeds into each well, and leave to germinate under a 14/10 h (light/dark) photoperiod, and at 28/22°C (day/night).
3. Harvest the fully expanded leaves from five seedlings of each mutant line (i.e. from each well), and place in 2-mL Safe-Lock microcentrifuge tubes (Eppendorf, Hamburg, Germany) (see Note 1). The fresh weight of each sample in the tubes is then 100–150 mg.
4. Put a 3-mm tungsten bead (Qiagen, Hilden, Germany) into the sample tube. The samples are then frozen by inserting the tubes into a 24-well adaptor rack (Qiagen) soaked in liquid nitrogen. The adaptor rack is then vibrated at 17 Hz for 2 min with a grinding apparatus (Model: TissueLyser II, Qiagen).
5. After grinding of the samples, add 800 μL of DNA extraction buffer to each tube.
6. Add 10 μL of RNaseA (10 mg/mL). After quick vortexing, incubate the samples at 65°C for 7 min.
7. Add 650 μL of chloroform (Junsei Chemical, Tokyo, Japan) and invert several times to mix.
8. Centrifuge the sample tubes at 13,000 ×*g* for 10 min at room temperature.
9. Remove the sample tubes carefully from the centrifuge and transfer the upper aqueous phase containing DNA to a new 1.5-mL Eppendorf tube (Sorenson, Salt Lake City, UT).
10. Add 0.7 volume of isopropanol (Junsei Chemical) and invert to mix.

11. The mixtures are then immediately centrifuged at 13,000 × *g* for 10 min at room temperature.
12. After centrifugation, discard the supernatant. Wash the DNA pellet with 1 mL of 70% (v/v) ethanol. Remove residual ethanol with a pipette after a quick spin, and dry the DNA pellet.
13. Dissolve the DNA pellet in 50 μL of 1× TE buffer (see Note 2). After this step, the DNAs can be stored for several months at −20°C.

3.2. Isolation of T-DNA Flanking Sequences by iPCR

1. A schematic depiction of the isolation of T-DNA flanking regions by iPCR is shown in Fig. 1. The choice of restriction endonuclease is important for the efficiency of this procedure (see Note 3). After the selection of restriction enzymes, design four sets of oligonucleotide PCR primers (about 22–24 nucleotides in length, and of about 50% GC) to amplify the flanking regions for both the T-DNA ends (see Note 4). The position and direction of the PCR primers we use are shown in Fig. 1.
2. Digest 1 μg of genomic DNA with 10 units of restriction endonuclease in a 50 μL volume for 10 h (see Note 5).
3. Directly add 50 μL of ligation mixture containing 15 units of T4 DNA ligase, 10 μL of 10× ligation buffer, and water to the digested DNA (see Note 6). After mixing by gentle agitation, incubate at 14°C for 10 h.
4. Prepare two distinct PCR mixtures for the amplification of both T-DNA ends as listed below. The amplification conditions are initial denaturation at 94°C for 5 min, followed by 35 cycles of 1 min at 94°C, 1 min at 58°C, and 4 min at 72°C, followed by a final step of 10 min at 72°C.

For amplification of right end region		For amplification of left end region	
Ligated DNA	2 μL	Ligated DNA	2 μL
Primer R1 (5 μM)	1 μL	Primer L1 (5 μM)	1 μL
Primer R2 (5 μM)	1 μL	Primer L2 (5 μM)	1 μL
dNTP mix (each 10 mM)	0.4 μL	dNTP mix (each 10 mM)	0.4 μL
EF-Taq buffer (10×)	2.5 μL	EF-Taq buffer (10×)	2.5 μL
Band Doctor™	3 μL	Band Doctor™	3 μL
EF-Taq DNA polymerase	0.1 μL	EF-Taq DNA polymerase	0.1 μL
Distilled water	15 μL	Distilled water	15 μL
Total	25 μL	Total	25 μL

5. Prepare nested PCR mixtures for right and left ends respectively, as outlined below. The amplification conditions are identical to the first PCR as described above.

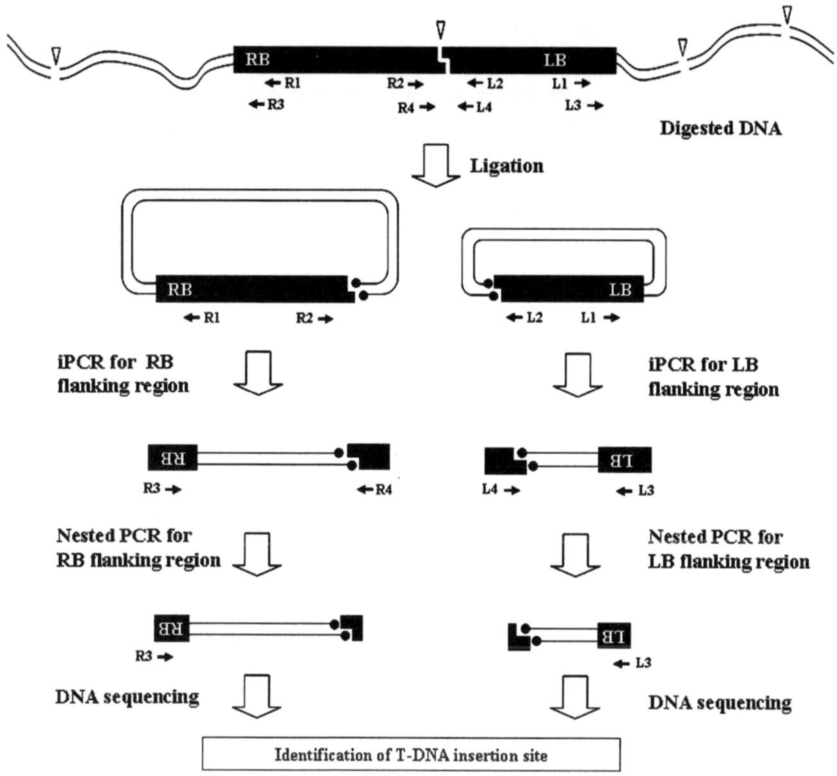

Fig. 1 Schematic representation of the iPCR procedure. Upon T-DNA integration into the plant genome, the right border (RB) and the left border (LB) of these inserts connect with unknown sequences. Genomic DNAs from the resulting insertional mutants are digested with the appropriate restriction endonucleases (*triangle*). The obtained monomeric DNA fragments (T-DNA segment with unknown flanking regions) are circularized by ligation (ligation sites are indicated by the *filled circles*). Only the target molecules are depicted here among the various forms of ligation products. The unknown T-DNA flanking regions are amplified by iPCR using T-DNA specific primers. For the specific amplification of target molecules to obtain a sufficient quantity of DNA, nested PCR is then performed. Finally, the T-DNA flanking regions are identified by sequencing of the nested PCR products using the R3 or L3 primers specific for the T-DNA ends. The annealing sites and direction of the oligonucleotide primers are denoted by *arrows*.

For amplification of right end region		For amplification of left end region	
The first PCR product	1 μL	The first PCR product	1 μL
Primer R3 (5 μM)	1 μL	Primer L3 (5 μM)	1 μL
Primer R4 (5 μM)	1 μL	Primer L4 (5 μM)	1 μL
dNTP mix (each 10 mM)	0.4 μL	dNTP mix (each 10 mM)	0.4 μL
EF-Taq buffer (10×)	2.5 μL	EF-Taq buffer (10×)	2.5 μL
Band Doctor™	3 μL	Band Doctor™	3 μL
EF-Taq DNA polymerase	0.1 μL	EF-Taq DNA polymerase	0.1 μL
Double distilled water	16 μL	Double distilled water	16 μL
Total	25 μL	Total	25 μL

6. Prepare a 1% (w/v) agarose gel in 0.5× TBE electrophoresis buffer.
7. Electrophorese 15 µL of the nested PCR products. Store the remaining mixtures at 4°C for subsequent DNA sequencing.
8. Prepare template DNA for DNA sequencing from total, or purified PCR product (see Note 7).

4. Notes

1. To remove cross contaminating genomic DNA, replace the disposable vinyl gloves after each sampling.
2. DNA concentration should be approximately 200 ng/µL.
3. Multiple complexes of transgenes at one locus are frequently observed in *Agrobacterium*-mediated transgenic plants. Thus, approximately 43% of the insertional mutant rice plants harbor T-DNA vector backbone junction structures (29). To prevent amplification of the vector backbone, avoid using restriction enzymes that cut within this region. We recommend a restriction endonuclease that cuts once within the T-DNA region through the tagging vector. This facilitates the isolation of the both T-DNA ends via one digestion-ligation reaction (Fig. 1).
4. Both ends of a T-DNA insert may be damaged during its transfer, and result in a shortened length after insertion into the genome. The right ends of the T-DNA are mostly well conserved with only small deletions (~30 nt), but the left ends can sometimes be deleted by up to 180 nt from the predicted border cleavage site (29). Thus, we recommend that the primers for T-DNA ends (i.e. R3 and L3 primers in Fig. 1) are designed to recognize the regions at about a 100 bp distance from the right border cleavage site, and 200 bp from the left border cleavage site.
5. For high-throughput analysis, digest the DNA in a 96-well plate (Simport, Beloeil, Canada) with sealing film (Simport). This assists with the transfer of the template DNAs from the 96-well plate after ligation, using a multi-channel pipette into PCR tubes.
6. In our old method, restriction enzymes were inactivated by heat treatment, and the digested DNA fragments were purified by ethanol precipitation. This simplified method omits these steps, and produces equivalent yields.
7. Preparations of template DNA from agarose gels for DNA sequencing are time consuming and costly. In our laboratory, we use PCR products without further purification as template

DNA, and approximately 0.5–2 μL (2–50 ng of DNA) of the nested PCR products are directly used in sequencing reaction. For a sample with multiple PCR products, extract the DNA from the each excised gel slice using a DNA purification kit (Promega). To identify the exact T-DNA insertion sites, R3 and L3 primers should be used for sequencing of the right and left border-plant DNA junctions, respectively.

Acknowledgments

The methods described in this report were developed with the support of the Crop Functional Genomic Center, the 21st Century Frontier Program (Grant CG1111); from the Biogreen 21 Program, Rural Development Administration (20070401-034-001-007-03-00); from the Korea Research Foundation Grant funded by the Korean Government (MOEHRD, Basic Research Promotion Fund, KRF-2007-341-C00028); and from Kyung Hee University.

References

1. Arabidopsis Genome Initiative. (2000) Analysis of the genome sequence of the flowering plant *Arabidopsis thaliana*. *Nature* **408**, 796–815.
2. Yu, J., Hu, S., Wang, J., Wong, G. K., Li, S., Liu, B., et al. (2002) A draft sequence of the rice genome (*Oryza sativa* L. ssp. *Indica*). *Science* **296**, 79–92.
3. International Rice Genome Sequencing Project. (2005) The map-based sequence of the rice genome. *Nature* **436**, 793–800.
4. Krysan, P. J., Young, J. C., and Sussman, M. R. (1999) T-DNA as an insertional mutagen in *Arabidopsis*. *Plant Cell* **11**, 2283–2290.
5. Weigel, D., Ahn, J. H, Blazquez, M. A., Borevitz, J. O., Christensen, S. K., Fankhauser, C., et al. (2000) Activation tagging in *Arabidopsis*. *Plant Physiol.* **122**, 1003–1013.
6. Jeon, J. -S., and An, G. (2001) Gene tagging in rice: a high throughput system for functional genomics. *Plant Sci.* **161**, 211–219.
7. Østergaard, L., and Yanofsky, M. F. (2004) Establishing gene function by mutagenesis in *Arabidopsis thaliana*. *Plant J.* **39**, 682–696.
8. Krishnan, A., Guiderdoni, E., An, G., Hsing, Y. I., Han, C. D., Lee, M. C., et al. (2009) Mutant resources in rice for functional genomics of the grasses. *Plant Physiol.* **149**, 165–170.
9. Ochman, H., Gerber, A. S., and Hartl, D. L. (1988) Genetic applications of an inverse polymerase chain reaction. *Genetics* **120**, 621–623.
10. Triglia, T., Peterson, M. J., and Kemp, D. J. (1988) A procedure for *in vitro* amplification of DNA segments that lie outside the boundaries of known sequences. *Nucleic Acids Res.* **16**, 81–86.
11. Liu, Y. G., and Whittier, R. (1995) Thermal asymmetric interlaced PCR: automatable amplification and sequencing of insert end fragments from P1 and YAC clones for chromosome walking. *Genomics* **25**, 674–681.
12. Liu, Y. G., Mitsukawa, N., Oosumi, T., and Whittier, R. (1995) Efficient isolation and mapping of *Arabidopsis thaliana* T-DNA insert junctions by thermal asymmetric interlaced PCR. *Plant J.* **8**, 457–463.
13. Balzergue, S., Dubreucq, B., Chauvin, S., Le-Clainche, I., Le Boulaire, F., de Rose, R., et al. (2001) Improved PCR-walking for large-scale isolation of plant T-DNA borders. *Biotechniques* **30**, 496–504.
14. Cottage, A., Yang, A., Maunders, H., de Lacy, R. C., and Ramsay, N. A. (2001) Identification of DNA sequences flanking T-DNA insertions by PCR-walking. *Plant Mol. Biol. Rep.* **19**, 321–327.

15. Parinov, S., Sevugan, M., De, Y., Yang, W. C., Kumaran, M., and Sundaresan, V. (1999) Analysis of flanking sequences from dissociation insertion lines. A database for reverse genetics in *Arabidopsis*. *Plant Cell* **11**, 2263–2270.
16. Hanley, S., Edwards, D., Stevenson, D., Haines, S., Hegarty, M., Schuch, W., et al. (2000) Identification of transposon-tagged genes by the random sequencing of mutator-tagged DNA fragments from *Zea mays*. *Plant J.* **23**, 557–566.
17. An, S., Park, S., Jeong, D. -H., Lee, D. -Y., Kang, H. -G., Yu, J. -H., et al. (2003) Generation and analysis of end sequence database for T-DNA tagging lines in rice. *Plant Physiol.* **133**, 2040–2047.
18. Alonso, J. M, Stepanova, A. N., Leisse, T. J., Kim, C. J., Chen, H., Shinn, P., et al. (2003) Genome-wide insertional mutagenesis of *Arabidopsis thaliana*. *Science* **301**, 653–657.
19. Yamada, K., Lim, J., Dale, J. M., Chen, H., Shinn, P., Palm, C. J., et al. (2003) Empirical analysis of transcriptional activity in the *Arabidopsis* genome. *Science* **302**, 842–846.
20. Hirochika, H., Guiderdoni, E., An, G., Hsing, Y. I., Eun, M. Y., Han, C. D., et al. (2004) Rice mutant resources for gene discovery. *Plant Mol. Biol.* **54**, 325–334.
21. Kolesnik, T., Szeverenyi, I., Bachmann, D., Kumar, C. S., Jiang, S., Ramamoorthy, R., et al. (2004) Establishing an efficient Ac/Ds tagging system in rice: large-scale analysis of Ds flanking sequences. *Plant J.* **37**, 301–314.
22. An, G., Jeong, D. H., Jung, K. H., and Lee, S. (2005) Reverse genetic approaches for functional genomics of rice. *Plant Mol. Biol.* **59**, 111–123.
23. Jeong, D. -H., An, S., Park, S., Kang, H. G., Park, G. G., Kim, S. R., et al. (2006) Generation of a flanking sequence-tag database for activation-tagging lines in japonica rice. *Plant J.* **45**, 123–132.
24. Chern, C. G., Fan, M. J., Yu, S. M., Hour, A. L., Lu, P. C., Lin, Y. C., et al. (2007) A rice phenomics study-phenotype scoring and seed propagation of a T-DNA insertion-induced rice mutant population. *Plant Mol. Biol.* **65**, 427–438.
25. Hsing, Y. I., Chern, C. G, Fan, M. J., Lu, P. C., Chen, K. T., Lo, S. F., et al. (2007) A rice gene activation/knockout mutant resource for high throughput functional genomics. *Plant Mol. Biol.* **63**, 351–364.
26. Kumar, C. S., Wing, R. A., and Sundaresan, V. (2008) Efficient insertional mutagenesis in rice using the maize En/Spm elements. *Plant J.* **44**, 879–892.
27. Wan, S., Wu, J., Zhang, Z., Sun, X., Lv, Y., Gao, C., et al. (2009) Activation tagging, an efficient tool for functional analysis of the rice genome. *Plant Mol. Biol.* **69**, 69–80.
28. Jeon, J. S., Lee, S., Jung, K. H., Jun, S. H., Jeong, D. H., Lee, J., et al. (2000) T-DNA insertional mutagenesis for functional genomics in rice. *Plant J.* **22**, 561–570.
29. Kim, S. R., Lee, J., Jun, S. H., Park, S., Kang, H. G., Kwon, S., et al. (2003) Transgene structures in T-DNA-inserted rice plants. *Plant Mol. Biol.* **52**, 761–773.

Chapter 12

Transposon Insertional Mutagenesis in Rice

Narayana M. Upadhyaya, Qian-Hao Zhu, and Ramesh S. Bhat

Abstract

Insertion mutants offer one of the direct ways to relate a gene to its function by employing forward or reverse genetics approaches. Both T-DNA and transposon insertional mutants are being produced in several crops, including rice, the first cereal with its complete genome sequenced. Transposons have several advantages over T-DNA including the ability to produce multiple independent insertion lines from individual starter lines, as well as producing revertants by remobilization. With our new gene constructs, and a two-component transposon *iAc/Ds* mutagenesis protocol, we have improved both gene trapping and screening efficiencies in rice.

Key words: Two-component *iAc/Ds* system, Flanking sequence tag, Rice, Transiently-expressed transposase system, Transposons

1. Introduction

One of the most direct approaches to determine gene function is insertional mutagenesis. Alterations in a plant phenotype, as a consequence of mutation, provide insight into the gene's function. The known DNA insertion sequence in the inactivated gene facilitates isolation of the "tagged" gene by various cloning and PCR-based strategies. Such tagging can be achieved by employing both non-transgenic and transgenic strategies. Endogenous transposons (both autonomous elements and their non-autonomous counterpart elements), such as *Activator* (*Ac*)/*Dissociation* (*Ds*), *Enhancer* (*En*)/*Inhibitor* (*I*) (also known as *Suppressor-Mutator* or *Spm*/*dSpm*), and *Mutator* (*MuDR*/*Mu*) in maize, or retrotransposons such as *Tos17* in rice, have been used to generate insertional mutants by non-transgenic means (1). Transgenic strategies include *Agrobacterium*-mediated T-DNA insertions (2–5) and heterologous transposons delivered through T-DNA (6, 7).

Andy Pereira (ed.), *Plant Reverse Genetics: Methods and Protocols*, Methods in Molecular Biology, vol. 678,
DOI 10.1007/978-1-60761-682-5_12, © Springer Science+Business Media, LLC 2011

In both the cases, plants can be initially screened for changes in phenotype. One can then clone part of the mutated/tagged gene (commonly referred to as flanking sequence tags or FSTs) using the inserted DNA tag as a reference point, and compare its sequence to sequences in the genome databases, thus linking the mutant phenotype with a known gene sequence.

With a population saturated with insertions, i.e., having at least one insertion in each gene, it is possible to apply both "forward genetics" and "reverse genetics" approaches to identify gene function (1). In the forward genetics approach, a mutant with a phenotype is first identified by screening the transposon tagged population, and sequences flanking the insert are then cloned and compared with database sequences to enable assignment of function to the mutated gene (see Fig. 1). In the reverse genetics approach, one starts with a computer predicted gene from the genome sequence and searches for an insertion mutant in that gene. Oligonucleotide primers from the insertional element and from the gene of interest are used for PCR amplification. Appropriately pooled DNA samples are used for high throughput screening for this often rare event in such populations. Once a mutation in the gene of interest has been identified, homozygotes are isolated and the phenotype confirmed.

To validate phenotype and gene sequence relationships, a number of strategies can be employed. Complementation experiments can be carried out by introducing the corresponding wild-type sequence into the mutant line as a transgene. The availability of multiple mutant alleles will also facilitate the validation. Alternatively, an RNAi system can be employed to determine whether the mutant phenotype can be mimicked by a targeted

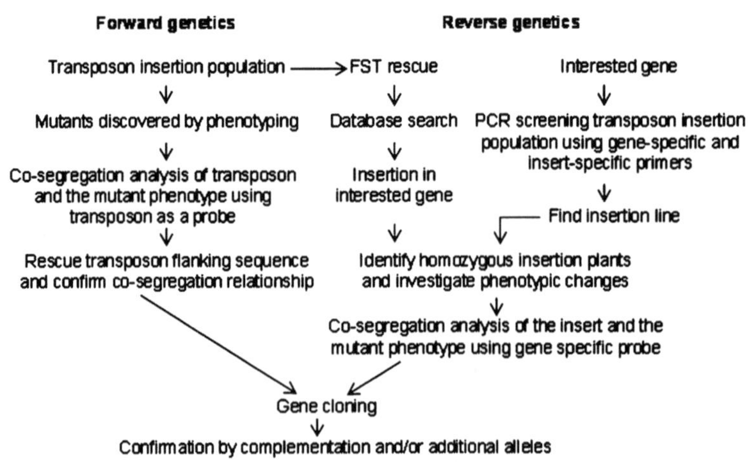

Fig. 1. Application of forward and reverse genetics strategies in gene identification using transposon mutagenized population.

gene expression knock-out. Finally, excision of the transposon resulting in restoration of the wild-type phenotype can confirm the gene function.

1.1. Ac Transposons as Insertion Mutagens

The *Ac* element can excise, and integrate randomly throughout the plant genome (although it tends to transpose close to its original position in the chromosome). A protein called transposase produced by *Ac* mediates these transpositions. A deletion derivative of *Ac* called *Dissociation* (*Ds*) is able to jump, but only in the presence of *Ac*, as it lacks the capacity to produce its own transposase. By removing certain terminal sequences from *Ac*, it can be made immobile (*iAc*), but still produce transposase. Several laboratories, including ours, have developed sophisticated transposon tagging systems with improved tagging and screening efficiencies in rice (8), which are primarily based on the two-component *iAc/Ds* (6) or *En/I* (7) system. In this system, essentially starter lines of *Ds* containing T-DNA (also referred to as *Ds*/T-DNA launch pads), and *iAc* containing T-DNA are produced by *Agrobacterium*-mediated transformation. *Ds*/T-DNA launch pad and *iAc*/T-DNA lines are then crossed to produce mutagenic F_1 plants. Mutagenic lines can also be produced by co-transformation or super-transformation of *Ds*/T-DNA and *iAc*/T-DNA constructs, or by having a construct with both *Ds* and *iAc*. In the mutagenic population, under the influence of a transposase produced by *iAc*, the normally-stable *Ds* element starts to jump or undergo transposition. This transposition will continue while the *iAc* is present. In subsequent generations, the *iAc* can be segregated from *Ds* elements that have integrated into new regions of the genome. Such stable insertion (SI) lines are then analyzed using forward or reverse genetics strategies as described above.

Transposons are preferred over T-DNA as insertional mutagens. Large-scale transposon mutagenised populations can be produced using a relatively small number of starter lines, as many independent insertions can be generated amongst the progeny of a single parental line. The tagged gene can be confirmed by revertants resulting from excision of the transposon. Transposons can also be remobilised to produce new insertion lines, in order to target genes in a specific chromosomal region, e.g., mapped quantitative trait loci (QTL). The *iAc/Ds*-based gene and enhancer trap systems in rice yield 5–10% unique stable insertion lines (8). The *Ds* re-insertions linked to the original location of *Ds* (the *Ds*/T-DNA launch pad) varied from 36 to 67%, with the majority being within 1 cM of the *Ds* launch pad (8).

1.2. Gene Traping and Enhancer Traping

Most of the tagging constructs used as insertional mutagens have gene trapping, or enhancer trapping facility. Use of a reporter gene with a minimal promoter (enhancer trap) or with intron splice acceptors (gene trap), linked to T-DNA or *Ds* transposon sequences, facilitate "trapping" of genetic regions which do not have a visible phenotype and will report on the expression of the chromosomal gene at the site of insertion (9, 10). For example, expression of the reporter gene in flowers will identify sequences related to flower development. Gene trapping efficiencies of ~6% have been reported for these constructs in rice (1). For efficient gene trapping either by T-DNA, or subsequently by *Ds*, the original T-DNA insertions need to be "clean", i.e., insertions devoid of direct or inverted T-DNA repeats, or of the incorporation of vector backbone (VB) sequences derived from outside the T-DNA borders (5).

1.3. Transiently-Expressed Transposase System

We have developed a novel method of producing stable *Ds* insertion lines using a transiently-expressed transposase (TET) system (11). We have developed constructs suited for high efficiency insertional mutagenesis in general, and the TET system in particular. By super infecting callus tissue from single-copy *Ds*/T-DNA lines, having both *Ds* excision and reinsertion markers, with *Agrobacterium* harboring *iAc* constructs containing a visual marker – *sgfpS65T*, we have been able to regenerate stable *Ds* insertion lines at a frequency of ~5% in addition to *iAc*/*Ds* double transformants. Mapped single-copy *Ds*/T-DNA launch pads produced using these constructs are highly suitable for efficient chromosomal region-directed insertion mutagenesis.

1.3.1. TET Construct Design Considerations

Three *Ds*/T-DNA constructs and two *iAc* constructs developed for TET system (11) are shown in Fig. 2. In the first construct pUR224NB (GenBank Acc. No. DQ225747), a CaMV35S promoter-driven *bar* gene cassette (12) serves as the *Ds* excision marker due to the placement of the gene trap *Ds* (*DsG*) cassette between the promoter and the *bar* gene. This construct has the

Fig. 2. (continued) (*D5′-Ds3′* orientation) is placed between the CaMV35S promoter and *hph(i)-nos*T cassette so that *hph* gene can serve as *Ds* excision marker. Orientations of the entire cassette within the T-DNA border sequences of these two constructs are shown; (**c**) pNU435 (Acc. No. DQ225750) has the promoter-*DsG*-excision marker cassette of pNU393A1. A specially designed cassette – maize Ubiquitin promoter-first exon – modified intron (with LB repeat sequences incorporated)-intron interrupted *barnase* [*bn(i)*] – *nos*T, to serve as vector backbone counter selector as well as dormant gene activator, and (**b**) a promoterless intron interrupted *barnase-nos*T cassette placed next to RB before the promoter-*DsG*-excision marker cassette to serve as T-DNA direct repeat (RB-LB-RB-LB) counter selector and T-DNA gene trap counter selector. This construct also does not have any pBR322 *ori* in the vector backbone; (**d**) *iAc* constructs pKU352NA (Acc. No. DQ225751) has the Ubi1 P(I)-Ω-*iAc*-T and CaMV35S P-*hph*-CaMV35S T cassettes from the construct pSK300 (13) and a Ubi1P-*sgfpS65T-nos*T cassette from pSK200 (13); (**e**) *iAc* constructs pKU400D (Acc. No. DQ225752) has the Ubi1 P(I)-Ω-*iAc*-T cassette from the construct pSK300 (13) and a Ubi1P-*sgfpS65T-nos*T cassette from pSK200 (13). The positions of recognition sites for *Sal* I (Sa), *Afl*II (Af), *Sph*I (Sp), *Hin*dIII (H), *Nhe*I (Nh), *Not* I (N) and *Xho*I (Xh) used in plasmid rescue and/or Southern blot analyses are indicated (*Reproduced from Upadhyaya et al (2006) Theor. Appl. Genet. 112:1326–1341 with permission from Springer*).

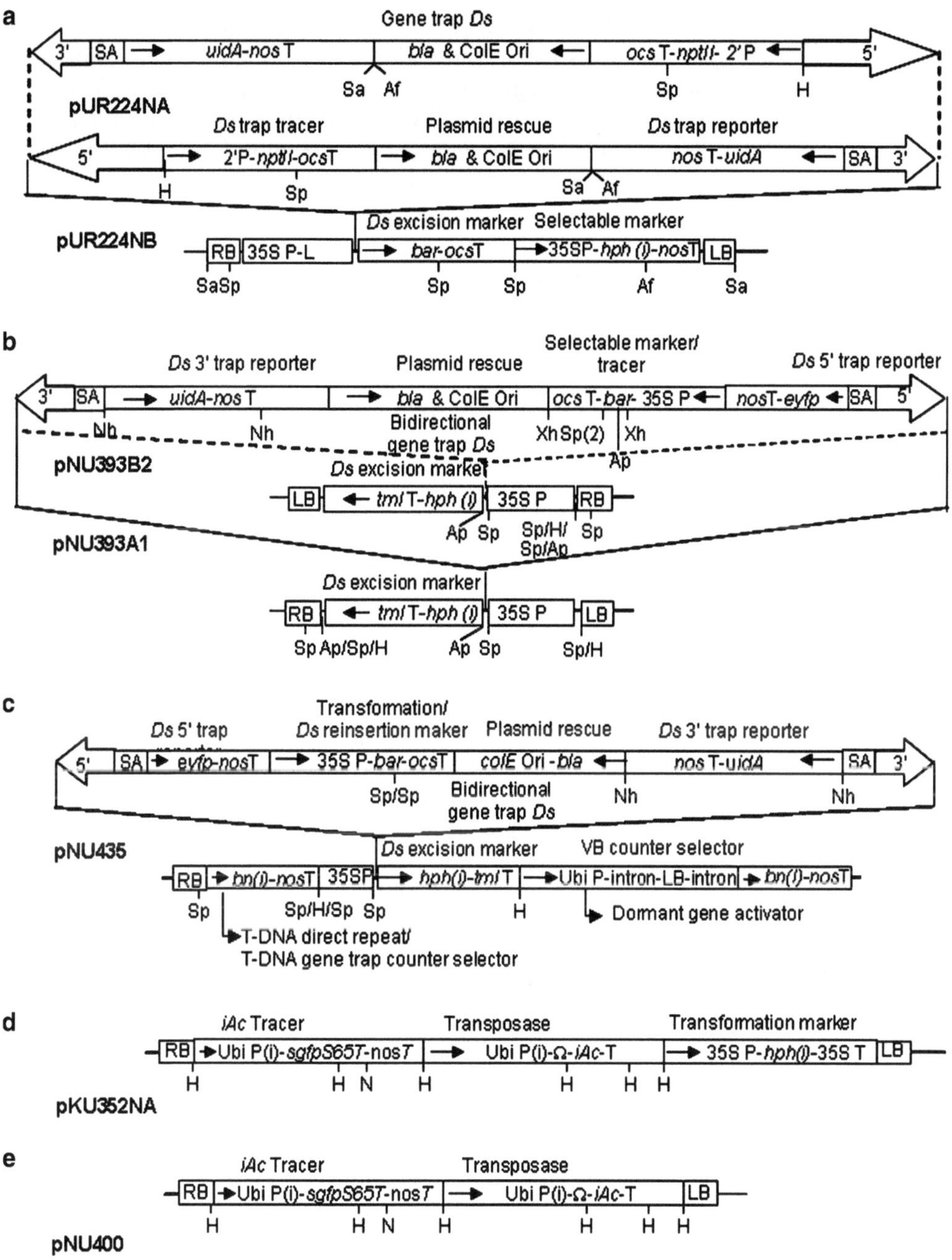

Fig. 2. Salient features of different tagging constructs. (**a**) Constructs pUR224NA (GENBANK. Acc. No. DQ225746) and pUR224NB (Acc. No. DQ225747) have a unidirectional gene trap *Ds* (*DsG*) with GPA1 intron-triple splice acceptor (SA)-*uidA* as trap reporter, *bla* and pBR322 *ori* as the flanking sequence tag (FST) recovery cassette, and 2′ promoter-*nptII-nos*T as *Ds* tracer, inserted between the CaMV35S promoter and *bar-nos*T cassette. Here the *bar* serves as the *Ds* excision marker. These constructs have a CaMV35 promoter driven, intron interrupted hygromycin phosphotransferase gene [*hph(i)*] as the selectable marker for transformation. Orientations of different components in these constructs indicated; (**b**) Constructs pNU393A1 (Acc. No. DQ225748) and pNU393B2 (Acc. No. DQ225749) have bidirectional gene trap *Ds* cassette (*Ds*3′-GPA1-SA-*uidA-nos*T and *Ds*5′-GPA1-SA-*eyfp-nos*T), also containing FST recovery cassette and CaMV35S P-*bar-ocs*T as the transformation selection marker and *Ds* reinsertion marker. The *Ds* cassette

preferred orientation of the unidirectional *DsG* cassette where the *Ds* 5′ end is facing the CaMV35S promoter, thus avoiding *cis*-activation of the gene trap reporter *uidA*. The unidirectional gene trap *Ds* cassette (with GPA1 intron-triple splice acceptor-*uidA* as trap reporter) in this construct was from pSK200 (13, 14), which also has 2′ promoter-*nptII-ocs*T as the *Ds* reinsertion marker or tracer, *bla* and pBR322 *ori* as the FST recovery cassette (plasmid rescue system). This also has the hygromycin resistance gene cassette (CaMV35S P-*hph*(*i*)-*nos*T) as a transformation marker (12) placed near the left-border (LB) end. The positioning of this selectable marker at the LB end is to avoid selection of transformants with truncated T-DNAs (3, 5, 15).

The constructs pNU393A1 (GenBank Acc. No. DQ225748) and pNU393B2 (GenBank Acc. No. DQ225749) have a previously proven (14, 15) bidirectional gene trap *Ds* cassette (*Ds*3′-GPA1-SA-*uidA-nos*T and *Ds*5′-GPA1-SA-*eyfp-nos*T). The *Ds* cassette also contains a CaMV35S P-*bar-ocs*T cassette (12) as the transformation selection marker and *Ds* reinsertion marker, and the FST recovery cassette. A previously tested CaMV35S promoter-driven intron-interrupted *hph-tml*T gene cassette (16) was used as the *Ds* excision marker by placing the *DsG* cassette (*Ds5′-Ds3′* orientation) between the promoter and the *hph(i)* gene cassette. The second, pBR322 origin of replication sequence was removed from both the binary vector backbones to increase the efficiency of FST recovery by plasmid rescue (14, 15). In our experience, when two copies of pBR322 *ori* are present in a construct, one gets mutated and becomes inactive. By removing the pBR322 *ori* in the vector backbone, we are ensuring the activity of the pBR322 *ori* in the *Ds* for future plasmid rescue. These features should improve the quality of the *Ds*/T-DNA launch pads which is very crucial for the TET system to work.

Further construct improvements were made to increase the frequency of selecting single-copy, clean T-DNA insertion lines. In the construct pNU435 (GenBank Acc. No. DQ225750), a maize Ubiquitin promoter (with its own first exon and the intron and LB sequence incorporated in the intronic region) driven, intron-interrupted *barnase* serves as a vector backbone (VB) counter selector, as transformed cell lines with VB-containing T-DNA inserts will be eliminated by the activity of *barnase* gene. Moreover, with a clean T-DNA insert, the Ubiquitin promoter near the LB has the potential to act as a dormant gene activator. A second copy of promoterless intron interrupted *barnase-nos*T (*bn(i)-nos*T) cassette placed within the T-DNA next to RB of the construct pNU435 (before the promoter-*DsG*-excision marker cassette) has the potential to serve as T-DNA direct repeat (RB-LB-RB-LB) counter selector, and as a T-DNA gene trap counter selector. A T-DNA direct repeat transgene will have the strong Ubiquitin promoter upstream of the RB side copy of the *barnase* gene and the resulting cell lines will be eliminated by the *barnase* gene activity. Insertion of this *Ds*/T-DNA within a gene with high constitutive expression has the potential to form an active gene-*barnase* fusion, thus enabling counter selection of such gene traps. pNU435 does not contain pBR322 *ori* in the vector backbone.

Two versions of *iAc* constructs were produced. A maize Ubiquitin promoter driven (also containing the first exon and intron and the Ω) *iAc* (13, 14) was used in these constructs. In the construct pKU352NA (GenBank Acc. No. DQ225751), a Ubi1P-*sgfpS65T-nos*T cassette is present to serve as a visual marker for *iAc* as we and others have shown that this expression cassette works very well in rice (14, 17). Consequently, this can be used effectively as a counter selector for *iAc* in the TET system of generating stable insertion lines. Another version of the *iAc* construct (pNU400, GenBank Acc. No. DQ225752) was made by removing the selectable marker gene cassette (CaMV35S P-*hph*-CaMV35S T) to make it compatible with the pNU393A1, pNU393B2, and pNU435 *Ds*/T-DNA constructs which have *hph* as excision marker.

Sources of various components of these constructs are: Binary vector backbone, T-DNA borders and selectable marker constructs – part of pCAMBIA1300 (http://www.cambia.org/daisy/cambialabs/materials.html), or pWBVec8 (16), or pMN-DRBBin1 (12); *Ds3* and GPA1 intron – pSK200 (13); *bla* and ColE Ori – pSP72 (Promega Corporation, Madison, USA); *Ds5* – A 378 bp end region of *Ds5′* from pEU334BN (15); *sgfpS65* – kindly supplied by J. Sheen (18); *Barnase* – PCR amplified from pMT416 (19) as described previously (15); CaMV35S P-*hph(i)*-*nos*T – from pWBVec8 (16).

The strategy for generating and screening stable insertion lines by the TET system, crossing and double transformation is illustrated in Fig. 3.

2. Materials

2.1. Seeds, Bacterial Strains, Growth Media, and Stock Solutions

1. Rice cultivar Nipponbare seed.
2. *Escherichia coli* strains DH5α (20) or XL-1 Blue MRF' (Stratagene, La Jolla, California) for cloning, and flanking sequence tag (FST) recovery by plasmid rescue.
3. *Agrobacterium* strain AGL1 (21).
4. Standard growth media Luria broth for *E. coli* strains and *Agrobacterium*, and antibiotics (ampicillin, spectinomycin, kanamycin and rifampicin).

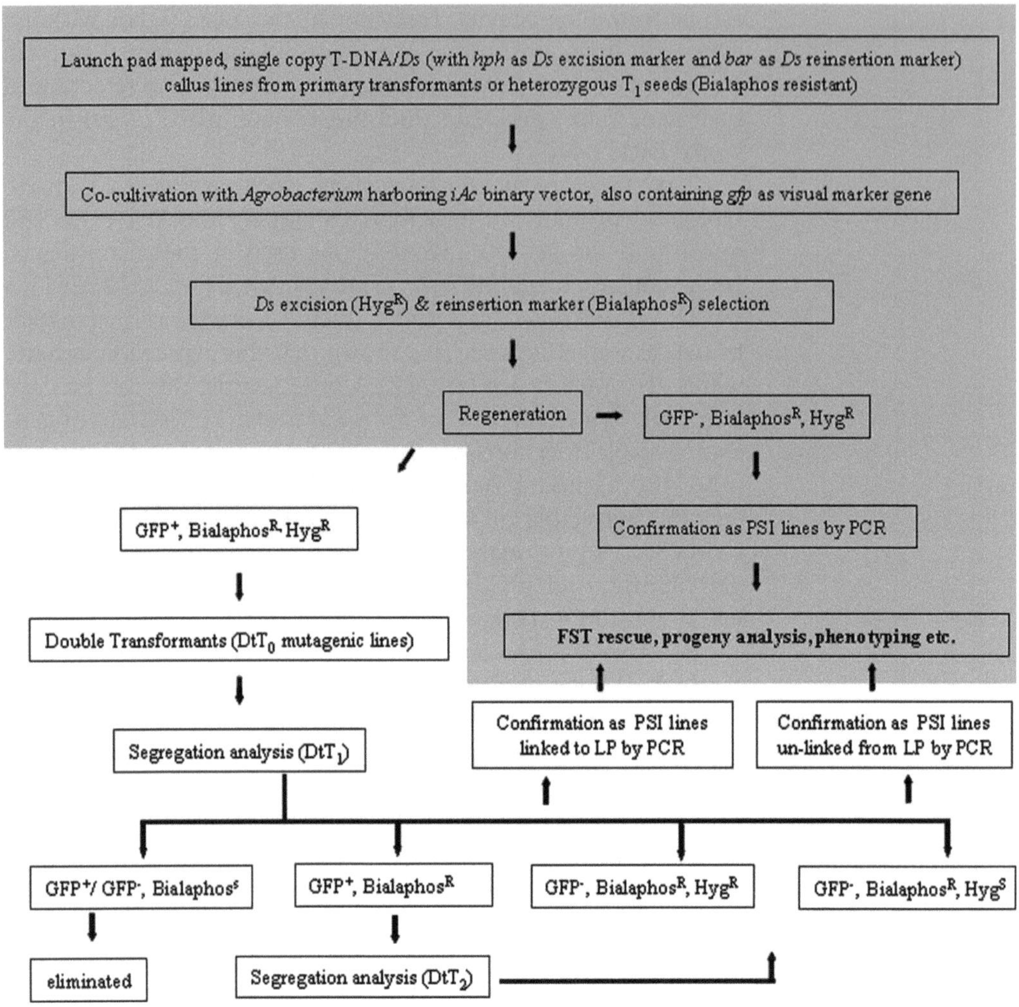

Fig. 3. Strategy for generating PSI lines using *Ds*/T-DNA launch pads either by the TET system (shaded) or by double transformant progeny screening (not shaded) (*Reproduced from Upadhyaya et al. (2006) Theor. Appl. Genet. 112:1326–1341 with permission from Springer*).

5. 20× N6 macronutrients, ingredients for 1 L stock: $(NH_4)_2SO_4$ (9.3 g), KNO_3 (56.6 g), KH_2PO_4 (8.0 g), $MgSO_4·7H_2O$ (3.7 g), and $CaCl_2·2H_2O$ (3.3 g).
6. 1,000× N6 micro, ingredients for 100 mL stock: $MnSO_4·4H_2O$ (440 mg), $ZnSO_4·7H_2O$ (150 mg), H_3BO_3 (160), and KI (80 mg).
7. 100× B5 micronutrients (100×), ingredients for 1,000 mL stock: $MnSO_4·4H_2O$ (1,000 mg), $Na_2MoO_4·2H_2O$ (25 mg), H_3BO_3 (300 mg), $ZnSO_4·7H_2O$ (200 mg), $CuSO_4·5H_2O$ (3.87 mg), $CoCl_2·6H_2O$ (2.5 mg), and KI (75 mg).

8. 1,000× N6 vitamins, ingredients for 100 mL stock: Glycine (200 mg), thiamine-HCl (100 mg), pyridoxine-HCl (50 mg), and nicotinic acid (50 mg).
9. 1,000× B5 vitamins: Commercial Gamborg's vitamin stock solution (Sigma).
10. 200× Fe-EDTA (ferric-sodium salt): 1.47 g for 200 mL stock.
11. 2,4-Dicholro-phenoxyacetic acid (2-4-D, 1 mg/mL): Dissolve 100 mg in 1 mL absolute ethanol, add 3 mL of 1 N KOH, adjust to pH 6.0 with 1 N HCl (very sensitive, adjust carefully after pH 9.0).
12. 6-Benzylaminopurine (BAP, 1 mg/mL): Commercially available solution (Sigma).
13. Naphthalene acetic acid (NAA, 1 mg/mL): Commercially available stock solution (Sigma).
14. Abscisic acid (ABA, 2.5 mg/mL): Dissolve 250 mg of ABA in 2 mL of 1 M NaOH and make to 100 mL with sterile distilled water.
15. Hygromycin B (50 mg/mL): Commercially available stock solution (Roche).
16. Timentin (150 mg/mL): Commercial mixture of 3 g ticarcillin sodium and 100 mg clavulanic acid as potassium clavulanate (Glaxo SmithKline Australia Pty Ltd). Dissolve the contents of the vial (3,100 mg) in 20.66 mL of sterile water, aliquot in 1.5 mL Eppendorf tubes and store at −20°C.
17. MS salts and vitamin mixture: Commercial Murashige minimal organic medium (Sigma).
18. Phytagel: (Sigma).
19. Bialaphos sodium salt: (Wako Pure Chemical Industries Ltd, Japan).
20. Wetting Agent 600: Commercial preparation of 600 g/L nonyl phenol ethylene oxide condensate non-ionic organic surfactant (David Gray and Co, WA, Australia) or AGRAL 60 (ICI Australia).
21. Callus induction medium N6D (22), Ingredients for 1 L: 10× N6 macronutrients (100 mL), 1,000× N6 micronutrients (1 mL), 1,000× N6 vitamins (1 mL), 200× MS Fe-EDTA (5 mL), myoinositol (100 mg), casamino acid (300 mg), proline (2.9 g), 1 mg/mL 2,4-D (2 mL), and sucrose (30 g). Adjust pH to 5.8 with 1 M KOH. Add 3 g phytagel/L and autoclave.
22. Sub culture medium NB, ingredients for making 1 L: 10× N6 macronutrients (100 mL), 1,000× B5 micronutrients (10 mL), 1,000× B5 vitamins (1 mL), 200× Fe-EDTA (5 mL), 1 mg/mL 2,4-D (2 mL), sucrose (30 g), proline

(500 mg), glutamine (500 mg), casein enzymatic hydrolysate (300 mg). Adjust pH to 5.8–5.85 with 1 M KOH, add 3 g phytagel/L and autoclave.

23. First selection medium NBTH30 or NBTB10 (see Note 1): NB medium plus 150 mg/L Timentin and 30 mg/L Hygromycin B, or 10 mg/L Bialaphos added after autoclaving just before pouring.
24. Second selection medium NBTH50, or NBTB10: NB medium plus appropriate selective agents (150 mg/mL Timentin and 50 mg/L Hygromycin B, or 10 mg/L Bialaphos).
25. Pre-regeneration medium PRTH50 or PRTB10: NB medium without 2,4-D, plus the following were added per liter after autoclaving: 1 mg/L BAP (2 mL), 1 mg/L NAA (1 mL), 2.5 mg/L ABA (2 mL), 50 mg/mL hygromycin (1 mL) or 10 mg/mL Bialaphos (1 mL), and 150 mg/mL Timentin (1 mL).
26. Regeneration medium RHT50: NB medium without 2,4-D, plus the following were added per liter after autoclaving: 1 mg/L BAP (1 mL), 1 mg/mL NAA (0.5 mL), 50 mg/mL Hygromycin B (1 mL) or 10 mg/mL Bialaphos (1 mL), and 150 mg/mL Timentin (1 mL).
27. Rooting medium, 0.5× MS medium; ingredients for making 1 L: MS salts and vitamin mixture (2.21 g), sucrose (10 g). Add 2.5 g phytagel, autoclave, and then add 1 mg/L NAA (50 μL), 50 mg/mL Hygromycin B (1 mL) or 10 mg/mL Bialaphos (1 mL), and 150 mg/L Timentin (1 mL) (see Note 2).
28. Plant propagation medium: Compost:perlite mix (4:1) supplemented with osmocote (1 g/L).
29. Hygromycin B (150 mg/L) and Basta (5 mL/L) in water containing 0.03% AGRAL 60 (ICI, Australia) as a wetting agent for seedling spray.
30. 0.08% of EDTA ferric-sodium salt with 0.03% AGRAL 60 for spraying seedlings showing iron deficiency.
31. Fertilizer pellets, nutrient solution, plant protection chemicals.

2.2. Bacterial Culture, Plant Tissue Culture, and Plant Propagation Facilities

1. Incubators (28 and 37°C) and Laminar flow cabinets.
2. Constant temperature (27°C) tissue culture growth rooms with 16 h light.
3. Misting facility in the glasshouse for hardening of tissue culture derived explants.
4. Glasshouse facilities for growing rice plants at 25 ± 3°C with light supplement (2–4 h) in winter months, benches, water tubes (for growing under submerged condition), Jiffy pots

(Jiffy International, Kristiansand, Norway), various sized pots and trays (with 24–96 well template for planting).

5. Growth cabinets for short day treatments (normally 16 h light and 8 h dark) of rice plants with both temperature and light controls.
6. Standard laboratory plastic and glassware supplies.
7. Micropore tape (3 M Health Care, Germany) for sealing the Petri dishes use in bacterial and tissue culture.

2.3. DNA or RNA Extraction (PureGene Protocol) Buffers, Reagents

1. Cell lysis solution: TE (Tris-HCl 10 mM, EDTA 1 mM, pH 8.0) with 1% SDS.
2. Protein precipitation solution: 6 M Ammonium Acetate.
3. Nucleic acid precipitation solution: Isopropanol.
4. Nucleic acid wash solution: 70% ethanol.
5. DNA resuspension buffer: Tris-HCl (10 mM) and EDTA(0.2 mM) at a pH 8.0.
6. RNA resuspension buffer: TE.
7. Bench top centrifuge, Eppendorff tubes, liquid nitrogen, and plant tissue grinding facility.

2.4. Agarose Gel Electrophoresis Reagents and Equipment

1. Agarose.
2. 1× TAE buffer: 0.04 M Tris-acetate and 1 mM EDTA.
3. Power pack (300 V, 400 mA).
4. Electrophoresis apparatus.
5. Nucleic acid stain: 10 mg/mL Ethidium bromide.
6. UV transilluminator.

2.5. DNA Blot Analysis

1. Amersham Hybond N^+ membrane.
2. 0.4 M NaOH transfer solvent.
3. 2× SSC (300 mM NaCl, 30 mM sodium citrate pH 7.0) buffer for membrane rinsing.
4. 3 mm filter paper for solvent wick.
5. Paper towel for capillary action.
6. Ultraviolet cross-linker (UVP, USA).

2.6. DNA Hybridisation

1. Hybridisation buffer (7% SDS, 1% BSA, 0.5 M $NaHPO_4$ at a pH 7.2, and 1 mM EDTA).
2. Washing buffers: Wash buffer 1 (2× SSC, 0.1% SDS), wash buffer 2 (1× SSC, 0.1% SDS), wash buffer 3 (0.5× SSC, 0.1%SDS).
3. 65°C hybridisation oven capable of sample rotation (ThermoHybaid).

4. Bottles for membrane incubation (Hybaid HB-OV-BM).
5. Shaking 65°C incubator for high stringency washing.
6. X-ray film and cassettes (Kodak Biomax), and X-ray film developing facilities (see Note 3).
7. Membrane stripping solution (5 mM Tris-HCl – pH 8.0, and 0.1% SDS).

2.7. Labeling of DNA Probes

1. ^{32}P-dCTP, Megaprime™ DNA labeling system (Amersham).
2. Sepharose G50 beads for separation of unincorporated nucleotides.
3. Appropriate facilities for handling and disposing of radioactive compounds.

2.8. Formaldehyde Gel Electrophoresis Reagents and Equipment (for RNA Blot)

1. 10× MOPS buffer: 0.2 M MOPS, 50 mM sodium acetate, 5 mM EDTA. Adjust pH to 7.0, aliquot, and store at –20°C.
2. Formaldehyde: Commercial 12.3 M (37%) preparation.
3. Formaldehyde agarose gel: 1.5% agarose, 1× MOPS buffer, 0.63 M Formaldehyde.
4. RNA sample preparation premix, ingredients in 775 μL: 10× MOPS (100 μL), formaldehyde (175 μL), and formamide (500 μL).
5. 10× RNA loading buffer: 50% glycerol, 1 mM EDTA, 0.4% bromophenol blue, 0.4% xylene cyanol, and 0.125% ethidium bromide. Stocks are stored at –20°C.
6. Gel and electrophoresis apparatus.

2.9. GUS Histochemical Staining

1. Staining buffer: 0.1 M $NaPO_4$ – pH 7.0, 10 mM EDTA, 10% methanol, 200 μM potassium ferricyanide, 200 μM potassium ferrocyanide, 0.75 mg/mL of X-gluc substrate, and 0.01% silwet.
2. Vacuum pump and bell jar for stain infiltration.
3. 25%, 50%, 70%, and 95% ethanol in water for tissue decolourisation (remove chlorophyll) following staining.

2.10. GFP Visualization

1. MZ6 stereomicroscope (Leica Microscopy and Scientific Instruments) illuminated by a 50 W HP mercury vapor lamp, and having a fluorescence GFP-Plus filter set (480/40 nm excitation filter, 505-nm LP dichromatic mirror and a 510 nm LP barrier filter).
2. S550/50 nm NP filter to minimize red chlorophyll autofluorescence in leaf tissues.

2.11. PCR Reactions

1. Thermostable DNA polymerase.
2. 10 × buffer supplied by the manufacturer.

3. 0.5 mM deoxyribonucleotide triphosphates (dATP, dCTP, dTTP, and dGTP).
4. Appropriate DNA primers.
5. PCR thermocycler.
6. Appropriate tubes.
7. PCR purification columns (Qiagen, Germany).

2.12. RT-PCR Reactions

1. One-Step RT-PCR Kit (Qiagen).
2. Appropriate DNA primers.
3. PCR thermocycler.

2.13. Molecular Biological Reagents for Cloning and Sequencing

1. ABI Prism BigDye termination cycle sequencing kit (PE Applied Biosystems, California, USA).
2. UltraClean™ 15 agroase gel DNA purification kit (Cambio, Cambridge, UK).
3. Restriction enzymes and buffers: From any commercial source (e.g., New England Biolabs, MA).
4. T4 DNA ligase and buffers: MBI Fermentas Inc, NY, USA.
5. DNase free RNase: From any commercial source. Make 10 mg/mL stock and store at –20°C.
6. DNase and buffer: RQ1 RNase-free DNase (1 U/μL) and 10× buffer (Promega Corporation, Madison).

3. Methods

3.1. Bacterial and Plasmid Culture

1. *Ds*/T-DNA or *iAc*/T-DNA binary vectors are introduced into the super virulent *Agrobacterium* strain AGL1 (21) by standard electroporation or triparental mating (23), and cultured at 28°C for 48 h on LB medium (24) containing the antibiotics rifampicin (50 mg/L) and kanamycin (25 mg/L).
2. Plasmid DNA isolations were performed using the QIAprep Spin Miniprep kit (QIAGEN Inc, California, USA) according to the manufacturer's instructions.

3.2. Rice Tissue Culture and Transformation

Any efficient tissue culture and transformation system can be used to generate the *Ds*/T-DNA and *iAc* starter lines. It is advisable to keep the tissue culture period to a minimum to reduce tissue-culture-induced somaclonal variations. Previously developed protocols of callus induction (25) in N6D solid medium (22), callus subculture in NB solid medium (26) and transformation (27) have been adapted successfully with cv. Nipponbare and several other *japonica* cultivars.

3.2.1. Callus Induction and Sub-culture

1. Remove the husk from rice seeds.
2. Soak dehusked seeds in 70% ethyl alcohol for 1 min in a 50 mL screw-cap plastic tube.
3. Rinse with sterile water three times.
4. Soak in 50% commercial bleach (see Note 4) for 30 min. A rotary shaker can be used for efficient sterilization.
5. Wash thoroughly with sterile water (perform under aseptic conditions).
6. Blot dry and place seeds on N6D media plates and seal with Micropore tape.
7. Leave plates under light at 28°C and look for any contamination, if so transfer to new plates (see Note 5). Wait for 6–8 weeks for callus production.
8. Pick healthy looking (small to medium sized creamy callus buds) and subculture in NB media, or use for transformation.
9. Repeat subculturing once every 2 weeks. For stable transformation, use calli subcultured for less than four to five times to minimize the extent of somaclonal variants in the regenerants.

3.2.2. Transformation

1. Pick healthy looking calli from subculture plates and transfer them onto new NB plates (25–30 calli/plate) (day 1). Pick ~2 mm sized calli, so that when they are used after 5 days they will be ~3 mm.
2. On the third day, spread 150–200 μL glycerol stock (thawed on ice) of *Agrobacterium* containing the binary construct on LB media with rifampicin and other appropriate antibiotics. Spread one plate for every 50 calli used. Incubate in the dark at 25°C for 48–72 h.
3. Gently, but thoroughly resuspend the plate-grown *Agrobacterium* culture in 5–10 mL of NB liquid media containing 100 μM acetosyringone and transfer into 50 mL tubes. Allow the culture to stand for 3 h at room temperature (25°C).
4. Immerse healthy embryogenic calli into *Agrobacterium* suspension and leave for 20–30 min.
5. Drain off the bacterial suspension and blot calli lightly on sterile filter paper, then place on NB plate containing 100 μM acetosyringone and incubate in the dark at 25°C for 3 days.
6. After co-cultivation, wash calli in sterile water containing 150 mg/L Timentin until the wash solution becomes clear. Calli may require five to eight washes. Be gentle, but through while washing.

7. Blot dry calli. If a visual marker is present in the construct check for the transient expression. At least 50% of the calli should show transient expression.
8. Place calli onto first selection media with appropriate antibiotics and culture in dark for 3–4 weeks. Watch for any *Agrobacterium* over-growth. If so, transfer the calli without overgrowth onto new plates (see Note 6).
9. Start observing for new resistant callus outgrowth (24–30 days). May have to wait for 40 days to see the resistant callus buds in case of Bialaphos selection (see Note 7).
10. Transfer resistant callus buds onto second selection media and culture for 14–21 days. Mark the callus lines picked in the original plate. You may find new callus lines (not picked earlier) producing resistant callus buds 7–14 days after the first pick.
11. Produce single callus descent from each resistant callus line, if planning to adapt the TET system for generating putative stable insertion (PSI) lines (see Note 8), else transfer resistant calli onto pre-regeneration plates and culture in dark for 8–12 days.
12. Transfer calli with signs of regeneration (see Note 9) to regeneration media plates and culture in light for 15–30 days.
13. Transfer the established plantlets to 0.5× MS pots (with appropriate selection, see Note 10) and culture in light for 10–14 days.

3.3. Plant Propagation and Maintenance

3.3.1. Primary Transformants

1. Transfer seedlings from MS pots to Jiffy pots with pre-wetted standard potting mix. Remove the adhering callus and Phytagel before planting in the Jiffy pots (see Note 11).
2. Transfer established seedlings to appropriately sized (normally 15 cm) pots or trays, and place then in water tubs (see Note 12).
3. Complete all the required transgene analyses as soon as possible (see Subheading 3.7).
4. Perform all the aftercare operations (see Note 13).
5. Harvest when a majority of the panicles are fully mature and dry.
6. Record in the respective datasheets and enter the data promptly into database.
7. Process the seeds after they are fully dried, weigh, and pack in storage boxes. Enter all the relevant information in the database.

3.3.2. Progeny Plants

1. Put the seeds to sprout in wetted filter paper in Petri dishes or in 24–48 well plates. If seeds have fungal contamination, surface sterilize (see Subheading 3.2.1).
2. Observe for GFP (see Subheading 3.7.5) after 36–48 h (not more than 48 h) and straight away plant in designated pots.
3. Pre-wet the potting mix (previously firmed and leveled) in the pots and make 0.5 cm hole, carefully place the germinating seeds, and cover with more potting mix and lightly firm the media.
4. Sprinkle water over the pot and cover with cardboard for 2–3 days. Sprinkle more water after 2 days (to keep the surface moist). Do not fill the tray with water.
5. Remove the cover after 3 days and keep the pots moist (sprinkle) for 7–10 days, or until all seedlings have emerged. Trays are then filled with water above the level of the potting mix in the pot (see Note 12).
6. Complete required analyses (see Subheading 3.7) and observations.
7. Perform all the aftercare operations (see Note 13).
8. If plant samples are to be collected, collect from 3–4 week-old seedlings.
9. Perform Hygromycin B and/or BASTA spray as required on 15–20 days seedlings (see Note 14).
10. If GFP segregation data needs to be collected, sacrifice one half of a matured panicle (say 24 seeds) for observation.
11. Perform Southern blot or plasmid rescue as required to determine the copy number in *Ds*/T-DNA and *iAc*/T-DNA (to select single copy lines) as detailed in Subheading 3.7.
12. Transgene copy/locus number can be estimated by segregation analyses by visual marker, spray assay, or gene specific PCR as detailed in Subheading 3.7. Use 12–24 progeny seeds for segregation analyses.

3.4. Transiently-Expressed Transposase System

1. Induce callus from heterozygous T_1 seeds of single-copy *Ds*/T-DNA lines as described in Subheading 3.2.1 (see Note 15), or use single callus descent lines with single *Ds*/T-DNA launch pad (see Note 8).
2. Regenerate a portion of the primary transgenic calli to plants and produce seeds for subsequent use.
3. Introduce *iAc* constructs (also contains *gfp*) by super-transformation (see Subheading 3.2.2). A transient GFP expression gives a measure of number of host cells infected with *Agrobacterium* which would also express transposase transiently.

4. Subsequent selection pressure for *Ds* and its excision, allows proliferation of callus cells with transposed *Ds*. Use Bialaphos as *Ds* excision marker selection for lines derived from pUR224NA, or pUR224NB, and pKU352NA. Use Hygromycin B as excision marker selection and Bialaphos as reinsertion marker selection for lines derived from the use of pNU393A1, or pNU393B2, and pNU400. In both cases, GFP serves as a visual marker for *iAc* to separate PSI lines (GFP^-) from possible double transformants (GFP^+).
5. Perform required transgene analysis and FST rescue from PSIs as detailed in Subheading 3.7.
6. Regenerate, propagate, and harvest as described in previous sections (see Subheadings 3.2.2 and 3.3).

3.5. Production of Mutagenic (Containing Both Ds LP and iAc) Lines

1. The double transformants obtained in the TET system can be used as mutagenic lines, and their progeny as screening populations for identifying stable *Ds* insertion lines (lines devoid of *iAc*).
2. Mutagenic lines are also produced by crossing *Ds*/T-DNA launch pad lines with *iAc* lines as described (28) previously (see Note 16). F_1 hybrids can easily be selected with GFP (specific to *iAc* construct), Basta spray, (*Ds* or T-DNA launch pad marker) and gene specific PCR.
3. It is better to use well characterized single copy *Ds*/T-DNA and single copy or single locus *iAc* lines combinations showing good transposition. This however has to be assessed by performing a pilot scale screening of the progeny seeds. The extent of transient expression spots (GUS) of the gene trap reporter gene (*uidA*) in a callus or plant line super transformed with *iAc* gives a measure of *iAc* activity.

3.6. High-Throughput Screening for Putative Stable Insertion Lines

1. As the *iAc* locus also contains *gfp*, it can be used as a counter selector for plants still containing *iAc* (means still mutagenic). Soak seeds in water for 2 days.
2. Observe GFP expression under an MZ6 stereomicroscope with fluorescence GFP-Plus filter set (480/40 nm excitation filter, 505-nm LP dichromatic mirror, and a 510 nm LP barrier filter).
3. Use S550/50 nm NP filter to minimize red chlorophyll autofluorescence in leaf tissues.
4. Imbibing the seeds in water for 1–2 days, allows this prescreening to be performed before planting (GFP^+ seeds are planted separately for producing more seeds for further screening, or discarded if there are no plans to produce more seeds for screening from the lines in question).
5. GFP^- seeds are then planted in a high-throughput set-up (24–96 well format in trays) in the glasshouse.

6. Perform *Ds* tracer (reinsertion marker) assay (either Basta or Hygromycin B spray depending on the *Ds*/T-DNA launch pad type) to eliminate plants without *Ds*.
7. If *hph* is the *Ds* excision marker and *bar* is the *Ds* tracer, hygromycin spray assay is done simultaneously to differentiate launch pad-linked (hyg^R) and launch-pad-unlinked (hyg^S) PSIs (see Note 16).
8. If *bar* is the *Ds* excision marker, do not use the spray assay (as it will kill PSIs unlinked to the launch pad), instead perform excision specific PCR (see Table 1) to differentiate linked and unlinked transpositions.

Table 1
Primers

Primer Name	Sequence	Target
Ac_1931$^+$	5′-CAGCTCCAAAGACAAAGACAAC-3′	*iAc* forward
Ac_2382$^-$	5′-TGCAGCAGCAATAACAGAGTC-3′	*iAc* reverse
bar5′_68$^+$	5′-CCATCGTCAACCACTACATC-3′	*bar* forward
bar3′_489$^-$	5′-AGAAACCCACGTCATGC-3′	*bar* reverse
35S_TAN	5′-GATCCGCAAGACCCTCC-3′	CaM35S minimal promoter forward
35S_AS	5′-AATACGCAAACCGCCTCTC-3′	CaMV35_reverse
35S3′	5′-GGGATGACGCACAATCCC-3′	CaMV35S promoter forward
Ds3_6587$^+$	5′-CCGTCCCGCAAGTTAAATATG-3′	Ds3 forward
Ds5_112$^-$	5′-ATCGGTTATACGATAACGGTC-3′	Ds5 reverse
GPAInt	5′-TCCAAGTCCACAAGGAAAATTG-3′	GPA1 intron
hph5′_5$^+$	5′-AAAAGCCTGAACTCACCGC-3′	*hph* forward
hph3′_515$^-$	5′-TCGTCCATCACAGTTTGCC-3′	*hph* reverse
hph5′_180$^-$	5′-GATCTTTGTAGAAACCATCGGC-3′	*hph reverse*
nos3′_48$^-$	5′-ATTCAATCTTAAGAAACTTTATTGCCA-3′	*nos* terminator
nptII_45$^-$	5′-TAGCCGAATAGCCTCTCCAC-3′	*nptII* reverse
sgfp_3F	5′-GAGCAAAGACCCCAACGAG-3′	*sgfPS65T* 3′ forward
sgfp_AS	5′-TCCTTGAAGAAGATGGTGCG-3′	*sgfpS65T* antisense
NUbPro_106	5′-TTTTTTAGCCCTGCCTTCATAC-3′	*Ubi1* Promoter
GUS_1346F	5′-TCACCGAAGTTCATGCCAGTCC-3′	*uidA* forward
GUS_313$^-$	5′-TCACTTCCTGATTATTGACCCAC-3′	*uidA* reverse

(continued)

Table 1 (continued)

Primer Name	Sequence	Target
GUS_1781R	5′-ACGCTCACACCGATACCATCAG-3′	*uidA* reverse
RB_TAIL1	5′-GCTGATAGTGACCTTAGGCGAC-3′	RB flank
RB_TAIL2	5′-CGTTGCGGTTCTGTCAGTTCC-3′	RB flank
RB_TAIL3	5′-CAAACGTAAAACGGCTTGTCCC-3′	RB flank
RB_TAIL4	5′-ATCAGATTGTCGTTTCCCGC-3′	RB flank
LB_TAIL1_393	5′-CGGCTAATCTATGTGATTGAGTGTGT-3′	LB flank_pNU393
LB_TAIL2_393	5′-TCTCTTTAATATCACCACGATCAGCTCG-3′	LB flank_pNU393
LB_TAIL2	5′-ATTAAAAACGTCCGCAATGTG-3′	LB flank
LB_TAIL3	5′-ACGTCCGCAATGTGTTATTAAG-3′	LB flank
LB_TAIL4	5′-GTGTTATTAAGTTGTCTAAGCGTC-3′	LB flank
AD2	5′-STTGNTASTNCTNTGC-3′	Degenerate primer for TAIL PCR
AD5	5′-RCAGNTGWTNGTNCTG-3′	Degenerate primer for TAIL PCR
AD7	5′-NTCGASTWTSGWGTT-3′	Degenerate primer for TAIL PCR
AD8	5′-NGTCGASWGANAWGAA-3′	Degenerate primer for TAIL PCR
Ds3_TAIL1	5′-ACCCGACCGGATCGTATCGGT-3′	Ds3′ flank
Ds3_TAIL2	5′-CCCGTCCGATTTCGACTTTAACCC-3′	Ds3′ flank
Ds3_TAIL3	5′ GTATTTATCCCGTTCGTTTTCGT-3′	Ds3′ flank
Ds3_TAIL4	5′-TATCCCGTTTTCGTTTCCGTCC-3′	Ds3′ flank
Ds5_TAIL1	5′-ACGGTCGGGAAACTAGCTCTAC-3′	Ds5′ flank
Ds5_TAIL2	5′-CCGTTTTGTATATCCCGTTTCCGT-3′	Ds5′ flank
Ds5_TAIL3	5′-TACCTCGGGTTCGAAATCGAT-3′	Ds5′ flank
Ds5_TAIL4	5′-TACGATAACGGTCGGTACGG-3′	Ds5′ flank
SS1_F+	5′-TGCCTTGATCGAAGCTGAC-3′	*OsRss1* forward
SS1_R	5′-AGCAAGGGGTAGAGGCTCTC-3′	*OsRss1* reverse

9. Perform FST rescue as detailed in Subheading 3.7.
10. Make regular observations for any visible mutant phenotypes such as seedling vigor, plant height, tiller number, flower morphology, and grain filling.

3.7. Transgene Analysis

3.7.1. Genomic DNA or RNA Extraction

Extract genomic DNA or RNA from *Ds*/T-DNA transformants using the PureGene total nucleic acid isolation kit according to the manufacturer's instructions as follows.

1. Collect samples and freeze with liquid nitrogen.
2. Crush ~20–40 mg frozen material (don't take more).
3. Add 600 μL of cell lysis solution, vortex and incubate at 65°C for 60 min.
4. Add 3 μL RNase A solution (for DNA extraction only), mix and incubate for 15–30 min at 37°C.
5. Spin at high speed for 3 min to pellet plant debris, and collect supernatant (see Note 17).
6. Cool sample to room temperature, then add 200 μL of protein precipitation solution, vortex vigorously for 20 s, and spin at high speed for 3 min.
7. Remove supernatant to new tube, add 600 μL isopropanol and mix thoroughly by inverting 50 times, and centrifuge at high speed for 5 min.
8. Pour off supernatant and wash pellet with 70% ethanol.
9. Air dry pellet and resuspend in 50 μL of 10 mM Tris-HCl/0.2 mM EDTA for DNA or TE for RNA (see Note 18).
10. For RNA extraction, centrifuge the suspension for 15–20 min (4°C) to remove any non-hydrated DNA, or cellular debris left over.
11. Then, add equal volume of 6 M LiCl, leave overnight at 4°C and precipitate bulk of the RNA by centrifugation for 15 min (leaving behind bulk of the DNA in solution).
12. If the pellet is visible, wash it carefully with 70% ethanol. Vacuum dry and resuspend in 20–50 μL of RNase free water.
13. If for RT-PCR or RNA Northern blot, treat RNA samples with RNase-free DNAse (5 μL of 1 U/μL) for 4 h before cleaning with phenol–chloroform extraction, and ethanol precipitation, and resuspension in water. Monitor DNA contamination using control RT-PCR with any plant gene specific primers binding to exons flanking intronic sequences such as rice sucrose phosphate synthase gene (see Note 19).

3.7.2. PCR Analysis for Detection of Transgenes, Ds insertion, Ds Excision, Launch Pad, and Zygosity

Following the initial selection and separation based on hygromycin and/or Basta selection and GFP visualization, confirm the presence or absence of respective transgenes by PCR (see Table 1 for primer list), and/or Southern blot hybridisation (see Subheading 3.7.3).

3.7.3. Southern Blot Hybridisation Analysis for Ds/T-DNA Copy Number Determination

1. Digest DNA from transformed callus or plant lines along with control Nipponbare DNA with appropriate restriction enzyme (see Table 2).

Table 2
Restriction enzymes, probes and primers used for plasmid rescue, Southern blot, TAIL PCR and FST sequencing for *Ds*/T-DNA (launch pad) and *Ds* lines derived from the use different constructs

Construct	Analysis/ rescue target	Restriction enzyme	Probe	Minimum fragment size (kb)	FST sequencing primer
pUR224NA	LB	*Sal* I	*hph* or *bar*	9.9	LB_TAIL3 or LB_TAIL4
	RB	*Hin*dIII	*uidA*	8.5	RB_TAIL3 or RB_TAIL4
	*Ds*3	*Hin*dIII *Sph*I	*uidA*	6.7 5.7	Ds3_6587
	*Ds*5	*Afl*II	*nptII*	6.3	Ds5_112
pUR224NB	LB	*Hin*dIII	*hph* or *bar* or *uidA*	10.3	LB_TAIL3 or LB_TAIL4
	RB	*Sal*II	*nptII*	8.2	RB_TAIL3 or RB_TAIL4
	*Ds*3	*Hin*dIII *Sph*I	*uidA*	6.7 5.7	Ds3_6587
	*Ds*5	*Sal*I	*nptII*	6.3	Ds5_112
pNU393A1	LB	*Nru*I	*bar* or *eyfp*	7.4	LB_TAIL3 or LB_TAIL4
	RB	*Xho*I	*bar* or *uidA*	7.9	RB_TAIL3 or RB_TAIL4
	*Ds*3	*Xho*I	*bar* or *uidA*	5.7	Ds3_6587
	*Ds*5	*Nru*I	*bar* or *eyfp*	6.5	Ds5_112
pNU393B2	LB	*Apa*I *Sph*I	*hphuidA* or *hph*	1.8[a] 5.4	LB_TAIL3 or LB_TAIL4
	RB	*Bgl*II *Nru*I	*eyfgbar* or *eyfp*	2.8[a] 7.4	RB_TAIL3 or RB_TAIL4
	*Ds*3	*Apa*I *Sph*I	*uidA*	5.4 5.2	Ds3_6587
	*Ds*5	*Bgl*II *Nru*I	*eyfpbar* or *eyfp*	1.9[a] 6.5	Ds5_112
pNU435	LB	*Sph*I	*hph* or *uidA*	8.0	LB_TAIL3 or LB_TAIL4
	RB	*Nru*I	*eyfp*	8.00	RB_TAIL3 or RB_TAIL4
	*Ds*3	*Sph*I	*uidA*	5.2	Ds3_6587
	*Ds*5	*Nru*I	*eyfp*	6.6	Ds5_112

2. Run digested DNA on 0.8% agarose gels using 1 × TAE buffer for 16 h at 40 V.
3. Transfer to a Hybond N^+ membrane by alkaline lysis as described by the manufacturer for 4–6 h.
4. Prepare a radioactively labeled probe using either *hph* or *uidA* gene sequences as template (see Note 20) by random hexamer priming using ^{32}P-dATP, and the Klenow fragment of DNA polymerase I according to the Megaprime™ DNA labelling kit protocol.
5. Remove un-incorporated nucleotides using Sepharose G50 columns.
6. Cross-link the transferred DNA and membranes using an ultraviolet cross-linker, and then rinse three times for 10 min in 2× SSC solution.
7. Pre-hybridize with hybridisation solution at 65°C for 1 h.
8. Boil the DNA probe for 3 min and add to the pre-hybridisation solution.
9. Hybridize overnight at 65°C.
10. Wash the membranes briefly in pre-heated (65°C) wash buffer 1, and perform each wash for 15 min at 65°C in wash buffer 1, 2, and 3, consecutively.
11. Air dry membranes, wrap in plastic bags, and expose to X-ray film (Biomax, Kodak) or Phosphor screens.
12. Number of hybridizing bands represent the copy number.

3.7.4. Northern Blot for Expression Analysis of Mutated Gene

1. Isolate RNA from mutant lines as well as wild-type and/or segregating nulls and adjust the concentration to 1 μg/μL.
2. Prepare the gel as follows: For making 100 mL, melt 1.5 g agarose in 85 mL of water, cool to 60°C, add 10 mL of 10× MOPS, and 5.1 mL 12.3 M formaldehyde (in the hood), mix and pour the gel into conveniently sized gel mould with combs. Gel sets in 10 min.
3. Prepare the RNA samples as follows: Add 35 μL of sample preparation premix for normal gel (8.75 μL for mini gel) to 10 μL of RNA sample (2 μL for mini gel), heat at 65°C for 10 min, cool on ice, spin down, and add 5 μL of loading buffer (1 μL for mini gel).
4. Transfer the gel to the electrophoretic apparatus with 1× MOPs running buffer filled just to the level of the gel (not submerged).
5. Fill the well with the running buffer and load RNA samples carefully.
6. Run at 200 V for 10 min, then at 100 V for 1.5–2.0 h until the bromophenol blue dye is two-thirds way down (run mini gels at 80 mA for 30 min).

7. Photograph the gel with a ruler.
8. Transfer to a Hybond N+ membrane by capillary blotting using neutral buffers as described by the manufacturer for 4–6 h.
9. Prepare probe, hybridize and autoradiograph as described in Subheading 3.7.3.

3.7.5. GFP Expression Visualization

1. Use segments of callus, leaf and root tips of regenerated plantlets and root, leaf, spikelet, floral parts, mature seeds, and germinating T_1 seed samples of primary transformants, or of subsequent progeny to screen for GFP expression.
2. Use a Leica MZ6 stereomicroscope with special emission and barrier filters (29) mentioned in Subheading 2.10.

3.7.6. GUS Visualization

Screening for GUS expression using 5-bromo-4-chloro-3-indolyl-β-D-glucuronide (X-gluc) was done according to Jefferson et al. (30), as detailed below. As with GFP, segments of callus, leaf, root tips, spikelets, floral parts, mature seeds, and germinating seeds can be stained to visualize GUS expression.

1. Harvest small segments of tissue (1–2 cm) in Eppendorf tubes.
2. Add 1 mL of GUS staining solution and place samples under vacuum for 2 min before releasing the vacuum.
3. Repeat this process five times. Release of the vacuum, forces staining solution into the leaf.
4. Incubate samples overnight at 37°C.
5. Remove staining solution and perform sequential washes (20 min each) in 25%, 50%, 70%, and 95% ethanol with low agitation at room temperature to remove plant pigments.
6. Tissue showing *uidA* activity can be then seen as blue GUS stained regions.

3.7.7. Recovery and Analysis of Ds/T-DNA Launch Pad or Ds Flanking Sequences

3.7.7.1. Plasmid Rescue

1. Digest 5 μg of genomic DNA with 10 U of appropriate restriction endonucleases (see Table 2 and Note 21).
2. Phenol/chloroform extract, precipitate, and resuspend in water.
3. Self-ligate DNA in 1,000 μL of ligation mix containing 7.5 Weiss units T4 DNA ligase at 16°C overnight.
4. Precipitate ligated DNA, resuspend in a small volume, and perform standard electroporation using electro-competent *E. coli* stain such as XL-1 Blue MRF.
5. Isolate plasmids (see Subheading 3.1) and perform diagnostic digestion with appropriate restriction enzyme.
6. Sequence selected clones using the reagents from the ABI Prism BigDye termination cycle sequencing kit with appropriate primers according to the manufacturer's instructions.

7. Sequence rescued plasmids showing the correct construct-derived "footprint," following cleavage with restriction endonuclease (*Apa*L1) using the appropriate primers and reagents from the ABI Prism BigDye Terminatior Cycle Sequencing Kit according to the manufacturer's instructions.
8. Use Programs from the University of Wisconsin Genetics Computer Group (GCG) for initial flanking sequence analysis (31). BLAST search all sequences against publicly available rice sequences from GenBank (http://www.ncbi.nlm.nih.gov/genome/seq/BlastGen/BlastGen.cgi?taxid=4530), China Rice GD (http://rise.genomics.org.cn/rice/index2.jsp b), or the Michigan State University Rice Genome Annotation Project (http://rice.plantbiology.msu.edu/blast.shtml), to map the FSTs.

3.7.7.2. TAIL-PCR

1. Perform TAIL-PCR to rescue the RB, LB, *Ds*3, or *Ds*5 flanking sequences of transformants according to previously reported protocol (32).
2. The primers to be used in the primary, secondary, and tertiary (if required) PCR reactions for rescues of RB, LB, *Ds5*, or *Ds3* flanking sequences are listed in Table 1. Respective RB_TAIL, LB_TAIL, Ds3_TAIL or Ds5_TAIL primers are used with one of the arbitrary degenerate primers (AD2, AD5, AD7 or AD8; see Note 22).
3. Load the secondary and tertiary PCR reactions on 1% agarose gels side by side. Excise specific bands with expected size difference from the gel, and elute using UltraClean™ 15 agroase gel DNA Purification Kit. Re-amplify the purified PCR products using the primer combinations used in the tertiary PCR before sequencing.

3.8. Mutant Identification: Forward Genetics

We have been concentrating on screening for stable insertion lines i.e., lines where *Ds* elements have transposed from its original T-DNA launch pad into a new location, and *iAc* has segregated away. The average frequencies of PSI lines observed in the F_2, DtT_1, F_3, and DtT_2 populations were 6.61%, 5.58%, 11.47%, and 7.05% respectively, with large variations in these frequencies in screening populations derived from different mutagenic lines (14). We have phenotyped ~1,500 stable *Ds* insertion lines under normal glasshouse conditions and observed altered phenotypes in 30% of the lines, including late germination, defective shoot apex formation, low seedling vigor, seedling lethality, dwarfism, variegated or twisted leaves, early or late flowering, partial or complete sterility, deformed spikelets, and small seeds. An analysis of 350 stable *Ds* insertion lines has shown that 15 and 70% of these lines expressed GUS reporter gene in leaves and spikelets, respectively.

Confirmation that a *Ds* insertion is responsible for a mutant phenotype has to be done using more than one approach (apart from initial segregation analysis), such as (1) identification of multiple alleles showing the same phenotype, (2) revertants with wild-type phenotype, (3) complementation, and (4) phenocopying with RNAi transgenes.

3.8.1. Identifying Reversion Events

It is possible to obtain revertants if *iAc* is still present in the mutant line which could induce retransposition. However, reversion to wild-type can occur only if the excision is clean (without leaving a footprint that affects gene function). In the case of stable insertion mutants (where *iAc* has already segregated away), it needs to be crossed with an *iAc* line and the resulting progeny is screened for revertants.

3.8.2. Complementation

1. Clone wild type sequence surrounding the insertion site into a suitable binary vector (see Note 23).
2. Transform the binary vector containing the wild type sequence into the mutant rice plant.
3. Observe for complementation in the resulting transgenic lines.

3.8.3. Phenocopy with RNAi

Using the RNAi or gene silencing technology, it is possible to phenocopy the mutant phenotype which is another strategy available for the confirmation of gene function.

1. Generate a silencing constructs (33) for the gene in question using appropriate plant promoter to drive its expression (like maize-ubiquitin promoter).
2. Insert this RNAi construct into a suitable rice transformation binary vector.
3. Generate transgenic rice plants containing the RNAi construct.
4. Examine transgenic lines for comparable mutant phenotype.
5. Confirm reduced levels of endogenous target gene expression by RNA Northern blot analysis, or RT-PCR.

3.9. Mutant Identification: Reverse Genetics

In plant populations saturated with multiple endogenous transposon copies, it is possible to identify knock-out mutations in a specific gene (34–38). Using T-DNA or *Ac/Ds*, it is theoretically possible to achieve this saturation with substantial numbers of tagged lines, from which DNA pools can be prepared. It is then possible to screen for mutations in a particular gene by polymerase chain reaction (PCR) analysis, using pooled DNA as a template and a gene-specific primer in combination with an insertion sequence-specific primer. Recovered mutants can then be subjected to custom screening for visible/obvious phenotypes.

However, the chance of recovering an insert in a target gene is dependent on the population size of the insertion lines, the size of the genome, and the size of the target gene. The number of independent insertion mutants needed to tag every gene in rice is estimated to be between 180,000 and 460,000 (1, 39).

4. Notes

1. Select pUR224NA and pUR224NB transformant calli with Hygromycin B (conferred by the CaMV35S promoter-driven *hph* chimeric gene) resistance and pNU393A1 and pNU393B2 transformants with Bialaphos (conferred by CaMV35S promoter-driven *bar* gene) resistance. We prefer using Bialaphos over Phosphinothricin (PPT) for rice, as in our hands, Bialaphos selection works better than PPT. Both selection media should have Timentin to suppress *Agrobacterium* growth.
2. It is better not to add any selection in the rooting media to obtain better rooting and growth especially when callus growth in selection media has been very good. Another benefit is the cost consideration.
3. Instead of X-ray films, phosphor screens can be used with appropriate imaging equipment such as FLA-5000 PhosphoImagers (Fuji, Japan).
4. Most of the commercial bleaches are 4.2% solutions of sodium hypochlorite with some other additives. Depending on the contamination load, you may use 1:2–1:4 diluted commercial formulations. Perform a pilot test to workout the optimum concentration of the bleach and duration of the treatment (30–60 min) as too much bleach treatment hinders callus induction.
5. Contamination of a few seeds can be managed by transferring the uncontaminated ones to a new plate. However, if too many seeds are contaminated, it is better to start again with a higher level of bleach treatment.
6. If only a few calli show *Agrobacterium* overgrowth, transfer the calli without any *Agrobacterium* over growth onto new plates. However, if many calli show *Agrobacterium* overgrowth, repeat the co-cultivation with fresh calli.
7. Bialaphos selection takes more time (40 days) to take effect than hygromycin in rice.
8. These single callus descent lines are numbered for tracking. Once sufficiently multiplied, split the calli from each single callus descent into three parts – one for regeneration, one for DNA extraction for copy number determination, and the

third for super-transformation with *iAc* (TET system, see Subheading 3.4).

9. Air dry calli after the transfer (dehydration effect) for better pre-regeneration. Do not overcrowd the pre-regeneration plates. Regenerable calli look whitish due to profuse production of leaf primordial within 4–7 days in pre-regeneration media.
10. At this stage, the selection pressure may be removed for better regeneration. Timentin may also be omitted if no prior *Agrobacterium* overgrowth was encountered.
11. Optionally, you may trim all the minor shoots and collect as samples for DNA extraction. Firm the mix with hand. Leave in the mister for 7–10 days (until the seedlings get hardened). If there is no mister facility (or running out of mister space), transfer the Jiffy pots into water tubs in the glasshouse. Make sure that the Jiffy pots are just submerged, which avoids any salt accumulation on the surface due to evaporation. Rice seedlings are very sensitive to salt and even a short exposure to salt will kill the seedlings. All the pre-screening (like Basta spray and/or hygromycin spray, sample collection for DNA extraction) can be done at this stage so that unwanted seedlings can be discarded. Spray with iron (0.08% EDTA ferric sodium salt with 300 μL/L Wetting Agent 600) if seedlings show iron deficiency. Take maximum care not to lose any seedlings. For single transformants, keep two to three plants per line. With super transformation experiments keep five plants for each putative double transformant line (GFP^+), and two to three for each putative stable insertion callus line (GFP^-).
12. For primary single transformants and their progenies, use 13–15 cm pots (medium density layout). For F_1 and T_0 double transformants, use bigger pots to increase seed yield. For screening population (F_2 and T_1 double transformants and subsequent screening populations), use high density planting in trays with 24–96 well planting setup. Assemble the pots in the water tub the same day of the transplanting. Ideally, it is better to remove the Jiffy pot and transplant sufficiently deep to have the original adhering potting mix completely covered in the new pot (to have good anchorage). Fill the tub at least to the level of potting mix. For the next 2 weeks, water from the top of the pot (to avoid salt accumulation). Alternatively, after 2–3 days, submerge the pots in water. For 15 cm pots, a side hole should be there at the potting mix surface level for submerging the pot surface to avoid salt accumulation. Inspect plants daily for at least a week to make sure that plants are establishing satisfactorily. If there are algal growths on the surface of the water, remove them to

avoid suffocation of seedlings. Once seedlings are established, any algal growth on the surface of the water will not effect the plant growth.

13. Make sure to take care of the seedlings throughout the experiment (especially initial water levels, iron spray, fertilizer tablets, pest control, sample collection, analyses, phenotypic observation, etc.). Make sure to have the observation sheets and harvest sheet well in advance of the actual operation. Place a slow release fertilizer tablet in each pot. Make sure to spray iron (0.08% of EDTA ferric-sodium salt with 300 μL/L Wetting Agent 600). It may require two sprays at an interval of 10 days. Afterwards, always keep the potting media submerged in water. Apply nutrient solutions fortnightly (replacing with normal watering). Observe the plants for any pests and diseases, and request for appropriate sprays. Drain the water few days before harvest (if no plans to re-pot).
14. Hygromycin B spray does not kill the rice seedlings but produces necrotic spots only whereas Basta spray will kill sensitive plants. Care should be taken while handling hygromycin as it is a very toxic antibiotic. Always wear protective gear while handling.
15. For the high-throughput screening procedures to work, callus lines should be from single copy *Ds*/T-DNA (*Ds* launch pad) lines, and should be heterozygous for *Ds*/T-DNA launch pad so that the excision marker selection can be applied as in the case of single callus descent lines (with single *Ds*/T-DNA copy) from primary *Ds*/T-DNA transformants. *Ds*/T-DNA copy number can be determined by Southern blot hybridization of appropriately digested DNA with radioactively labeled *hph* probe as described (see Subheading 3.7.3), FST rescue (see Subheading 3.7.7), or by observed selectable marker-gene segregation ratio.
16. Depending on the sun light intensity, anthesis commences after 10 a.m. and may continue up to mid-day. On cloudy days, anthesis gets delayed. Anthesis and fertilization is not synchronous across the rice panicle. Therefore, individual spikelets are identified and selected according to their developmental stage by monitoring accurately the beginning of anthesis, which can be recognized by the opening of the spikelet. One can pick the set of near-matured spikelets in the panicle for emasculation the day before crossing. In rice, pollen remains viable for 5–10 min and so it is important to collect the anthers from the donor plants just before anthesis, and sprinkle pollens over the emasculated spikelets.
17. If plant debris precipitation is not achieved, it means that too much proteins are there, some of which still bound to DNA. If so, restart the process with less plant material.

18. With DNA, make sure it is dissolved well. Preferably, leave samples at 37°C overnight. Run 1 µL on an agarose gel to estimate the yield (if required). For PCR, use 0.2–1 µL in a 20 µL reaction mix.
19. Rice sucrose synthase gene has several introns and is a single copy gene. Primers which anneal to exons flanking introns are very useful in monitoring the DNA contamination. With the primers SS1_F and SS1_R (see Table 1), if there is genomic DNA contamination, two PCR products of size 442 and 266 bp will be produced. If there is no DNA contamination, only one band of 266 bp will be produced.
20. It is advisable to use template DNA excised from a simple cloning vector containing just the gene of interest. Using an eluted fragment from a binary vector used for transformation will give non specific hybridization bands due to template contamination.
21. While selecting the restriction enzyme for copy number determination, use a restriction enzyme for which there is a single recognition site preferably closer to the left border sequence, and for which there are frequent recognition sites in the rice genome sequence. This will give better resolution of the hybridizing band. It is important to note that the template sequences used for probe preparation should be entirely within the RE recognition site and the left border.
22. Suitable primer combinations for primary, secondary, and tertiary TAIL PCR need to be determined for each type of rescue (RB, LB Ds3 or Ds5 flanks). Sequences of various primers are provided in Table 1.
23. Ideally, it is better to use the wild-type gene's own promoter and terminator sequences for complementation.

References

1. Hirochika, H., Guiderdoni, E., An, G., Hsing, Y. I., Eun, M. Y., Han, C. D., Upadhyaya, N., Ramachandran, S., Zhang, Q., Pereira, A., Sundaresan, V., and Leung, H. (2004) Rice mutant resources for gene discovery. *Plant. Mol. Biol.* **54**, 325–334.
2. An, G., Lee, S., Kim, S. H., and Kim, S. R. (2005) Molecular genetics using T-DNA in rice. *Plant Cell Physiol.* **46**, 14–22.
3. An, S., Park, S., Jeong, D. H., Lee, D. Y., Kang, H. G., Yu, J. H., Hur, J., Kim, S. R., Kim, Y. H., Lee, M., Han, S., Kim, S. J., Yang, J., Kim, E., Wi, S. J., Chung, H. S., Hong, J. P., Choe, V., Lee, H. K., Choi, J. H., Nam, J., Park, P. B., Park, K. Y., Kim, W. T., Choe, S., Lee, C. B., and An, G. (2003) Generation and analysis of end sequence database for T-DNA tagging lines in rice. *Plant Physiol.* **133**, 2040–2047.
4. Chen, S., Jin, W., Wang, M., Zhang, F., Zhou, J., Jia, Q., Wu, Y., Liu, F., and Wu, P. (2003) Distribution and characterization of over 1000 T-DNA tags in rice genome. *Plant J.* **36**, 105–113.
5. Sallaud, C., Gay, C., Larmande, P., Bes, M., Piffanelli, P., Piegu, B., Droc, G., Regad, F., Bourgeois, E., Meynard, D., Perin, C., Sabau, X., Ghesquiere, A., Glaszmann, J. C., Delseny, M., and Guiderdoni, E. (2004) High throughput T-DNA insertion mutagenesis in rice: a first step towards *in silico* reverse genetics. *Plant J.* **39**, 450–464.

6. Chin, H. G., Choe, M. S., Lee, S. -H., Park, S. H., Koo, J., Kim, N. Y., Lee, J. J., Oh, B. G., Yi, G. H., Kim, S. C., Choi, H. C., Cho, M. J., and Han, C. -D. (1999) Molecular analysis of rice plants harboring an *Ac/Ds* transposable element-mediated gene trapping system. *Plant J.* **19**, 616–623.
7. Greco, R., Ouwerkerk, P. B., Taal, A. J., Sallaud, C., Guiderdoni, E., Meijer, A. H., Hoge, J. H., and Pereira, A. (2004) Transcription and somatic transposition of the maize *En/Spm* transposon system in rice. *Mol. Genet. Genomics* **270**, 514–523.
8. Zhu, Q. -H., Eun, M. Y., Han, C. -D., Kumar, C. S., Pereira, A., Ramachandran, S., Sundaresan, V., Eamens, A. L., Upadhyaya, N. M., and Wu, R. (2007) Transposon insertional mutants: a resource for rice functional genomics, in *Rice Functional Genomics – Challenges, Progress and Prospects* (Upadhyaya, N. M., Ed.), Springer, New York, pp. 223–271.
9. Springer, P. S. (2000) Gene traps: tools for plant development and genomics. *Plant Cell* **12**, 1007–1020.
10. Sundaresan, V., Springer, P., Volpe, T., Haward, S., Jones, J. D., Dean, C., Ma, H., and Martienssen, R. (1995) Patterns of gene action in plant development revealed by enhancer trap and gene trap transposable elements. *Genes Dev.* **9**, 1797–1810.
11. Upadhyaya, N. M., Zhu, Q. H., Zhou, X. R., Eamens, A. L., Hoque, M. S., Ramm, K., Shivakkumar, R., Smith, K. F., Pan, S. T., Li, S., Peng, K., Kim, S. J., and Dennis, E. S. (2006) *Dissociation* (*Ds*) constructs, mapped *Ds* launch pads and a transiently-expressed transposase system suitable for localized insertional mutagenesis in rice. *Theor. Appl. Genet.* **112**, 1326–1341.
12. Lu, H. -J., Zhou, X. -R., Gong, Z. -X., and Upadhyaya, N. M. (2001) Generation of selectable marker-free transgenic rice using double right-border (DRB) binary vectors. *Aust. J. Plant. Physiol.* **28**, 241–248.
13. Kumar, S. C., and Narayanan, K. K. (1997) Gene and enhancer trap constructs for isolating genetic regions from rice. *Rice Biotechnol. Q.* **31**, 17–18.
14. Upadhyaya, N. M., Zhou, X. -R., Ramm, K., Zhu, Q. -H., Wu, L. -M., Eamens, A., Sivakumar, R., Kato, T., Yun, D. -W., Kumar, S., Narayanan, K. K., Thomas, G., Peacock, W. J., and Dennis, E. S. (2002) An *iAc/Ds* gene and enhancer trapping system for insertional mutagenesis in rice. *Funct. Plant Biol.* **29**, 547–559.
15. Eamens, A. L., Blanchard, C. L., Dennis, E. S., and Upadhyaya, N. M. (2004) A bidirectional gene trap construct for T-DNA and *Ds* mediated insertional mutagenesis in rice (*Oryza sativa* L.). *Plant Biotechnol. J.* **2**, 367–380.
16. Wang, M., Upadhyaya, N. M., Brettell, R. I. S., and Waterhouse, P. M. (1997) Intron-mediated improvement of a selectable marker gene for plant transformation using *Agrobacterium tumefaciens. J. Genet. Breed.* **51**, 325–334.
17. Kolesnik, T., Szeverenyi, I., Bachmann, D., Kumar, C. S., Jiang, S., Ramamoorthy, R., Cai, M., Ma, Z. G., Sundaresan, V., and Ramachandran, S. (2004) Establishing an efficient *Ac/Ds* tagging system in rice: large-scale analysis of *Ds* flanking sequences. *Plant J.* **37**, 301–314.
18. Chiu, W. -L., Niwa, Y., Zeng, W., Hirano, T., Kobayashi, H., and Sheen, J. (1996) Engineered GFP as a vital reporter for plants. *Curr. Biol.* **6**, 325–330.
19. Hartley, R. W. (1988) Barnase and barstar. Expression of its cloned inhibitor permits expression of a cloned ribonuclease. *J. Mol. Biol.* **202**, 913–915.
20. Hanahan, D. (1983) Studies on transformation of *Escherichia coli* with plasmids. *J. Mol. Biol.* **166**, 557.
21. Lazo, G. R., Stein, P. A., and Ludwig, R. A. (1991) A DNA transformation-competent *Arabidopsis* genomic library in *Agrobacterium. Biotechnology* **9**, 963–967.
22. Toki, S. (1997) Rapid and efficient *Agrobacterium*-meditated transformation in rice. *Plant Mol. Biol. Rep.* **15**, 16–21.
23. Ditta, G., Stanfield, S., Corbin, D., and Helinski, D. R. (1980) Broad host range DNA cloning system for gram-negative bacteria: construction of a gene bank of *Rhizobium meliloti. Proc. Natl. Acad. Sci. U. S. A.* **77**, 7347–7351.
24. Miller, J. H. (1972) *Experiments in Molecular Genetics*, Cold Spring Harbor Laboratory: Cold Spring Harbor, New York.
25. Thompson, J. A., Abdullah, R., and Cocking, H. (1986) Protoplast culture of rice (*Oryza sativa* L.) using media solidified with agarose. *Plant Sci.* **47**, 123–133.
26. Li, L., Qu, R., Kochko, A. D., Fauquet, C., and Beachy, R. N. (1993) An improved rice transformation system using the biolistic method. *Plant Cell Rep.* **12**, 250–255.
27. Upadhyaya, N. M., Surin, B., Schünmann, P., Ramm, K., Gaudron, J., Taylor, W. C., and Waterhouse, P. M. (2000) *Agrobacterium*-mediated transformation of Australian rice cultivars Jarrah and Amaroo with modified

promoters and selectable markers. *Aust. J. Plant. Physiol.* **27**, 201–210.

28. Coffman, W. R., and Herrera, R. M. (1980) Rice, in *Hybridization of Crop Plants* (Fehr, W. R., and Hadley, H. H., Eds.), ASA and CSSA, Madison, pp. 511–522.
29. Kurup, S., Runions, J., Kohler, U., Laplaze, L., Hodge, S., and Haseloff, J. (2005) Marking cell lineages in living tissues. *Plant J.* **42**, 444–453.
30. Jefferson, R. A., Kavanagh, T. A., and Bevan, M. W. (1987) GUS fusions: beta-glucuronidase as a sensitive and versatile gene fusion marker in higher plants. *EMBO J.* **6**, 3901-3907.
31. Devereux, J., Haeberli, P., and Smithies, O. (1984) A comprehensive set of sequence analysis programs for the VAX. *Nucleic Acids Res.* **12**, 387–395.
32. Liu, Y. G., Mitsukawa, N., Oosumi, T., and Whittier, R. F. (1995) Efficient isolation and mapping of *Arabidopsis thaliana* T-DNA insert junctions by thermal asymmetric interlaced PCR. *Plant J.* **8**, 457–463.
33. Helliwell, C. A., and Waterhouse, P. M. (2005) Constructs and methods for hairpin RNA-mediated gene silencing in plants. *Meth. Enzymol.* **392**, 24–35.
34. Coen, E. S., Robbins, T. P., Almeida, J., Hudson, A., and Carpenter, R. (Eds.) (1989) *Consequences and Mechanisms of Transposition in Antirrhinum majus*, American Society of Microbiology, Washington DC, USA.
35. Das, L., and Martienssen, R. (1995) Site-selected transposon mutagenesis at the hcf106 locus in maize. *Plant Cell* 7, 287–294.
36. Gerats, A. G., Huits, H., Vrijlandt, E., Marana, C., Souer, E., and Beld, M. (1990) Molecular characterization of a nonautonomous transposable element (*dTph1*) of petunia. *Plant Cell* **2**, 1121–1128.
37. Koes, R., Souer, E., van Houwelingen, A., Mur, L., Spelt, C., Quattrocchio, F., Wing, J., Oppedijk, B., Ahmed, S., Maes, T., and et al. (1995) Targeted gene inactivation in petunia by PCR-based selection of transposon insertion mutants. *Proc. Natl. Acad. Sci. U. S. A.* **92**, 8149–8153.
38. Walbot, V. (1992) Strategies for mutagenesis and gene cloning using transposon tagging and T-DNA insertional mutagenesis. *Annu. Rev. Plant Physiol. Plant Mol. Biol.* **43**, 49–82.
39. Krishnan, A., Guiderdoni, E., An, G., Hsing, Y. I., Han, C. D., Lee, M. C., Yu, S. M., Upadhyaya, N., Ramachandran, S., Zhang, Q., Sundaresan, V., Hirochika, H., Leung, H., and Pereira, A. (2009) Mutant resources in rice for functional genomics of the grasses. *Plant Physiol.* **149**, 165–170.

Chapter 13

Reverse Genetics in *Medicago truncatula* Using *Tnt1* Insertion Mutants

Xiaofei Cheng, Jiangqi Wen, Million Tadege, Pascal Ratet, and Kirankumar S. Mysore

Abstract

Medicago truncatula has been chosen as one of the two model species for legume molecular genetics and functional genomics studies. With the imminent completion of *M. truncatula* genome sequencing, availability of large-scale mutant populations becomes a priority. Over the last 5 years, nearly 12,000 insertion lines, which represent approximately 300,000 insertions, have been generated at the Samuel Roberts Noble Foundation using the tobacco retrotransposon *Tnt1*. Individual genomic DNA was isolated from each insertion line and pooled into four levels with the super-pool containing 500 lines. Using *Tnt1*-specific and gene-specific primers, a PCR-based efficient reverse screening strategy has been developed. Amplified PCR products are purified and sequenced to identify the exact insertion locations. Overall, approximately 90% of genes screened were found to have one or more *Tnt1* insertions. Therefore, this PCR-based reverse screening is a rapid way of identifying knock-out mutants for specific genes in *Tnt1*-tagged population of *M. truncatula*. In addition to the DNA pool screening, a web-based database with more than 13,000 flanking sequence tags (FSTs) has also been set up. One can search the database to find an insertion line for the gene of interest.

Key words: *Medicago truncatula*, *Tnt1*, Reverse genetics, PCR-based screening, Mutants

1. Introduction

Legumes are second only to grasses in importance to humans and agriculture. *Medicago truncatula* is emerging as one of the model legume species. Apart from nodulation and nitrogen fixation, *M. truncatula* is a good model system to study compound-leaf development, flowering time, and plants' natural products. More importantly, the availability of various resources for *M. truncatula* makes it feasible for studies in genetics, genomics, proteomics, and metabolomics.

Andy Pereira (ed.), *Plant Reverse Genetics: Methods and Protocols*, Methods in Molecular Biology, vol. 678, DOI 10.1007/978-1-60761-682-5_13,

Mutant collections are important for both forward and reverse genetics. Insertion mutagenesis is a widely used powerful tool in gene discovery and functional characterization in plants (1). T-DNA was successfully used as an insertional mutagen in the model plant *Arabidopsis thaliana* (2). However, due to the limitation of high throughput *in planta* transformation efficiency, large-scale mutant generation by T-DNA insertional mutagenesis (as done in *Arabidopsis*) is not practical for *M. truncatula*. Retrotransposons (class I transposable elements) have been used successfully as insertional mutagens in plants, such as *Tos17* in rice, *Tnt1,* and *TtoI* in *Arabidopsis* and rice, respectively (3–5). Unlike DNA transposons, retrotransposons do not transpose in the vicinity of their original locations, but rather randomly disperse in the genome. Thus, retrotransposons are good candidates for gene tagging in leguminous plants.

Tnt1 was originally isolated from tobacco and it is one of the well characterized retrotransposons (6). We have previously demonstrated that *Tnt1* actively transposes during tissue culture in *M. truncatula* R108 and Jemalong lines (7, 8). The high efficiency of *Tnt1* transposition during tissue culture results in multiple *Tnt1* inserts in regenerated *M. truncatula* lines (from 4 to 50 insertions per genome) (8). More importantly, the inserts in the regenerated lines are stable during seed propagation. In addition, *Tnt1* insertions in regenerated *M. truncatula* lines are, in most cases, independent and can be segregated by genetic crossing. Several developmental and symbiotic mutants (7–11) we have characterized in the *Tnt1* insertion population indicate that the mutant phenotypes were caused by the *Tnt1* insertion. Over the last 5 years, nearly 12,000 *Tnt1* insertion lines, which represent approximately 300,000 insertions, have been regenerated at the Samuel Roberts Noble Foundation. In this chapter, we will describe the application of these *Tnt1* lines for reverse genetics in *M. truncatula*.

The *Tnt1*-tagged *M. truncatula* mutant population generated at the Samuel Roberts Noble Foundation will support legume research just like the SALK T-DNA collections have functioned in *Arabidopsis.* The currently available 12,000 lines represent over 300,000 *Tnt1* inserts, most of which are independently distributed in the gene-rich regions of the genome. From the efficiency of reverse screening and the genome saturation probability, the current 12,000 lines may cover approximately 85% of the *M. truncatula* genome. However, the FST database currently hosts only about ~13,000 FSTs from ~1,100 lines. Direct BLAST-searching the FST database has only a small chance to find an insertion in a gene-of-interest. This is where the reverse screening of DNA pools comes into play. Two gene-specific primers in combination with two *Tnt1*-specific primers will enable us to find an insertion in most genes in the *M. truncatula* genome. Once a *Tnt1*-tagged line is identified, further characterization of the mutant including growing under permissive conditions, and genotyping the segregating population for a possible phenotype is necessary to understand the gene function. When large-scale FST sequencing is performed on these collections, it will be possible to map most of the *Tnt1* inserts in the *Medicago* genome (www.medicago.org). Finding a mutant in your favorite gene will then be a matter of checking the web site (http://bioinfo4.noble.org/mutant/) and ordering the seeds from stock centers analogous to the SALK T-DNA lines. The development of this FST database corresponding to the majority of the *Tnt1* inserts in the population will thus represent a very valuable tool for the scientific research community.

2. Materials

2.1. Regeneration of Tnt1 Mutant Lines

Tnt1 insertion mutants were regenerated by tissue culture. Originally, *Tnt1* was introduced into *M. truncatula* (R108) through *Agrobacterium*-mediated transformation (7). The original line that we used for large-scale regeneration was tnk88-7-7. Tissue culture and plant regeneration were carried out as previously described (7, 8).

2.2. Plant Genomic DNA Isolation

1. Extraction buffer: 100 mM Tris-HCl at pH 8.0, 50 mM EDTA-Na2 at pH 8.0, 500 mM NaCl, 2-mercaptoethanol (350 μl/500 ml). Before using, mix 9.35 ml of above buffer with 0.625 ml of 20% SDS to make the working solution.
2. 3 M potassium acetate.
3. Chloroform.
4. Isopropanol.
5. 75% ethanol.

2.3. PCR Reactions and PCR Product Purification

Ex Taq™ (Takara Bio Inc.) was used for PCR amplification following the manufacturer's protocol. QIAquick PCR Purification Kit (Qiagen) was used for PCR product purification following the manufacturer's protocol, except that the products were eluted with water. The concentration of the PCR product was measured by Nanodrop spectrometer (Nanodrop Technologies, Inc). The products were sequenced using Tnt1-F2 or Tnt1-R2 primer depending on the primers used for PCR reactions.

3. Methods

3.1. Screening DNA Pools

We use a combination of one gene-specific primer and one *Tnt1*-specific primer to selectively amplify the *Tnt1*-tagged gene-of-interest from large DNA pools. Genomic DNA is extracted from individual lines and pooled together so that fewer subsequent PCR reactions are required to screen the entire population. Since we currently have approximately 12,000 lines and we expect to saturate 90% of the *M. truncatula* genome with less than 20,000 *Tnt1* lines, we employ a simple and efficient one-dimensional pooling strategy. Every 500 independent lines make one super-pool. This strategy allows screening of 10,000 lines in 20 super-pools with 80 PCR reactions. This turns out to be quite successful with efficiency of ~90% for more than 200 genes tested so far.

3.1.1. Preparation for PCR Screening

3.1.1.1. Genomic DNA Isolation

1. Approximately 0.3 g of fresh leaf tissue from each regenerated plant (R0) is collected in 2 ml Eppendorf tubes, frozen in liquid nitrogen, and ground to fine power with glass beads.
2. Add 0.5 ml extraction buffer (working solution) into each tube, mix well. Heat the samples in 65°C water bath for 15 min. Mix two to three times during incubation.
3. Add 200 μl of 3 M potassium acetate, invert to mix thoroughly, and set on ice for 10–15 min. Add 200 μl chloroform and mix well.
4. Spin at 17,000 × *g* for 10 min at room temperature.
5. Transfer clear supernatant to a new tube containing 400 μl isopropanol, and invert to mix well.
6. Place the tubes in −80°C freezer for 15–20 min.
7. Spin at 17,000 × *g* for 15 min at 4°C.
8. Pour off the liquid and add 1 ml of 75% ethanol to wash the pellet.
9. Air dry the pellet for 20 min.
10. Dissolve the pellet in 250 μl of distilled, de-ionized H_2O.

3.1.1.2. Genomic DNA Pooling

Figure 1 shows the schematic diagram of DNA pooling strategy.

1. Mini-pool (M): Take 100 μl of genomic DNA from each of ten individual samples, put into one Eppendorf, and invert to mix well to make a 1 ml mini-pool.
2. P-pool (P): Take 500 μl of genomic DNA from each of ten mini-pools, invert to mix in one tube to make a 5-ml pool.
3. Super-pool (S): Take 2 ml of genomic DNA from each of five P-pools; mix in one tube for a 10 ml super-pool.
4. Storage: Aliquot P-pools and super-pools into 500 μl each. Keep one aliquot of each P-pool and super-pool at 4°C for use, and store the rest aliquots, mini-pools, and individual lines at −20°C.

3.1.1.3. Primer Design

Tnt1 Primer Design: Three forward primers, Tnt1-F (5′-ACAGTG CTACCTCCTCT GGATG-3′), Tnt1-F1 (5′-TCCTTGTTG GA

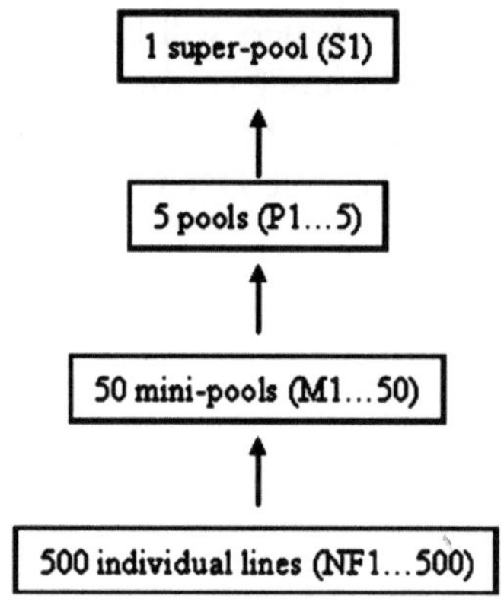

Fig. 1. Genomic DNA pooling chart of the Tnt1 insertion lines.

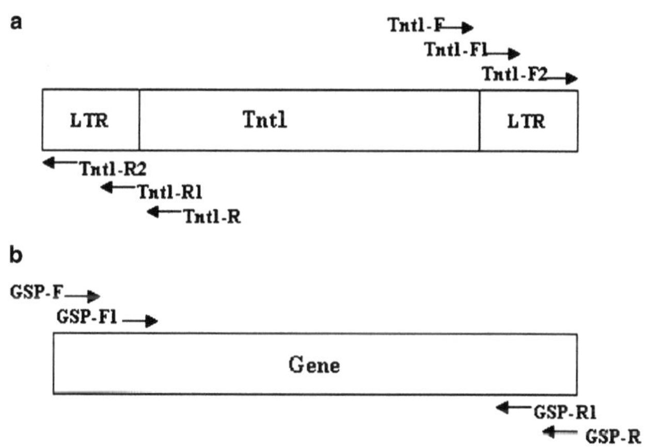

Fig. 2. Primer design for *Tnt1* and specific gene. (**a**) The location and direction of the *Tnt1* forward and reverse primers. (**b**) The location and direction of the gene-specific primers.

TTGGTAGCCAACTTTGTTG-3′), and Tntl-F2(5′-TCTTGT TAATTACCGTATCT CGGTGCTACA-3′); and three reverse primers, Tntl-R (5′-CAGTGAACGAGCAGAAC CTG TG-3′), Tntl-Rl (5′-TGTAGCACCGAGATACGGTAATTAACAAGA-3′), and Tntl-R2 (5′-AGTTGGCTACCAATCCAACAAGGA-3′), are designed from both ends of *Tntl* (Fig. 2a). Primers Tntl-F, Tntl-Fl and Tntl-R, Tntl-Rl are used for PCR screening, whereas primers Tntl-F2 and Tntl-R2 are used for PCR product sequencing.

Gene-Specific Primer Design: Two pairs of gene-specific primers are designed based on the genomic sequence of the gene (Fig. 2b). Primer length is 22–24 bp with 9–11 G/C to match the melting temperatures of *Tntl* primers. The forward primers are located close to the start codon region, and the reverse primers are close to the stop codon region. If the gene sequence is larger than 5 kb, it can be split into two fragments, and then two pairs of primers will be designed for each fragment. The amplification efficiency of the gene-specific primers should be tested by PCR using both wild-type A17 (the reference genome) and R108 genomic DNA as templates.

3.1.2. PCR-Based Reverse Screening in *Tnt1* Insertion Population

3.1.2.1. Screening for *Tnt1* Insertions of Specific Gene(s) in Super-Pools

The schematic screening procedure is shown as a flow chart (Fig. 3). The screening is started with PCR in super-pools. There are four primer combinations for each gene-of-interest: Gene-specific forward (GSP-F) with Tnt1-F, GSP-F with Tnt1-R, GSP-R with Tnt1-F, and GSP-R with Tnt1-R. The combinations of GSP-F with Tnt1-F or Tnt1-R are first used for the screening (see Note 1).

Primary PCR (*first PCR*): Ex Taq™ is used for all PCR reactions. PCR master mixture is prepared according to the product protocol with 1 μM of GSP primer (GSP-F or GSP-R) and 0.25 μM of *Tnt1* primer (Tnt1-F or Tnt1-R). Aliquot 37 μl of the mixture into each PCR tubes and add 3 μl of super-pool DNA into each tube. Perform PCR using a touchdown program: 95°C for 5 min; 94 for 30 s, 60°C for 30 s, and 72°C for 2.5 min, five cycles; 94°C 30 s, 57.5°C 30 s, and 72°C 2.5 min for five cycles; 94°C for 30 s, 55°C for 30 s, and 72°C 2.5 min for 25 cycles; 72°C for 5 min and stored at 10°C.

Secondary PCR (*Nested PCR*): After the first PCR, take 2 μl of each PCR reaction into 98 μl of H_2O and mix well to make 50 times dilution of the first PCR products, and then take 2 μl of diluted first PCR products into PCR tubes as the template for the nested PCR. Prepare nested-PCR reaction mixture with 0.25 μM of GSP (GSP-F1 or GSP-R1) and *Tnt1* nested primers (Tnt1-F1

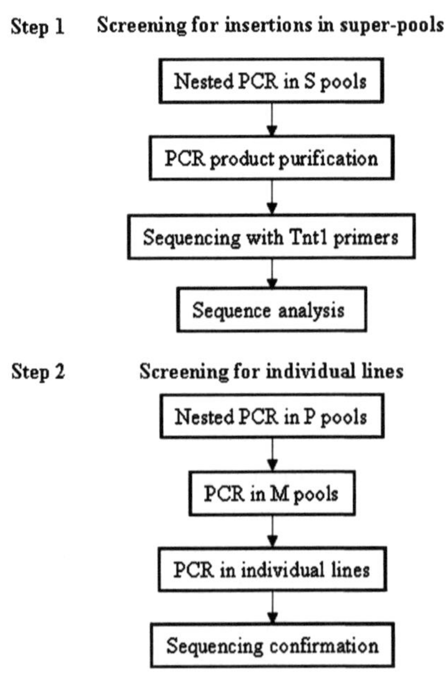

Fig. 3. PCR-based reverse screening flow chart.

or Tnt1-R1). Aliquot 38 μl of the PCR mixture into each tube. Perform PCR under the same program as the first round.

3.1.2.2. Electrophoresis of PCR Products

Take 10 μl of each PCR reaction products from first and second round, add 2 μl 5× loading dye, and mix well, and load on 1% agarose gel side by side. The PCR product should be subject to electrophoresis, and then visualize and capture the image under UV (Fig. 4a). Select the PCR reactions that show bright significant bands on the gel (see Note 2).

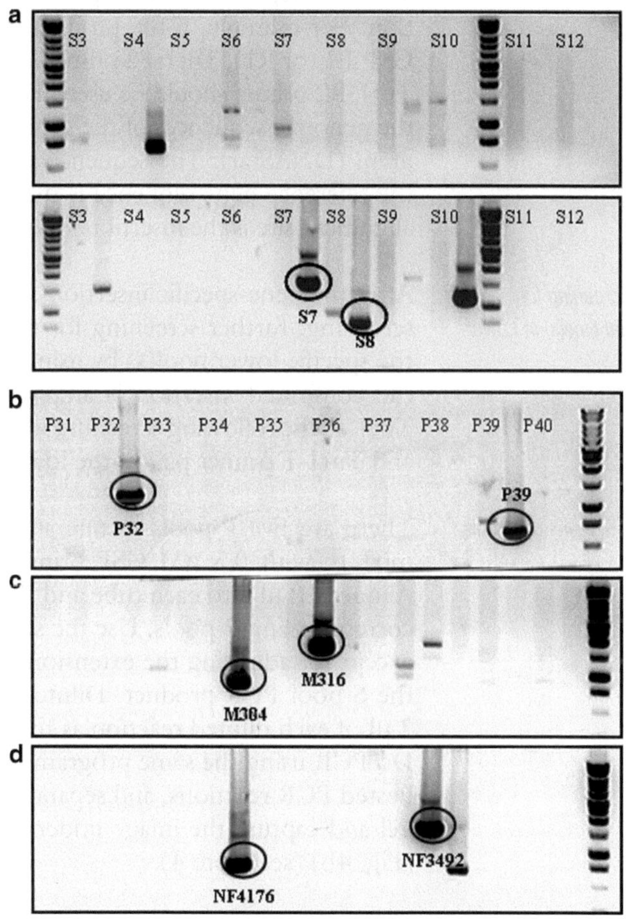

Fig. 4. PCR results of *Tnt1* insertion screening for one gene. (**a**) PCR results in ten super-pools. Two lanes for each super-pool: first lane for first PCR and next lane for the nested PCR. *Upper panel* showing the results with GSP-F and Tnt1-F primer pair. *Lower panel* showing the results of the primer combination of GSP-F and Tnt1-R. One significant band obtained in super-pool 4 with GSP-F and Tnt1-F primers; three significant bands in S7, S8 and S10 with GSP-F and Tnt1-R (see Note 3). (**b**) PCR results in selected pools. S7 includes P31 to P35, S8 includes P36 to P40. The product in P32 showed similar size to that in S7, and the same for the product in P39 and S8. (**c**) PCR results in selected mini-pools. P32 includes M311 to M320 and P39 includesM381 to M390. The expected bands of P32 and P39 are obtained in M316 and M384, respectively. (**d**) PCR results from selected individual lines. M316 includes NF3485 to NF3494 and M384 includes NF4273 to NF4282. The expected bands of M316 and M384 are obtained in individual line NF3492 and NF4276 respectively.

3.1.2.3. PCR Product Purification

The selected PCR products, which have 30 μl of reaction products, should be further purified. QIAquick PCR purification kit (QIAGEN) can be used, and purification should be performed by following the kit protocol, except that the product should be eluted in 30 μl of H_2O. The concentration of the PCR product can be measured using Nanodrop spectrometer (Nanodrop Technologies, Inc).

3.1.2.4. Sequence Analysis

The purified PCR products are sequenced using primer Tnt1-F2 or Tnt1-R2 depending on the primers used for the nested PCR reaction. For example, if the product is amplified with Tnt1-F1 and GSP-F1 (or R1), Tnt1-F2 primer is used for sequencing; otherwise Tnt1-R2 primer should be used. The sequences are compared with the genomic sequences of the gene-of-interest using SeqMan program (DNAStar). The sequence(s) which forms one contig with the reference gene sequence is the gene-specific insertion(s). The alignment site is the insertion location (see Note 3).

3.1.3. Screening for Individual Line(s) in Lower Pools

After the gene-specific insertion(s) is confirmed from super-pool screening, further screening for the insertion(s) is performed in the specific lower pool(s) by using one primer pair. For example, the confirmed insertion is amplified from S1 with GSP-F and Tnt1-F; the following screening will be only preceded with GSP-F and Tnt1-F primer pair in the lower pools of S1.

3.1.3.1. Screening P-Pools

There are five P-pools in one super-pool. Prepare PCR master mixture with 0.5 μM GSP primer and 0.25 μM *Tnt1* primer. Aliquot 28 μl into each tube and add 2 μl of genomic DNA from corresponding P-pools. Use the same touchdown PCR program, except for adjusting the extension time depending on the size of the S-pool PCR product. Dilute the reaction for 50× and take 2 μl of each diluted reaction as the template for the nested PCR. Do PCR using the same program. Take 10 μl from each first and nested PCR reactions, and separate the products on a 1% agarose gel and capture the image under UV light as described in 3.1.2 (Fig. 4b) (see Note 4).

3.1.3.2. Screening Mini-pools

After the product is obtained in a specific P-pool, the screening is followed in the mini-pools. There are ten mini-pools in one P-pool. Prepare PCR master mixture with 0.25 μM GSP and *Tnt1* primers, aliquot 28.5 μl into PCR tubes, and add 1.5 μl of corresponding mini-pool DNA as template. Perform the first PCR as described in Subheading 3.1.2. Separate and visualize the PCR product in agarose gel as above, and check for the same size PCR product (Fig. 4c) (see Note 5).

3.1.3.3. Screening for Individual Lines

After the product is obtained in a specific mini-pool, the screening is followed in the individual lines. There are ten individual lines in one mini-pool. Prepare PCR master mixture with 0.25 μM of GSP and *Tnt1* primers, aliquot 29 μl into PCR tubes, and add 1 μl of individual line DNA as template. Perform first PCR and check the PCR product as described above (Fig. 4d). After the same size product is obtained, purify and sequence the PCR product as described in Subheadings 3.1.2 and 3.1.3. Compare the sequence with the gene-of-interest sequence to confirm that the correct line is identified.

3.2. Searching Flanking Sequence Tags Database

The *M .truncatula Tnt1*-tagged mutant population is a very useful resource for discovering gene function in legumes. If you have a gene of interest for study, especially if you have the genomic sequence of the gene, you may just simply go to our web-based FSTs database (http://bioinfo4.noble.org/mutant/). You can use the sequence of your gene of interest to BLAST search the FST database. Currently, the database includes approximately 13,000 FSTs from the regenerated *Tnt1* lines. Most of the 13,000 FSTs were recovered by thermal asymmetric interlaced (TAIL)-PCR (12, 13). We are in the process of exploring high throughput sequencing approach to recover more FSTs. Once the approach becomes practical, the number of total FSTs will be dramatically increased in the next few years. If you find an FST that matches your gene sequence, you can order seeds online from the same website. Once you receive seeds from us, you may need to genotype the progeny to confirm the insertion and/or identify homozygous plants.

3.3. Progeny Genotyping

3.3.1. Seed Germination

The seeds are treated in concentrated sulfuric acid for 8 min, washed in H2O for two to three times, sterilized for 8–10 min in 30% commercial bleach with 0.01% Tween 20, and then washed in sterilized H_2O for three to five times. Place the seeds on MS media and store in 4°C in dark. After 10-day cold treatment, transfer the seeds to a culture room. When the cotyledons are open, take one cotyledon from each seedling for genomic DNA extraction following the method described in Subheading 3.1.1, with scaled down solutions. Dissolve the pellet in 30 μl of H_2O. The genotyping can also be done at a later stage, when green house plants have only few leaves.

3.3.2. PCR-Based Genotyping in Seedlings

Since the seeds from R1 plants will be segregating for the mutation, it is necessary to confirm which seedlings contain the gene-specific *Tnt1* insertion. In order to address this, prepare PCR mixture with the primer pairs of GSP and *Tnt1* from the screening, add 1–2 μl of genomic DNA from each seedling as the template. Use the PCR program as described in the previous screening. Check the PCR results by separating the products on 1% agarose gel and take a photograph (an example is shown in Fig. 5a). Mark the seedlings which contain the gene-specific *Tnt1* insertion for further homozygosity examination.

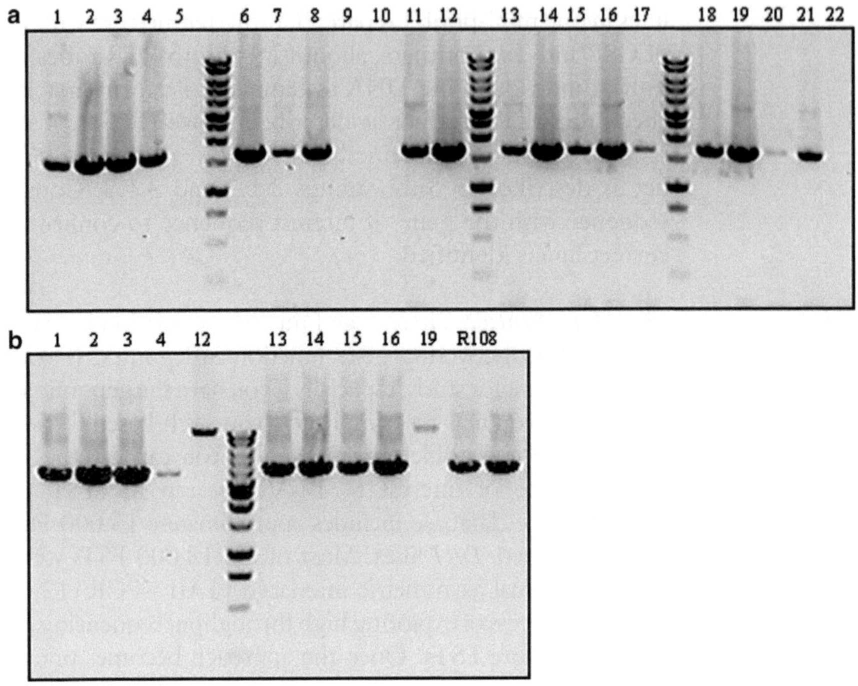

Fig. 5. PCR-based genotyping for the progenies of the specific *Tnt1* insertion line. (**a**) *Tnt1* insertion detection by PCR with GSP-F and Tnt1-R in 22 seedlings. The amplified band indicated that the seedling contained the expected *Tnt1* insertion. (**b**) PCR detection for the insertion homozygote with GSP-F and GSP-R primer pair. The small product (~4.5 kb) represents the gene fragment and the large product represents the gene fragment plus *Tnt1* insertion (~10 kb). Among ten seedlings with the *Tnt1* insertion, eight are heterozygous and two are homozygous for the insertion.

To check seedlings that are homozygous for the specific *Tnt1* insertion, prepare PCR mixture with gene-specific primer pairs flanking the insertion site, add 1–2 μl of seedling genomic DNA. Do PCR with longer extension time, depending on the size of the gene. The expected PCR product for the homozygous is the gene size between the primer pair plus 5.3 kb of *Tnt1* insertion (sample number 12 and number 19 in Fig. 5b) (see Note 6). Transfer the homozygous seedlings into soil for phenotype observation and seed collection (see Note 7).

4. Notes

1. Theoretically, any insertions in the gene should be recovered by using these two primer combinations. If no insertion is recovered in the PCR, the primer combinations of GSP-R with Tnt1-F or Tnt1-R will be used for re-screening from super-pools.

2. Usually, the significant band indicates true insertion in the specific gene. Therefore, choose the strong PCR products for sequencing.
3. If the genomic sequence is not available, and the primers are designed based on the cDNA or EST, the insertion may fall into the intron region and not form one contig with the gene sequence. To prevent missing the insertion, sequence alignment should be analyzed manually. Another problem is that the primers designed from cDNA or EST may span the exon/intron junction and fail to amplify the genomic fragment.
4. Compare the size of the PCR product with the previous one and ensure they are the same size.
5. If no expected PCR product is obtained, dilute the PCR reaction as described above and proceed with the nested PCR. Ensure that similar sized PCR product is obtained.
6. If the gene size is large, it is difficult to amplify the fragment of the gene plus 5.3 Kb *Tnt1* insertion. New primer pairs may be designed flanking the *Tnt1* insertion.
7. Since each line contains on an average 25 insertions, there may be more than one phenotype in the segregating progeny. To clarify whether the phenotype is caused by *Tnt1* insertion in the specific gene, the following two conditions must be met: (1) All plants with the same specific phenotype should be homozygous for the gene-specific insertion. (2) All plants that are homozygous for the gene-specific insertion should display same specific phenotype. Only when both the conditions are met, one can claim association of the specific phenotype with the specific insertion. Normally, the more the plants are examined, the more reliable the results will be.

Acknowledgements

This work was supported by the Samuel Roberts Noble Foundation, in part by NSF plant genome grant (DBI 0703285), and by the European Union (EU FP6-GLIP project FOOD-CT-2004-506223).

References

1. Benetzen J.L. (2000) Transposable element contributions to plant gene and genome evolution. *Plant Mol. Biol.*, 42, 251–269.
2. Alonso J.M., Stepanova A.N., Leisse T.J., Kim C.J., Chen H., Shinn P., Stevenson D.K., Zimmerman J., Barajas P., Cheuk R., Gadrinab C., Heller C., Jeske A., Koesema E., Meyers C.C., Parker H., Prednis L., Ansari Y., Choy N., Deen H., Geralt M., Hazari N., Hom E., Karnes M., Mulholland C., Ndubaku R., Schmidt I., Guzman P., Aguilar-Henonin L., Schmid M., Weigel D., Carter D.E., Marchand

T., Risseeuw E., Brogden D., Zeko A., Crosby W.L., Berry C.C. and Ecker J.R. (2003) Genome-wide insertional mutagenesis of *Arabidopsis thaliana*. *Science*, 301, 653–657.

3. Okamoto H. and Hirochika H. (2000) Efficient insertion mutagenesis of *Arabidopsis* by tissue culture-induced activation of the tobacco retrotransposon *Tto1*. *Plant J.*, 23, 291–304.
4. Courtial B., Feuerbach F., Eberhard S., Rohmer L., Chiapello H., Camilleri C. and Lucas, H. (2001) *Tnt1* transposition events are induced by in vitro transformation of *Arabidopsis thaliana*, and transposed copies integrate into genes. *Mol Genet Genomics*, 265, 32–42.
5. Yamazaki M., Tsugawa H., Miyao A., Yano M., Wu J., Yamamoto S., Matsumoto T., Sasaki T. and Hirochika H. (2001) The rice retrotransposon *Tos17* prefers low-copy-number sequences as integration targets. *Mol Genet Genomics*, 265, 336–344.
6. Grandbastien M.A., Spielmann A. and Caboche M. (1989) *Tnt1*, a mobile retroviral-like transposable element of tobacco isolated by plant cell genetics. *Nature*, 337, 376–80.
7. d'Erfurth I., Cosson V., Eschstruth, A., Lucas, H., Kondorosi A. and Ratet P. (2003) Efficient transposition of the *Tnt1* tobacco retrotransposon in the model legume *Medicago truncatula*. *Plant J.*, 34, 95–106.
8. Tadege M., Wen J., He J., Tu H., Kwak Y., Eschstruth A., Cayrel A., Endre G., Zhao P.X., Chabaud M., Ratet P. and Mysore K.S. (2008) Large-scale insertional mutagenesis using the *Tnt1* retrotransposon in the model legume *Medicago truncatula*. *Plant J.*, 54(2), 335–47.
9. Benlloch R., d'Erfurth I., Ferrandiz C., Cosson V., Beltran J.P., Canas L.A., Kondorosi A., Madueno F. and Ratet P. (2006) Isolation of *mtpim* proves *Tnt1* a useful reverse genetics tool in *Medicago truncatula* and uncovers new aspects of AP1-like functions in Legumes. *Plant Physiol.*, 142, 972–983.
10. Marsh J.F., Rakocevic A., Mitra R.M., Brocard L., Sun J., Eschstruth A., Long S.R., Schultze M., Ratet P. and Oldroyd G.E.D. (2007) *Medicago truncatula* NIN is essential for Rhizobial-independent nodule organogenesis induced by autoactive calcium/calmodulin-dependent protein kinase. *Plant Physiol.*, 144, 324–335.
11. Wang H., Chen J., Wen J., Tadege M., Li G., Liu Y., Mysore K.S., Ratet P. and Chen R. (2008) Control of compound leaf development by FLO/LFY ortholog Single Leaflet1 (SGL1) in *Medicago truncatula*. *Plant Physiol.*, 146, 1759–1772.
12. Liu Y.G., Mitsukawa N., Oosumi T. and Whittier R.F. (1995) Efficient isolation and mapping of *Arabidopsis thaliana* T-DNA insert junctions by thermal asymmetric interlaced PCR. *Plant J.*, 8, 457–463.
13. Liu Y.G., Chen Y. and Zhang Q. (2005) Amplification of genomic sequences flanking T-DNA insertions by thermal asymmetric interlaced polymerase chain reaction. *Methods Mol. Biol.*, 286, 341–348.

Chapter 14

Screening *Arabidopsis* Genotypes for Drought Stress Resistance

Amal Harb and Andy Pereira

Abstract

A high throughput drought screen is described for *Arabidopsis* that is based on a gravimetric method to monitor and control water content of the soil. To screen for plant growth under mild drought conditions, 30% of field capacity can be used, which is equal to 2 g H_2O/g dry soil. The screen allows the testing of a large number of plants of different sizes to the same level of soil water. Therefore, response to drought of different genotypes can be compared. This method can be used for knockout or overexpression genotypes, which are evaluated for their drought response in terms of their growth measured by change in biomass.

Key words: Abiotic stress, Drought, Gravimetric method, *Arabidopsis*, Reverse genetics

1. Introduction

Drought is a major abiotic stress, which causes detrimental losses in plants' biomass and yield. Much molecular data has been obtained on the response and adjustment of plants to drought stress (1–6). But, still there is a necessity for a comprehensive understanding of plants' response to such stress. With the sequencing of plant genomes and the identification of genes expressed under specific stress conditions by microarray analysis, it is of growing interest to find out the function of such genes. Reverse genetics is a well known strategy to discover the function of genes, and how they are involved in the different biological and physiological processes (7). Drought, in terms of soil water content means there is insufficient water to meet the demands for normal growth of plants. Under such water deficit, plants have different mechanisms to save and wisely allocate their resources. This is manifested as retardation in plants' growth, and consequently

Andy Pereira (ed.), *Plant Reverse Genetics: Methods and Protocols*, Methods in Molecular Biology, vol. 678, DOI 10.1007/978-1-60761-682-5_14, © Springer Science+Business Media, LLC 2011

reduction in their biomass. Therefore, change in biomass can be used as a criterion to test the differential response of different genotypes to drought stress (8). A drought screen that simulates to some extent the natural drought was designed. This screen is based on the exposure of different genotypes to the same soil-water level, so that their performance (response) can be compared.

1.1. Overview of the Drought Screen

1. Preparation of planting trays.
2. Sowing and cold treatment.
3. Plant growth for 3 weeks.
4. Drought treatment; dry down for 7–8 days, then mild drought for 5–10 days.
5. Harvest of plants' shoots, and biomass measurement.

2. Materials

2.1. Sowing and Cold Treatment

1. Peat pellets (Jiffy-7, Jiffy Products Ltd, Shippagan, Canada).
2. Trays: TOP press fill tray of 2.5 in. SVD pots (32 pots/tray) (TO Plastics, Clearwater, MN).
3. Flats: 1,020 per box, flat no holes (Dillen, Middlefield, OH).
4. Humidity domes (Dillen, Middlefield, OH).
5. Steel shelf storage trucks (http://www.globalindustrial.com).
6. Marker: Red color is recommended.
7. Labels.
8. Mutant seeds (*Arabidopsis* stock seed center).
9. Needle for sowing.
10. Smooth surface paper for sowing.
11. Cold room set at 4°C.

2.2. Mild Drought Treatment

1. Balance (AND GF-1000, A&D Company, Limited, Tokyo, Japan).
2. Syringes (5 and 10 ml) to adjust the water content.
3. Beakers.

2.3. Harvest and Biomass Measurement

1. Blades.
2. Analytical balance (AND HR-60, A&D Company, Limited, Tokyo, Japan).
3. Glassine bags (1 OZ. FLAT GLASSINE BAG (2¾ × 3¼), Paper Mart, Los Angeles, CA).

4. Card boxes.
5. Oven set at 75°C.

3. Methods

Applying the following method, one can screen large number of genotypes in a batch (about 100 with replications), requiring about six to eight racks available as Steel shelf-storage trucks (www.Globalindustrial.com) in a temperature controlled growth room. Each rack has four shelves, with 40 cm distance between them, and each shelf is provided with six fluorescent lamps giving light intensity of 90 μmol/s/m^2.

3.1. Preparation of Planting Trays

1. Planting trays that contain positions for 32 pellets were used (see Note 1). Choose good completely wrapped pellets (see Note 2), and label each pellet with its position in the planting tray using a marker.
2. Use a balance that is connected to the computer, weigh the labeled pellets, and print the weights directly on an Excel sheet (see Note 3).
3. Take some representative samples to dry, and calculate the dry weight of the soil at the beginning of the experiment.
4. After labeling and weighing, the pellets are ready to be soaked with tap water. Usually they need 30–60 min to be ready for sowing.

3.2. Sowing Process

1. The wet soaked pellets are ready for sowing. One can sow blocks of four genotypes in each tray with eight replications per genotype. Two trays are used with the same genotypes; one is drought-treated, and the other is the well-watered control.
2. After sowing, cover the planting trays with transparent plastic domes and keep them at 4°C for 2 days.

3.3. Plant Growth

1. After 2 days of cold treatment, transfer the planting trays to the growth room at 22°C.
2. Plants germinate on the second day of transfer from the cold treatment and so the plastic domes can be removed, and the plants be kept to grow for 25 days (see Note 4). During this growth period, plants require watering once or twice a week to field capacity.

3.4. Drought Treatment

1. After 25 days of growth to six-leaf stage (9), stop watering and let the pellets dry down. This usually takes 7–8 days (see Note 5).

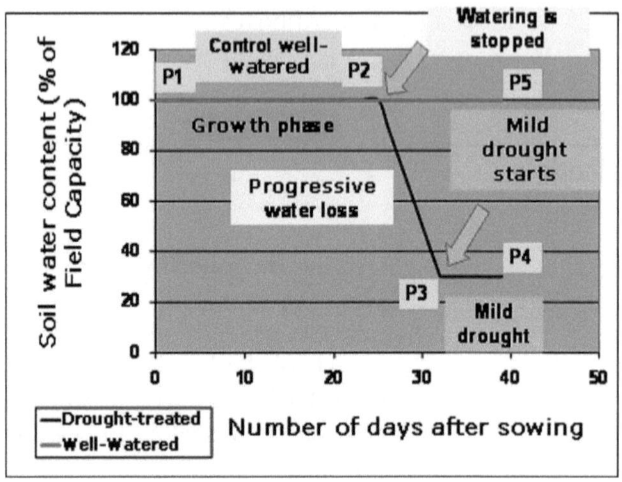

Fig. 1. Schematic illustration of drought treatments. The plant growth starts at point 1 (P1), with two treatments: well-watered control and drought-treated. At P2, watering is stopped for the drought-treated plants, while it is maintained for the well-watered control plants. At P3, mild drought starts, with the soil water content of 2 g H_2O/g dry soil. The duration of mild drought treatment is from 5 to 10 days. After that, the plants are harvested and dried for biomass measurement at P4 for the drought-treated plants, and at P5 for the well-watered plants.

2. When the water content in the pellets reaches 30% of field capacity (see Notes 6 and 10), mild drought stress treatment begins.

3.5. Mild Drought Treatment

1. Once the water content of the pellets reaches 2 g H_2O/g dry soil (30% of field capacity), mild drought treatment starts. Figure 1 shows the water loss from the peat pellets during the duration of the experiment.
2. Prepare an Excel sheet with equations to calculate the required amount of water to keep the water content constant to 2 g H_2O/g of dry soil. Shown below are the equations used to calculate the water content in the pellet, the target pellet weight, and the amount of water needed to reach the target weight (TW).

Water content of the pellet = (Final Pellet Weight– Dry Weight of the Pellet)/ Dry Weight of the Pellet × 100 (1)

Target Weight (to reach mild drought water content (TW)) = 3 × times Dry Weight of the Pellet (2)

Amount of water to reach mild drougnt level(ml) = TW–Final Pellet Weight (3)

3. When the excel sheet is ready, and the balance is connected to the computer, the balance software is activated and ready for use. Weigh the pellets and print (transfer) their final weight to the excel sheet, and the equations will be solved to give the amount of water to be added.

Fig. 2. Representative wild type Columbia plants showing the difference between drought-treated (*right*) and well-watered plants (*left*).

4. Use a syringe of suitable size to add the amount of water shown on the excel sheet (see Note 7 and Fig. 2), so that, each pellet will have the same water content.
5. This adjustment process has to be done on a daily basis. You do not have to write the equations every day, but can copy and paste the form from the previous day.
6. You can expose your plants to mild drought for 5–10 days (see Note 8).

3.6. Harvest and Measurement of Biomass

1. After mild drought treatment, arrange plants as follows: For each genotype, all the replications of the drought-treated and the well-watered control are arranged in one tray and ready to harvest.
2. Take pictures of the plants, which show the difference between drought-treated and well-watered control.
3. Cut the whole rosette of each plant with a blade, and using a weighing boat and a sensitive analytical balance measure the fresh weight.
4. The weighed plants are kept individually in glassine bags (see Note 9). The bags are labeled with the plant position in the tray, its genotype, and the treatment (drought or well-watered).
5. The glassine bags containing the plants are then packaged in card boxes (see Note 10), and kept in the oven at 75°C for 2 days.

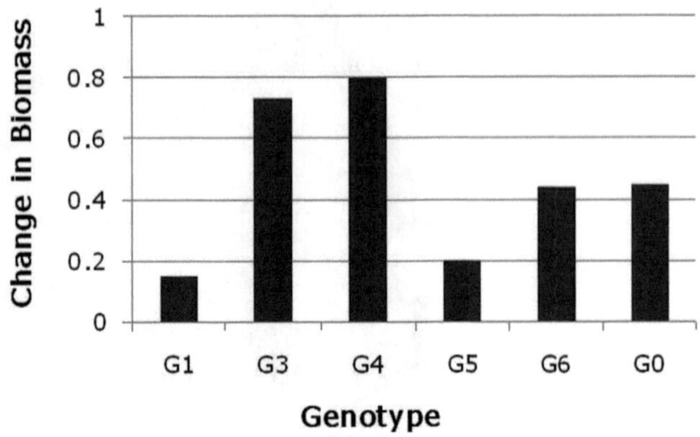

Fig. 3. The change in biomass under mild drought among different genotypes compared to the wild type (G0). Some genotypes (G1and G5) that show less reduction in biomass compared to the wild type (G0) are considered drought resistant genotypes. On the other hand, other genotypes (G3 and G4) are drought sensitive, showing more reduction in biomass compared to the wild type (G0). In addition, there are genotypes (G6) that respond to the mild drought the same as the wild type plants (G0).

6. After drying the plant samples, measure the dry weight (biomass).
7. Calculate the reduction in biomass as shown in the following equation:

   Reduction in Biomass (RB) = {(Biomass of Well Watered Control) – (Biomass of Drought Trated)} / (Biomass of Well Watered Control)
8. Use T-test ($p < 0.05$) to test the significance in the reduction of biomass between the wild type and the other genotypes (Fig. 3).
9. Another way to show the results is to plot biomass under drought and well-watered conditions for each genotype.
10. Other physiological parameters can be measured: Stomatal conductance, transpiration, and CO_2 fixation (see Note 11).

4. Notes

1. The planting-trays with 32 positions are labeled as follows: the rows are labeled with letters, and the columns are labeled with numbers. So, we have eight rows and four columns. In one tray, the rows are A–H, columns are 1–4, and the positions are A1, A2, …, H4. Similar numbering can be used for trays with other formats.

2. Usually, the peat pellets are wrapped with a biodegradable net. But, in a box of Jiffy pellets, you can find some bad, inappropriately wrapped pellets. So, you have to be careful and choose the completely wrapped pellets for your drought test.
3. The software that comes with the balance, transfers the data from the balance to the computer, and is activated during the weighing process. The balance has a print button, which when pressed, transfers the weight data to an active Excel sheet.
4. One can stop watering starting from 25 to 30 days after sowing, and plants are still sensitive to drought stress. At later developmental stages, the plants are not actively growing, and they lose sensitivity to drought in terms of reduction in the biomass.
5. To synchronize the time needed to reach mild drought level, it is recommended that you check the water content of the pellets 5 days after you stop watering, to ensure that all plants have almost the same water level, and remedial watering of some plants might be necessary. At this time, the water level is high enough, and plants do not face any drought stress yet, so you will not interfere with the physiology of the plants.
6. Based on literature and an exploratory drought test, the level of mild drought was determined (10). When plants were exposed to water level of 30% of field capacity, they showed reduction in their biomass compared to the well-watered control.
7. Add the water close to the root growth area in the pellet, but do not harm the roots with the syringe. In addition, be consistent and add the water to the same place for all genotypes, in order to reduce differences in water availability among the different plants.
8. To determine how long the mild drought treatment should continue, we first tested 10 days duration, to make sure that we will get a significant change in biomass. After many experiments, we have seen that at least 5 days of treatment can give a significant reduction in the biomass, which is measurable in all genotypes.
9. It is recommended to oven dry the glassine bags at 75°C overnight, or 1 day before the harvest day. Non-dry bags tend to stick to the plant samples, and thus cause some loss in the measured biomass.
10. Packaging of glassine bags in card boxes is very useful for high throughput screens. This saves space, and makes it feasible to dry large number of plant samples at the same time depending on the size of your oven.

11. During the course of mild drought treatment, plants seem to be acclimated, so that there is no difference in the photosynthesis parameters between drought-treated and well-watered plants. We tested measuring the photosynthesis of wilting plants (progressive drought), and this gave significant results for comparison within and between genotypes.

References

1. Ingram, J. and Bartels, D. (1996) The molecular basis of dehydration tolerance in plants. *Annu. Rev. Plant Physiol. Plant Mol. Biol.* **47**, 377–403.
2. Bray, E. (1997) Plant responses to water deficit. *Trends Plant Sci.* **2**, 48–54.
3. Shinozaki, K., Yamaguchi-Shinozaki, K., Mizoguchi, T., Urao, T., Katagiri, T., Nakashima, K., et al. (1998) Molecular responses to water stress in *Arabidopsis thaliana*. *J. Plant Res.* **111**, 345–351.
4. Ramanjulu, S. and Bartels, D. (2002) Drought- and desiccation-induced modulation of gene expression in plants. *Plant Cell Environ.* **25**, 141–151.
5. Chaves, M., Maroco, J. and Pereira, J. (2003) Understanding plant responses to drought – from genes to the whole plant. *Funct. Plant Biol.* **30**, 239–264.
6. Shinozaki, K. and Yamaguchi-Shinozaki, K. (2007) Gene networks involved in drought stress response and tolerance. *J. Exp. Bot.* **58**, 221–227.
7. Pereira, A. (2001) Genetic dissection of plant stress responses, in *Molecular Analysis of Plant Adaptation to the Environment* (Malcolm J Hawkesford, and Peter Bucner eds.), Kluwer Academic Publishers, Norwell, MA, pp. 17–42.
8. Granier, C., Aguirrezabal, L., Chenu, K., Cookson, S. J., Dauzat, M., Hamard, P., et al. (2005) PHENOPSIS, an automated platform for reproducible phenotyping of plant responses to soil water deficit in *Arabidopsis thaliana* permitted the identification of an accession with low sensitivity to soil water deficit. *New Phytol.* **169**, 623–635.
9. Boyes, D. C., Zayed, A. M., Ascenzi, R., McCaskill A. J., Hoffman, N. E., Davis, K. R., et al. (2001) Growth stage-based phenotypic analysis of *Arabidopsis*: a model for high throughput functional genomics in plants. *Plant Cell* **13**, 1499–1510.
10. Bouchabke-Coussa, O., Quashie, M. L., Seoane-Redondo, J., Fortabat, M. N., Gery, C., Yu, A., et al. (2008) *ESKIMO1* is a key gene involved in water economy as well as cold acclimation and salt tolerance. *BMC Plant Biol.* **8**, 1–27.

Chapter 15

Protein Tagging for Chromatin Immunoprecipitation from *Arabidopsis*

Stefan de Folter

Abstract

A powerful method to identify binding sites in target genes is chromatin immunoprecipitation (ChIP), which allows the purification of in vivo formed complexes of a DNA-binding protein and associated DNA. Briefly, the method involves the fixation of plant tissue and the isolation of the total protein-DNA mixture, followed by an immunoprecipitation step with an antibody directed against the protein of interest and, subsequently, the DNA can be purified. Finally, the DNA can be analyzed by PCR for the enrichment of specific regions.

A drawback of ChIP is that for each protein another antibody is needed. To overcome this, a generic strategy is possible using tags fused to the protein of interest. In this case, only antibody is needed against the tag. This protocol describes the tagging of proteins and how to perform ChIP.

Key words: ChIP, Chromatin immunoprecipitation, Protein-DNA complex, Target genes, Binding sites, Transcription factors

1. Introduction

Most transcription factors fulfill their role in complexes and regulate their target genes upon binding to DNA motifs located in upstream regions or introns. To date, knowledge about transcription factor target genes and their corresponding transcription factor binding sites are still very limited (e.g. (1, 2)).

A powerful method to identify target sites is chromatin immunoprecipitation (ChIP), which allows the purification of in vivo formed complexes of a DNA-binding protein and associated DNA (reviewed in: (3)). In short, the method involves the fixation of plant tissue and the isolation of the total protein-DNA mixture, followed by an immunoprecipitation step with an antibody directed against the protein of interest and, subsequently,

Andy Pereira (ed.), *Plant Reverse Genetics: Methods and Protocols*, Methods in Molecular Biology, vol. 678,
DOI 10.1007/978-1-60761-682-5_15, © Springer Science+Business Media, LLC 2011

the DNA can be purified. Finally, the DNA can be analyzed by PCR for the enrichment of specific regions. Alternatively, the precipitated DNA can be amplified and used to hybridize to microarrays containing promoter elements or the entire genome as tiled oligonucleotides (ChIP-on-chip; (4–6)), or, nowadays coming more in reach, next-generation sequencing technology may be used to sequence the precipitated DNA (ChIP-seq; (7, 8)).

However, one drawback of ChIP is the need of an antibody for each specific protein. One approach to overcome this problem is the use of a tag fused to the protein of interest. A specific tag can be used as a kind of generic tagging approach for ChIP experiments. The best location of the tag, N- or C-terminally fused to the tag, depends on the protein of interest and many times has to be experimentally verified. Furthermore, the gene that encodes for the protein of interest can be expressed under a strong constitutive promoter, e.g., CaMV 35S (9, 10), or under its own promoter. An advantage of a constitutive promoter is that high expression in almost all tissues of the fusion protein can be expected, though, this can also have negative effects on the development of the plant. For instance, it has been shown that some transcription factors (e.g., MADS-box family) are very sensitive to protein fusions and to expression under a strong constitutive CaMV 35S promoter (11).

Many different tags that can be used for affinity purification of a protein-DNA complexes are available (e.g., (12–14)). The GREEN FLUORESCENT PROTEIN (GFP) has shown to be very useful in scientific studies over the last 15 years (15–17). Besides the ability to observe the expression of a fused protein to GFP (e.g., (11, 18)), there are also commercially available antibodies. Therefore, we focus in this protocol on the use of GFP as a generic tag and how to perform the ChIP procedure (see Fig. 1). The basis of this protocol is work from Ito et al. (19) and Wang et al. (20). Examples of the use of, specifically, this protocol are work published by, e.g., Gómez-Mena et al. (21) and de Folter et al. (11).

2. Materials

2.1. Tagging Vector

1. PCR primers to amplify a genomic fragment of the gene of interest.
2. PCR reagents: proofreading DNA polymerase, dNTPs, buffer, $MgCl_2$.
3. PCR purification kit.
4. pENTR/D TOPO cloning kit (Invitrogen; Gateway™ cloning technology).

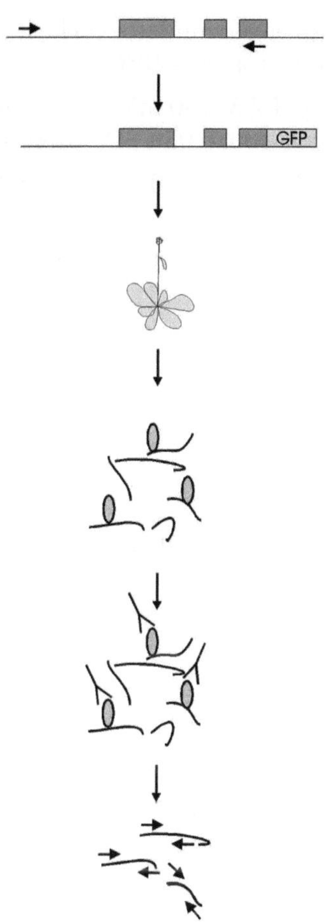

Fig. 1. Schematic overview of making a tagging construct and the ChIP procedure.

5. Suitable binary tagging vector that is Gateway™ compatible (pMDC204, contains a GFP-tag) (22).

2.2. Chromatin Immunoprecipitation

1. MC buffer (fresh): 10 mM sodium phosphate pH 7.0, 50 mM NaCl, 0.1 M sucrose.
2. M1 buffer (fresh): 10 mM sodium phosphate pH 7.0, 0.1 M NaCl, 1 M 2-methyl 2,4-pentanediol (hexylene glycol; Sigma), 10 mM β-mercaptoethanol, Complete™ Protease Inhibitor Cocktail (Roche Diagnostics GmbH, Mannhein, Germany) (protease inhibitor cocktail is added just before use; one tablet per 50 ml).
3. M2 buffer (fresh): M1 buffer with 10 mM $MgCl_2$, 0.5% Triton X-100.
4. M3 buffer (fresh): M1 buffer without 2-methyl 2,4-pentanediol.
5. Sonic buffer: 10 mM sodium phosphate pH 7.0, 0.1 M NaCl, 0.5% Sarkosyl, 10 mM EDTA, Complete™ Protease Inhibitor

Cocktail, 1 mM PMSF (protease inhibitor cocktail and PMSF are added just before use) (store at –20°C).

6. IP buffer: 50 mM Hepes pH 7.5, 150 mM KCl, 5 mM $MgCl_2$, 10 μM $ZnSO_4$, 1% Triton X-100, 0.05% SDS (store at –20°C).
7. Elution buffer (fresh): 50 mM Tris pH 8.0, 10 mM EDTA, 1% SDS.
8. Wash buffer to prepare beads: 10 mM Tris pH 7.5, 150 mM NaCl.
9. Formaldehyde (37% stock).
10. Glycine 1.25 M (store at 4°C).
11. 0.1 M PMSF (phenylmethylsulfonyl fluoride) (in ethanol; store at –20°C).
12. RNase A (10 mg/ml; store at –20°C).
13. Proteinase K (20 mg/ml; store at –20°C).
14. Phenol/chloroform (50/50).
15. Chloroform.
16. 100% ethanol.
17. 70% ethanol.
18. 3 M NaAc pH 5.4.
19. Glycogen (20 μg/μl; store at –20°C).
20. MilliQ water with 10 mM Tris pH 8.0.
21. Complete™ Protease Inhibitor Cocktail (Roche Diagnostics GmbH), 50× concentrated stock (one tablet in 1 ml water; ~12 weeks stable at –20°C).
22. Antibody.
23. Blocked protein A/G agarose beads for IP.

2.3. DNA Enrichment Analysis by PCR

1. PCR primers to amplify the (expected) binding site in a possible target gene and control PCR primers for a region lacking the binding site.
2. PCR reagents: Taq DNA polymerase, dNTPs, buffer, $MgCl_2$.

3. Methods

3.1. Tagging Vector

1. Amplify a genomic fragment of your gene of interest with specific primers that will include the native promoter till the last codon before the stop codon of the gene by PCR on genomic DNA (see Note 1), in the case of C-terminal fusions.

Verify the PCR product on gel and when a single fragment is obtained, purify the PCR reaction with a PCR purification kit (see Note 2). Clone the gene in the pENTR/D vector according to manufacture's instructions (Invitrogen) (see Note 3) and verify the integrity of the genomic fragment by sequencing. Subsequently, recombine the obtained Gateway™ ENTRY clone using LR clonase (Invitrogen) in a Gateway compatible binary vector (pMDC204; see Note 4) that contains a C-terminal tag of choice, which will be essential to be able to perform ChIP (see Note 5). Transform the binary tagging vector to *Agrobacterium tumefaciens*, followed by transformation *Arabidopsis* plants using the floral dip method (23).

3.2. Chromatin Immunoprecipitation

Day 1

1. Collect plant material, normally 1 g of Arabidopsis tissue (see Note 6).
2. Fixate (crosslink) tissue in 50 ml tube in MC buffer (25 ml) with 1% formaldehyde (676 μl of 37% stock). Place the 50 ml tube (without the cap) on ice and apply vacuum for 15 min (see Note 7).
3. Stop the fixation by adding glycine to a final concentration of 0.125 M (2.5 ml of 1.25 M stock) and apply again vacuum for 5 min.
4. Wash three times with cold MC buffer (afterward dip the tissue dry on a paper) (see Note 8).
5. Grind tissue in a mortar with liquid nitrogen till a fine powder is obtained (do not let it melt) (see Note 9).
6. Transfer the powder to a 50 ml tube with 25 ml cold M1 buffer and mix by shaking.
7. Filter the slurry through a 55 μm nylon mesh and collect the flow-through in a clean 50 ml tube, which is standing on ice (see Note 10).
8. Centrifuge the filtrate for 10 min at 1,000×*g* at 4°C (see Note 11).
9. Wash the nuclear pellet five times with 5 ml cold M2 buffer, and centrifuge as in step 8 each time (see Note 12).
10. Wash once with 5 ml cold M3 buffer and centrifuge as in step 8 (see Note 13).
11. Resuspend the crude nuclear pellet in 1 ml Sonic buffer (including PMSF (10 μl 0.1 M) and protease inhibitor cocktail (20 μl 50×)) and transfer to a 2 ml eppendorf tube (see Note 14).

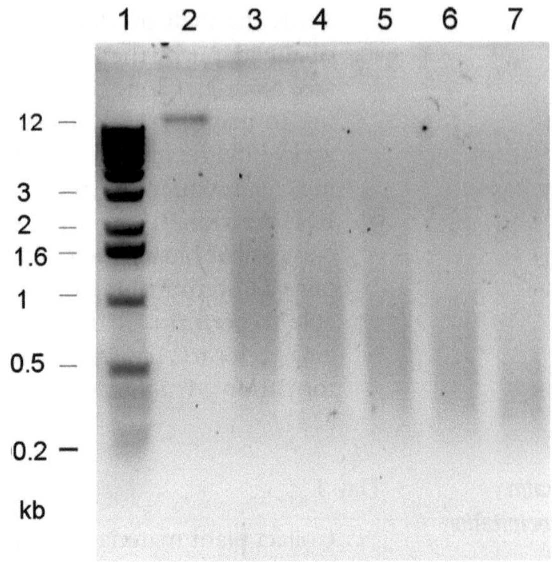

Fig. 2. Agarose gel analysis of sonicated ChIP samples of a sonication efficiency test. A probe sonicator MSE Soniprep 150 was used at 50% power output and different sonication rounds were performed (see Subheading 3.2 step 12 and Note 15). From each sample, 20 μl out of 1 ml was analyzed on a 0.8% agarose gel (lanes 2–7). Lane 1, 1 kb marker; lane 2, no sonication; lane 3, one round of 15 s sonication; lane 4, two rounds of 15 s sonication; lane 5, three rounds of 15 s sonication; lane 6, four rounds of 15 s sonication; lane 7, eight rounds of 15 s sonication.

12. Sonicate the chromatin on ice with a probe sonicator (50% power output, MSE Soniprep 150) three times for 15 s with 30 s break between sonication rounds (see Note 15; see Fig. 2). Leave the eppendorf tube 5 min on ice after sonication.
13. Centrifuge the suspension in a microcentrifuge for 5 min maximum speed at 4°C.
14. Transfer the supernatant to a new 2 ml eppendorf tube and mix with 1 ml IP buffer.
15. Set aside 50 μl to serve as "Input DNA" control in the PCR and to check the degree of sonication. Store sample at –20°C, continue with this sample in step 24 (see Note 16).
16. To preclear chromatin, add 50 μl protein-A/G beads (washed) and incubate on a rotating wheel for 30–60 min at 4°C (see Note 17).
17. Centrifuge in a microcentrifuge for 2 min at 1,000 × g at 4°C.
18. Divide the supernatant equally over two 1.5 ml eppendorf tubes, one that serves as the IP-sample and the other as the negative control.
19. Add to the IP-sample 2.5 μl antibody (depending on concentration antibody, typically 0.5 μg). Incubate for 60 min on a

rotating wheel at 4°C. To the negative control no antibody is added (see Note 18).

Day 2

20. Add 20 µl protein-A/G beads to each eppendorf tube and incubate overnight on a rotating wheel at 4°C.
21. Centrifuge for 2 min at 1,000 × *g* at 4°C and discard supernatant.
22. Wash the agarose beads five times with 1 ml IP buffer (without protease inhibitors). Incubate beads each time 5 min on a rotating wheel at RT, followed by centrifugation for 1 min 1,000 × *g* at RT.
23. Elute the beads by adding 50 µl Elution buffer and incubate the samples for 10 min at 65°C shaking, followed by centrifugation for 30 s 2,000 × *g*. Transfer 40 µl of the supernatant to a new 1.5 ml eppendorf tube.
24. Add 120 µl Elution buffer to the eluate. Also include the "Input DNA" sample (from step 15) and add 200 µl Elution buffer. Incubate samples overnight at 65°C (reverse-crosslinking).

Day 3

25. Add 1 µl RNase A (10 mg/ml) and incubate 30 min at 37°C.
26. Add 10 µl proteinase K (20 mg/ml) and incubate for 2 h at 37°C.
27. Extract DNA by adding 1 vol of phenol/chloroform (50/50), followed by centrifugation for 10 min at maximum speed (see Note 19).
28. Add 1 vol of chloroform, followed by centrifugation for 5 min at maximum speed.
29. Precipitate DNA with 2.5 vol 100% ethanol, 1/10 vol 3 M NaAc pH 5.4, and 1 µl glycogen (20 µg/µl overnight at −20°C or 1 h at −80°C).
30. Centrifuge 30 min at maximum speed at 4°C.
31. Wash the DNA with 500 µl 70% ethanol, followed by centrifugation for 5 min at maximum speed at 4°C.
32. Resuspend DNA in 25 µl milliQ water (with 10 mM Tris, pH 8.0). Samples can be stored at −20°C.

3.3. DNA Enrichment Analysis by PCR

ChIP PCRs can be performed to reveal if a specific DNA fragment is enriched in the IP-sample compared to the negative-control sample. Primers should be designed around the (expected) binding site, and control primers are made for a region lacking the binding site (see Note 20).

Template ChIP DNA can be used undiluted or diluted (IP and negative-control sample 1–1/10; "Input DNA" 1/100) (see Note 21), amplified by PCR, and analyzed typically after 35–40 cycles (see Note 22) on a 1–1.5% agarose electrophorese gel. Perform each PCR reaction in 25 μl: 15 μl DNA mix and 10 μl primer mix (make a matrix, see below).

DNA-mixtures		
DNA (1/100 input, IP, or negative)	1 μl	Multiply by "number of primer primers + 1"
PCR buffer 10×	2.5 μl	
10 mM dNTPs	0.5 μl	
Taq polymerase	0.2 μl	
MQ water	10.8 μl	
Primer mixtures		
Primer 1 (10 μM)	1 μl	Multiply by "number of DNA samples + 1"
Primer 2 (10 μM)	1 μl	
MQ water	8 μl	

4. Notes

1. A genomic fragment of the gene of interest means the native promoter and the gene (exons and all introns). It is difficult to predict the complete size of the promoter of a gene. A general rule that works normally well is to take 2.5 kb upstream of the start codon. When there is another gene in this upstream region, amplify a smaller region that excludes the other gene. The use of a constitutive promoter, like CaMV 35S, might work well, though, it is more artificial. The risk is higher to obtain secondary effects related to plant development because of a change in expression pattern and the level of expression. To predict the best location of the tag, N- or C-terminal, is not so easy. Mostly, this is trial and error. In many cases, C-terminal is working well. When using a C-terminal tag fused to the gene product, it means that the stop codon of the gene should not be present. The easiest is to design the reverse oligo in the C-terminal part of the gene and stopping exactly before the stop codon of the gene. Very important is to maintain the reading frame of the fused gene with the tag. The tagging approach can be used in a wild-type background, though there is the competition with the native protein lacking the tag. An option is to use the mutant back-

ground for the corresponding gene. This will also reveal whether the fused product is fully functional by observing if the mutant can be complemented.

2. The PCR fragment may also be isolated from gel.
3. The use of the Gateway™ system (Invitrogen) makes cloning procedures more easy and quick, however, other vector systems can be used.
4. There are many binary vectors available with different tags (22, 24–27). These references, among others, are examples of publicly available collections of tagging vectors.
5. Nowadays, for many tags commercial antibodies are available. A few examples of some companies that have ChIP grade antibodies are: Abcam, Upstate/Millipore, and Santa Cruz, among others. Verify which species is used to produce the antibody (see Note 17).
6. The type of tissue depends on where your protein of interest is expressed. The amount of tissue necessary depends on the level of protein expression and whether a wild-type background or a complemented mutant background is used. Though, a good start is 1 g of tissue.
7. Time needed for fixation may differ for different tissues or species. Too little fixation will result in the loss of protein-DNA complexes and overfixation will yield in less soluble chromatin.
8. The purpose of dipping the tissue dry on a paper is avoiding a big clomp of ice when liquid nitrogen is added. The grinding will be more difficult.
9. Instead of a mortar and liquid nitrogen a blender may be used. In this case, process tissue directly in 25 ml M1 buffer.
10. The amount of nuclei may be checked by DAPI staining and the use of a UV-microscope. When large amounts of nuclei remain on the nylon mesh, the supernatant after step 8 can be reapplied on the nylon mesh. Collect the flow-through in the same tube that contains the pellet. Repeat centrifugation step 8.
11. Should give a thick greenish pellet (chloroplasts).
12. The pellet gradually becomes whiter (mostly starch).
13. This wash step will remove the triton X-100. Triton will otherwise cause a lot of foaming during sonication, which results in less efficient sonication.
14. Use a 2 ml eppendorf tube. In this tube, the probe of the sonicator will be put, which results in rising of the solution.
15. The sonication step is to shear the chromatin to obtain DNA fragments with an average of 500 bp. This step is critical. The time of sonication depends on the type of sonicator, material,

and time of fixation. The eppendorf tube is the whole time on ice avoiding heating of the sample. Put the probe of the sonicator in the solution so that it does not touch the walls of the eppendorf tube and is ~0.5 cm above the bottom of the tube. When a lot of foam is formed during sonication, lower the output of the sonicator. Foam also lowers the sonication efficiency, but the protocol can be continued. Most part of the shearing happens during the first seconds of the first sonication round. When starting for the first time using a sonicator, it should be tested for efficiency. To do so, perform steps 1–10 as written (protease inhibitor cocktail can be left out for this test, also the Sonic buffer can be without). Resuspend the nuclei pellet in step 11, e.g., 6 ml Sonic buffer and divide over six eppendorf tubes (1 ml each). Change the parameters of sonication for each eppendorf tube (increase the volume of Sonic buffer and divide over more eppendorf tubes to be able to test more parameters). After step 13, transfer supernatant to new 1.5 ml eppendorf tubes and incubate overnight at 65°C, followed by steps 25 and 26. Check sonication efficiency by running 20 μl of each sample on a 1.2% agarose gel. Alternatively, when almost nothing is visible on gel, the samples can be precipitated. See Fig. 2 for an example of a sonicator test. Furthermore, chromatin samples can be stored for several months at −80°C.

16. The "Input DNA" serves as a control to check sonication efficiency and as a positive control in the PCRs. When you have a running ChIP protocol, the check for sonication efficiency is more to confirm in the end that the starting material was good (run, e.g., 3 μl from the 25 μl on an agarose gel). When it is desirable to check for the presence of the fusion protein by Western blot, another 50 μl can be stored.

17. The type of beads to be used depends on the origin of serum. Use for rabbit-derived antibodies protein-A agarose beads, and for goat-derived antibodies protein-G agarose beads. For other species derived antibodies, verify information about the efficiency of binding to the different beads. Examples of sources of beads are: Abcam, Upstate/Millipore, and Santa Cruz, among others. The beads should be washed before using them. For one experiment (one IP and one negative control sample), take 100 μl slurry (normally beads are as 50% slurry) and put in a clean 1.5 ml eppendorf tube. Add 500 μl Wash buffer beads (10 mM Tris pH 7.5, 150 mM NaCl) and mix. Centrifuge for 2 min at 1,000 × *g* at 4°C, remove supernatant and repeat once more. Resuspend the beads again as 50% slurry by adding Wash buffer beads. Store the beads at 4°C. Perform this washing step on the day that the beads are used.

18. Alternatively, to the negative control 2.5 μl preimmune serum or an unrelated antibody may be added when available. Another possibility for a negative control is using a wild-type sample that does not contain the fusion protein. In this case, also add the antibody to this sample.
19. After step 26, DNA may also be directly purified by using a PCR purification kit (e.g., Qiagen).
20. Design primers to amplify a relatively small region, like 200–400 bp. PCR conditions will have to be optimized depending on the material and primers.
21. Samples can be used directly or diluted in PCR. A good start is to use the IP and negative-control sample undiluted (alternatively, 10× dilution), and the "Input DNA" sample as a 100× dilution (alternatively, 1,000× dilution).
22. Optionally, remove after 35 cycles 10 μl sample of each tube and continue till 40 cycles with remaining part of the sample. Run 10 μl of each sample on a 1.2% agarose gel. Finally, the PCR result of the IP sample is compared with the result of the negative-control sample. Enrichment can be calculated based on intensity difference. Alternatively, real-time PCR reactions can be performed. Always repeat PCRs to confirm the results. For more detailed information on data analysis and background information about the different steps during the ChIP protocol, see ref. (28).

Acknowledgments

This work was previously financed by the Netherlands Proteomics Centre (NPC) and now the work in de Folter laboratory is financed by the Mexican Science Council (CONACyT 82826) and CONCyTEG (08-03-K662-116).

References

1. de Folter, S., and Angenent, G. C. (2006) *Trans* meets *cis* in MADS science. *Trends Plant Sci* **11**, 224–31.
2. Hannenhalli, S. (2008) Eukaryotic transcription factor binding sites – modeling and integrative search methods. *Bioinformatics* 24, 1325–31.
3. Orlando, V. (2000) Mapping chromosomal proteins in vivo by formaldehyde-crosslinked-chromatin immunoprecipitation. *Trends Biochem Sci* **25**, 99–104.
4. Wu, J., Smith, L. T., Plass, C., and Huang, T. H. (2006) ChIP-chip comes of age for genome-wide functional analysis. *Cancer Res* **66**, 6899–902.
5. Buck, M. J., and Lieb, J. D. (2004) ChIP-chip: considerations for the design, analysis, and application of genome-wide chromatin immunoprecipitation experiments. *Genomics* **83**, 349–60.
6. Mockler, T. C., and Ecker, J. R. (2005) Applications of DNA tiling arrays for whole-genome analysis. *Genomics* **85**, 1–15.
7. Wold, B., and Myers, R. M. (2008) Sequence census methods for functional genomics. *Nat Methods* **5**, 19–21.

8. Mardis, E. R. (2007) ChIP-seq: welcome to the new frontier. *Nat Methods* **4**, 613–4.
9. Odell, J. T., Nagy, F., and Chua, N. H. (1985) Identification of DNA sequences required for activity of the cauliflower mosaic virus 35S promoter. *Nature* **313**, 810–12.
10. Kay, R., Chan, A., Daly, M., and McPherson, J. (1987) Duplication of CaMV 35S promoter sequences creates a strong enhancer for plant genes. *Science* **236**, 1299–302.
11. de Folter, S., Urbanus, S. L., van Zuijlen, L. G., Kaufmann, K., and Angenent, G. C. (2007) Tagging of MADS domain proteins for chromatin immunoprecipitation. *BMC Plant Biol* 7, 47.
12. Hearn, M. T. W., and Acosta, D. (2001) Applications of novel affinity cassette methods: use of peptide fusion handles for the purification of recombinant proteins. *J Mol Recognit* **14**, 323–69.
13. Lichty, J. J., Malecki, J. L., Agnew, H. D., Michelson-Horowitz, D. J., and Tan, S. (2005) Comparison of affinity tags for protein purification. *Protein Expr Purif* **41**, 98–105.
14. Terpe, K. (2003) Overview of tag protein fusions: from molecular and biochemical fundamentals to commercial systems. *Appl Microbiol Biotechnol* **60**, 523–33.
15. Chalfie, M., Tu, Y., Euskirchen, G., Ward, W. W., and Prasher, D. C. (1994) Green fluorescent protein as a marker for gene expression. *Science* **263**, 802–5.
16. Chiu, W., Niwa, Y., Zeng, W., Hirano, T., Kobayashi, H., and Sheen, J. (1996) Engineered GFP as a vital reporter in plants. *Curr Biol* **6**, 325–30.
17. Schmid, J. A., and Neumeier, H. (2005) Evolutions in science triggered by green fluorescent protein (GFP). *Chembiochem* **6**, 1149–56.
18. Urbanus, S., de Folter, S., Shchennikova, A., Kaufmann, K., Immink, R., and Angenent, G. (2009) In planta localisation patterns of MADS domain proteins during floral development in *Arabidopsis thaliana*. *BMC Plant Biol* **9**, 5.
19. Ito, T., Takahashi, N., Shimura, Y., and Okada, K. (1997) A serine/threonine protein kinase gene isolated by an in vivo binding procedure using the *Arabidopsis* floral homeotic gene product, AGAMOUS. *Plant Cell Physiol* **38**, 248–58.
20. Wang, H., Tang, W., Zhu, C., and Perry, S. E. (2002) A chromatin immunoprecipitation (ChIP) approach to isolate genes regulated by AGL15, a MADS domain protein that preferentially accumulates in embryos. *Plant J* **32**, 831–43.
21. Gómez-Mena, C., de Folter, S., Costa, M. M. R., Angenent, G. C., and Sablowski, R. (2005) Transcriptional program controlled by the floral homeotic gene *AGAMOUS* during early organogenesis. *Development* **132**, 429–38.
22. Curtis, M. D., and Grossniklaus, U. (2003) A gateway cloning vector set for high-throughput functional analysis of genes in planta. *Plant Physiol* **133**, 462–9.
23. Clough, S. J., and Bent, A. F. (1998) Floral dip: a simplified method for *Agrobacterium*-mediated transformation of *Arabidopsis thaliana*. *Plant J* **16**, 735–43.
24. Karimi, M., Inze, D., and Depicker, A. (2002) GATEWAY(TM) vectors for *Agrobacterium*-mediated plant transformation. *Trends Plant Sci* 7, 193–95.
25. Karimi, M., Depicker, A., and Hilson, P. (2007) Recombinational cloning with plant gateway vectors. *Plant Physiol* **145**, 1144–54.
26. Earley, K. W., Haag, J. R., Pontes, O., Opper, K., Juehne, T., Song, K., and Pikaard, C. S. (2006) Gateway-compatible vectors for plant functional genomics and proteomics. *Plant J* **45**, 616–29.
27. Nakagawa, T., Suzuki, T., Murata, S., Nakamura, S., Hino, T., Maeo, K., Tabata, R., Kawai, T., Tanaka, K., Niwa, Y., Watanabe, Y., Nakamura, K., Kimura, T., and Ishiguro, S. (2007) Improved Gateway binary vectors: high-performance vectors for creation of fusion constructs in transgenic analysis of plants. *Biosci Biotechnol Biochem* **71**, 2095–100.
28. Aparicio, O., Geisberg, J. V., and Struhl, K. (2004) Chromatin immunoprecipitation for determining the association of proteins with specific genomic sequences in vivo. *Curr Protoc Cell Biol* Chapter **17**, Unit 17.7.

Chapter 16

Yeast One-Hybrid Screens for Detection of Transcription Factor DNA Interactions

Pieter B.F. Ouwerkerk and Annemarie H. Meijer

Abstract

The yeast one-hybrid system is widely recognized as a valuable and straightforward technique to study interactions between transcription factors and DNA. By means of one-hybrid screens, transcription factors or other DNA-binding proteins, expressed from cDNA expression libraries, can be identified due to the interactions with a DNA sequence-of-interest that is linked to a reporter gene, such as the yeast *HIS3* gene. Usually, the library is constructed in an *E. coli*-yeast shuttle vector designed for production of hybrid proteins consisting of a library protein and the *trans*-activating domain (AD) from the yeast GAL4 transcription factor. Here, we describe an optimized system of vectors for one-hybrid screenings together with detailed step-wise protocols, an elaborate trouble-shooting guide and many technical tips to conduct successful screenings. This system and other yeast genetic selection procedures derived from one-hybrid methodology proved highly useful to help understanding the regulatory networks controlling expression of the genome.

Key words: Yeast one-hybrid, Transcription factor, Regulatory sequence, pINT1 integration vector

1. Introduction

The yeast one-hybrid screening system is a versatile and efficient method to identify transcription factors (or other DNA-binding proteins) that can bind and regulate a given gene-of-interest. The screen is based on genetic selection in yeast for the specific interaction between a transcription factor expressed from a cDNA expression library and a DNA sequence (hereafter named bait sequence) resulting in activation of a reporter gene that can be selected for. A schematic overview of the concept of the yeast one-hybrid screening is presented in Fig 1. The yeast *HIS3* gene is a widely used reporter in yeast one-hybrid screens as well as in other related yeast screening techniques such as two-hybrid (1).

Andy Pereira (ed.), *Plant Reverse Genetics: Methods and Protocols*, Methods in Molecular Biology, vol. 678, DOI 10.1007/978-1-60761-682-5_16, © Springer Science+Business Media, LLC 2011

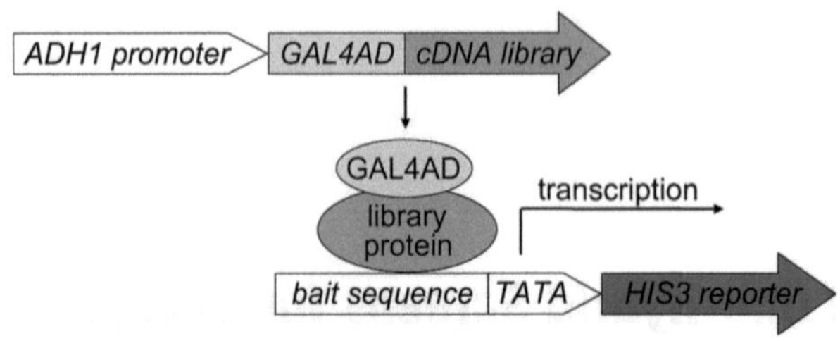

Fig. 1. Schematic overview of a yeast one-hybrid screening procedure to isolate new transcription factors or other DNA-binding proteins from cDNA expression libraries. In this situation, the cDNA is expressed from an *E. coli*-yeast shuttle vector by the constitutive yeast *ADH1* promotor and is fused to the activation domain (AD) sequence of the *GAL4* transcription factor gene from yeast. Detection of binding of the library protein to the bait sequence results in activation of expression the *HIS3* growth marker which complements for a chromosomal *his3* mutation thereby allowing colony formation on histidine-deficient SD medium. A frequently used cDNA library vector is pACTII that contains a *LEU2* selection marker complementing for a chromosomal *leu2* mutation.

In the yeast one-hybrid approach, a strain carrying a mutation (*his3*) in the histidine biosynthesis pathway that can be complemented by activation of the *HIS3* reporter is used, thereby allowing the strain to grow on histidine-deficient medium and form colonies on solid medium. Preferably, the *HIS3* reporter construct is stably integrated in the genome, thereby ensuring that the interaction between the transcription factor and bait sequence occurs in a chromosomal context. The yeast vector for expressing the cDNA library is usually designed to express the library proteins as translational fusions, with a strong transcription activation domain (AD) (hence the name one-hybrid), which can either be derived from the yeast transcription factor GAL4, or the *Herpes simplex* VP16 transcriptional activator (for useful vectors see for example Durfee et al. (2), Legrain et al. (3), or the websites of Invitrogen, Stratagene or BD Biosciences). This approach ensures that library proteins can also be cloned that do not contain a functional transcription activation domain.

For designing a yeast one-hybrid screening strategy, we recommend to use the pHIS3/pINT1 reporter cassettes described by Meijer et al. (4) (see also Table 1). Reporter gene constructs can be made in pHIS3NB or pHIS3NX, and subsequently transferred to the yeast integrative vector pINT1. This vector is designed for specific integration via double cross-over on the non-essential *PDC6* locus (YGR087c, CHr. VII). Selection for integration is via the dominant marker *APT1* that confers resistance to the antibiotic G418 (also named Gentamycine G418 or Geneticine). An advantage of this system is that, no other auxotrophic markers are used, so that these can be used for expression of other constructs in yeast. Furthermore, integration of a pHIS3/pINT1 construct is not dependent on leaky *HIS3* expression as required with some other yeast one-hybrid vectors. pINT1-HIS3NB is a follow-up of pINT1 and already has the *HIS3* gene cloned into pINT1, leaving three unique restriction sites (*Not*I, *Spe*I, and *Xba*I) available for insertional cloning of a bait sequence (Fig. 2). All Genbank Accession codes and vector details are summarized in Table 1. Note that, apart from being used for integration of *HIS3* reporter constructs in the *PDC6* locus, pINT1 can be used to integrate any other gene construct into the yeast genome (5).

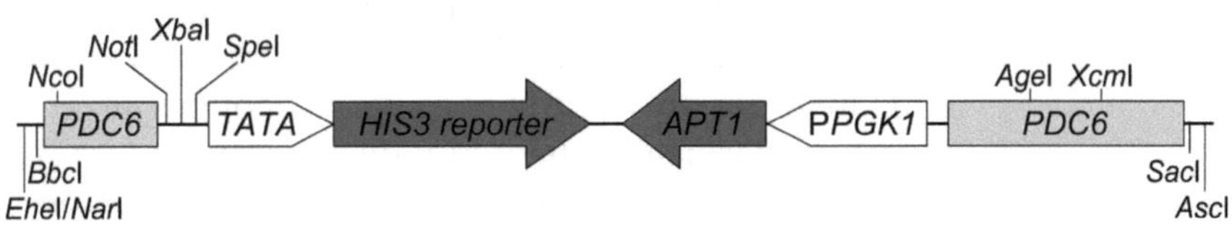

Fig. 2. Schematic representation of the *PDC6* integration fragment from pINT1-HIS3NB (Genbank Accession AY061966) after excision from its pUC29 backbone, using either one of the restriction sites *Nco*I, *Bbe*I or *Ehe*I on the *left side* and *Asc*I, *Sac*I, *Xcm*I or *Age*I on the *right side*. Essentially the same configuration is reached when pHIS3NB or pHIS3NX fragments are cloned into pINT1. The *HIS3* gene is preceded by its own minimal promoter including a TATA box and transcription start site, and a polylinker with unique restriction sites *Not*I, *Xba*I and *Spe*I sites for insertion cloning of the bait sequence. Inserts can be sequenced using the *HIS3* reverse primer (TACTAGGGCTTTCTGCTCTG). Selection for integration via double cross-over in the *PDC6* locus (YGR087c, CHr. VII) is via the dominant marker *APT1* conferring resistance in yeast to the antibiotic G418 (31). Expression of the *APT1* gene is controlled by the yeast *PGK1* promoter (P*PGK1*) and the *CYC1* terminator.

Table 1
Overview of vectors of the pINT1 system for yeast reporter strain construction. Vectors pINT1, pHIS3NB and pHIS3NX are described by Meijer et al. (4). Vector pINT1-HIS3NB was made by inserting the *HIS3* gene from pHIS3NB between restriction sites *Not*I and *Bam*HI of pINT1. The pHIS3/pINT1 vectors are annotated in full detail in NCBI/Genbank. All plasmids are derivatives of pUC-series vectors or SK Bluescript II (Strategene, Palo Alto, CA, USA) and replicate high-copy in *E. coli* and confer resistance towards 200 mg/l carbenicillin. Because of the presence of the *APT1* dominant marker in pINT1/pINT1-HIS3NB, *E. coli* cells harbouring these plasmids are also resistant towards low levels of kanamycin (25 mg/l medium). Integration of pINT1/pINT1-HIS3NB derivatives in yeast cells is selected for with 150 mg/l G418 in YAPD plates. All plasmids can be obtained from P.B.F. Ouwerkerk (p.b.f.ouwerkerk.2@gmail.com) or A.H. Meijer (a.h.meijer@biology.leidenuniv.nl)

Vector code	Genbank accession	Yeast markers	Vector type	*E. coli* selection
pINT1-HIS3NB	AY061966	*HIS3, APT1*	*PDC6* integrative	Cb^r, Km^r
pHIS3NB	AF275029	*HIS3*	Precursor for pINT1	Cb^r
pHIS3NX	AF275030	*HIS3*	Precursor for pINT1	Cb^r
pINT1	AF289993	*APT1*	*PDC6* integrative	Cb^r, Km^r
YCpHIS3B'	–	*HIS3, TRP1*	*ARS-CEN*	Cb^r
YCpHIS3BS'	–	*HIS3, TRP1*	*ARS-CEN*	Cb^r

Since the introduction of pINT1 and the two pHIS3 reporter vectors by Meijer et al. (4), this yeast one-hybrid vector system has been used in a number of successful screens for identification of novel transcription factors from plants, interacting with promoter sequences (6–13). In addition to using the yeast one-hybrid approach to screen for novel factors, it can also be used to verify interactions of known transcription factors, or other DNA binding proteins with target DNA sequences (4, 14–17). In such experiments, mutant derivatives of the bait sequence are developed and cloned into the pINT1 vector system, in order to identify specific nucleotides in a *cis*-acting sequence to which the transcription factor binds (see also (18)). We also developed two *ARS-CEN/TRP1* vectors named YCpHIS3B' and YCpHIS3BS' (4) that can be used for testing bait sequences and mutant derivatives. In contrast to pINT1/pINT1-HIS3NB, these vectors do not integrate in the yeast genome, but maintain stably under tryptophan selection via ARS-CEN replication and contain a multiple cloning site for insertion of *cis*-acting sequences in front of the *HIS3* reporter gene. Most GAL4 AD vectors for cDNA expression (such as pACT2 and pACTII) are LEU2-based and can be used in combination with YCpHIS3B' and YCpHIS3BS' or with the pHIS3/pINT1/pINT1-HIS3NB system vectors.

In this chapter, we describe the sequence of events leading to a successful yeast one-hybrid screen. The procedures described in this chapter are based on the use of yeast strains containing pHIS3/pINT1/pINT1-HIS3NB vectors and a pACTII cDNA library. The methods described include the construction of the reporter strains, yeast transformation, the library screening, and extraction of putative cDNA clones from yeast and re-introduction into *E. coli*. Bait sequence constructs and cDNA expression libraries in *E. coli*/yeast shuttle vectors are made with standard molecular biology protocols, and it is beyond the scope of this chapter to describe these in full detail. Instead, we present an outline, the requirements of which the bait sequence construct and cDNA expression library should meet. First, a strategy to clone the bait sequence into the pHIS3/pINT1/pINT1-HIS3NB vectors should be designed. A summary of details of these vectors is presented in Table 1, including Genbank Accession codes. The *HIS3* gene is already preceded by a minimal promoter containing the TATA box, transcription start site, and 5′ untranslated region and thus, the reporter construct design should be based on transcriptional fusion. Transformation of the pHIS3/pINT1/pINT1-HIS3NB vectors into yeast is described in Subheading 3.1.

The choice of the bait sequence can be based on gain-of-function and loss-of-function analysis studies with promoters. In many cases, this leads to the identification of rather short (10–50 bp) sequences that simply can be ordered as oligonucleotides and directly cloned into pINT1-HIS3NB using *Not*I and *Spe*I sites attached to the oligonucleotide (e.g. (13)). Obviously, also longer promoter fragments can be used, leading to successful screens (12). We have used promoter fragments up to 1 kb for yeast one-hybrid screens in successful screens. However, we have noted that the binding sites for transcription factors resulting from our screens are always located within 400 bp from the insertion cloning site in front of the *HIS3* reporter gene (unpublished results, P.B.F. Ouwerkerk and A.H. Meijer). Thus, it seems likely that there is a maximum distance over which *trans*-activation of the *HIS3* reporter by a transcription factor can occur in this type of screen.

To determine whether the reporter strain shows background growth on histidine-lacking medium, a titration is performed on 3-amino-1,2,4,-triazole (3-AT) (see Subheading 3.2). Background growth (also named leaky expression) is the result of activation of the *HIS3* reporter construct by interaction of endogenous yeast factors with the bait sequence. 3-AT is an herbicide and one of its actions is that it works as a competitive inhibitor of HIS3 enzyme activity and therefore, addition to the SD plates will reduce background growth. Once the required 3-AT concentration is determined, the reporter strain is ready for use in library screens. In some cases, endogenous activation is so strong that addition up to 75 mM 3-AT is not sufficient to inhibit background growth. In that case, it will be better to design a different bait sequence construct. If the bait sequence contains a well known *cis*-regulatory element in yeast genes, one can try to use a mutant for the corresponding yeast transcription factor in order to overcome the problem of background growth. This strategy was used to detect interaction of the homeodomain protein Oskn2 from rice with a downstream target binding site in a one-hybrid strategy using a yeast *GCN4* mutant ((19) and Table 2).

Obviously, the cDNA expression library should be made from a relevant source of mRNA. Ideally, the library is based on a convenient bacteriophage λ vector such as λACTII (20), which allows efficient cDNA library cloning and amplification. This vector contains an *E. coli*-yeast shuttle vector between the phage arms with Cre/loxP-mediated automatic sub-cloning properties, allowing excision of the plasmid library in *E. coli*. The resulting vector is named pACTII (2) and is maintained in yeast with *LEU2* selection and *2 μm*-based replication. The cDNA cloned in pACTII is driven by the yeast *ADH1* promoter. Several libraries in pACTII, using cDNA from rice, *Catharanthus roseus* and Arabidopsis, gave good results in yeast one-hybrid screens (6–8, 10, 12). λACTII/ pACTII libraries are fully compatible with yeast two-hybrid or other screens in yeast. The methods described in this chapter are based on the use of λACTII/pACTII-based libraries and *LEU2* selection for the pACTII/GAL4 AD vector.

Table 2
Overview of useful yeast strains for one-hybrid screens. Strains YPO101 and YPO102 are derived from Y187 and YM4271 respectively and contain the integration fragment from pINT1-HIS3NB lacking an upstream bait sequence. These strains serve as controls for the analysis of library clones. YPO8472 is derived from EE7 (30) and has extra deletions in transcription factor GCN4 and in the ADE2 and ADE5 genes

Strain	Genotype	References
Y187	*MATα ura3-52 his3-Δ200 ade2-101 trp1-901 leu2-3,112 met⁻ gal4-Δ512 gal80-Δ538 URA3::GAL1$_{UAS}$-GAL1$_{TATA}$-lacZ MEL1*	(27)
YPO101	*Y187 PDC6::pINT1-HIS3NB*	(23)
YM4271	*MATa ura3-52 his3-Δ200 ade2-101 lys2-801 trp1-901 leu2-3, 112 gal4-Δ512 gal80-Δ538 tyr1-501 ade5::hisG*	(28)
YPO102	*YM4271 PDC6::pINT1-HIS3NB*	This chapter
YPH500	*MATα ura3-52 lys2-801 ade2-101 trp1-Δ63 his3-Δ200 leu2-Δ1*	(29)
YPO8472	*MATα leu2-3,112 ura3-52 his3-Δ200 trp1-901 lys2-801 suc2-Δ9, Mel- ade2::PGK1-APT1 ade5::hisG gcn4-Δ1::hisG*	(19)

In the phase after the library screening (Subheading 3.3), false positives need to be discriminated from true positives. We recommend to PCR-amplify cDNA inserts directly from the re-streaked positive yeast colonies (Subheading 3.4), sequence the products, and determine their identity with BLAST searches. Common artefacts are ribosomal proteins, histones, nucleoside transporters, and RNA binding proteins. These proteins have functions in interacting with nucleic acids and because there they are fused with the GAL4 AD they turn into artificial transcription factors. Another possible artefact is the plant *HIS3* gene that is able to complement the *his3* mutation in yeast. This artefact turns up only occasionally, probably because the abundance in the libraries is low. Other artefacts are components of signal transduction pathways like phosphatases, kinases, and GTPases, or proteins that have enzymatic functions in metabolism. These proteins probably do not directly interact with the bait sequence, but perhaps they detoxify the 3-AT or interfere with endogenous pathways leading to activation of *HIS3* expression. If a well characterized bait sequence is used, one may consider to make a mutant bait sequence construct with an alternative reporter such as *LacZ*, *MEL1*, or *GFP*. In this approach, a dual reporter strain is made and tested during the screen for expression of both reporters, thus allowing direct elimination of most types of false positives. It is not possible to predict on forehand how many true positives may come up from a screen. This depends on the abundance of the targeted transcription factor in the library and the affinity of the library protein for the bait sequence. A true positive library clone can either be in frame with the GAL4 AD or, unexpectedly, out of frame and still activate reporter gene expression. We have previously reported on a screen with a bait-sequence for homeodomain-leucinezipper proteins from rice where members two different families (HD-Zip I and II) were identified (7). All HD-Zip family I members were out of frame with the GAL4 AD, while the HD-Zip family II members were in frame. Apparently, the HD-Zip I proteins contained an intrinsic activation domain. Because only out-of-frame fusions were obtained, in-frame fusions between the GAL4 AD and HD-Zip I proteins might have been toxic or nonfunctional.

The protocol for recovery of library plasmids from yeast and retransformation into *E. coli* for follow-up studies is based on a method described by Singh et al. (21) and outmatches other methods that we have used and described in previous papers (22, 23). The basic problem in retransformation of plasmid DNA from yeast to *E. coli* was the low transformation efficiency, which is caused by the presence of an unknown toxic factor that is co-isolated with the plasmid from yeast. The protocol described by Singh and summarized in Subheading 3.5 makes use of yeast spheroplasts prepared by lyticase treatment, from which DNA is extracted using a QIAprep kit from Qiagen. Plasmid DNA isolated from yeast using this method gives a high retransformation frequency.

Transcription factor-DNA interactions identified in yeast one-hybrid screens can be further verified using in vitro Electrophoretic Mobility Shift Assay (EMSA) studies (24) or in plants using protoplast transformations in transient expression studies, (25) followed by gain-of-function (overexpression) and loss of function studies (mutants) in plants. Follow-up experiments using other yeast strategies may include studies to map activation domains in the newly identified transcription factors. For this, protein–protein fusions can be made in, for example, the GAL4 BD fusion vector pAS2-1 (BD Biosciences, San Jose, CA). Proteins expressed from this vector are linked to the binding domain (BD) of GAL4. Such constructs are analyzed in reporter strains containing a construct with a binding site for GAL4 BD, linked to a reporter gene. For example, strain Y187 (Table 2) has a *GAL1* promoter (containing the GAL4 BD) fused to the LacZ gene. If the transcription factor-of-interest has an activation domain, fusion to the GAL4 BD will lead to reporter gene activation, if not, there will be no activation. However, one should be aware that activation in this or other assays does not necessarily mean that the factor has a function as transcriptional activator in plants. For example, the homeodomain protein Oshox1 from rice is able to activate expression of a reporter gene in yeast without the help of an extra activation domain (4), but in rice suspension cells, it down-regulates a reporter construct (26). Thus, in plant functions of transcription factors may be dependent on the actions of other transcription factors and cofactors.

2. Materials

2.1. Yeast Culture and Transformation

1. YPD: 2% (w/v) Bacto-peptone, 1% (w/v) yeast extract, and 2% (w/v) d-Glucose. YPD can also be obtained as a ready-to-use powder mix (Duchefa, Haarlem, The Netherlands). For solid plates, 1.5% (w/v) Bacto agar or micro-agar (Duchefa, Haarlem, The Netherlands) is used (see Note 1). Yeast Peptone Dextrose complete medium (YPD) supplemented with adenine hemisulphate (see Note 2) is named YAPD.
2. Minimal medium (SD): 0.17% (w/v) yeast nitrogen base without ammonium sulphate and aminoacids (Difco, Detroit, USA), 0.5% (w/v) ammonium sulphate, and 2% (w/v) D-glucose. For solid plates, 2% (w/v) Bacto agar or micro agar (Duchefa, Haarlem, The Netherlands) is added (see Note 1).
3. SD supplements: The following supplements (100× stocks) should be added to minimal SD medium, depending on the genotype of the yeast strain used, and the markers of the plasmids being selected for. Adenine hemisulphate (2 g), uracil (2 g/l), l-histidine (2 g/l), l-leucine (3 g/l), l-lysine (3 g/l), l-methionine (2 g/l), and l-tryptophan (2 g/l) (see Note 3).
4. 1 M 3-amino-1,2,4,-triazole (3-AT; e.g. from Sigma or Fluka), stored at +4°C (see Note 4).
5. 1,000× G418 stock: 150 mg/ml G418 in water (Duchefa, Haarlem, The Netherlands), stored as aliquots at –80°C (see Note 5).
6. 50% (w/v) polyethyleneglycol (PEG) stored at room temperature (see Note 6).
7. 10× LiAc stock: 1 M lithium acetate at pH 7.5 (HAc).
8. 10× TE stock: 100 mM Tris-HCl pH 7.5, and 10 mM EDTA.
9. Yeastmaker carrier DNA (10 µg/ml) (BD Biosciences, San Jose, CA, USA) (see Note 7).
10. Disposable inoculation loops.

2.2. Assessment of the pINT1/pINT1-HIS3NB Bait Sequence Strains

1. SD His+ and SD His– plates (standard Petri dishes of 9 cm diameter), and SD His– plates with a range of 3-AT concentrations (5, 10, 25 to 50 mM).

2.3. cDNA Library Transformation

1. SD plates (standard Petri dishes with 9 cm diameter) without leucine (SD Leu-) for determining transformation efficiency. Large SD plates (Petri dishes of 15 cm diameter) without leucine, without histidine (SD Leu- His-), but with the appropriate concentration of 3-AT determined in Subheading 3.2 (see Note 8).

2.4. Direct PCR on Yeast

1. Phusion DNA polymerase kit from Finnzymes (Finland) for PCR reactions.
2. 5′ and 3′ primers for cDNA amplification from the library plasmids (see Note 9).

2.5. Recovery of cDNA Clones from Putative Positive Yeast Colonies

1. Solution of 5 mg/mL lyticase (L-4025, Sigma) in 1.2 M sorbitol, 0.1 M sodium phosphate buffer at pH 7.4 (see Note 10).
2. QIAprep kit (Qiagen, Venlo, The Netherlands) with QIAprep columns and buffers P1, P2, N3, PB, PE, and EB.

3. Methods

3.1. Preparation of pINT1/pINT1-HIS3NB Bait Sequence Strains by LiAc-Mediated Transformation

1. Depending on the bait-sequence, choose a pair of restriction enzymes flanking the PDC6 sequences in pINT1/pINT1-HIS3NB (see Fig. 2; (4)), and isolate 100–500 ng of the restriction fragment from pINT1 or pINT1-HIS3NB derivatives containing the bait sequence fused to the *HIS3* reporter, the *APT1* dominant marker gene, and the flanking *PDC6* sequences (see Note 11).
2. Select a suitable yeast strain (e.g. Y187 or YM4271, see Note 12) and streak a single colony on a YAPD plate and incubate at 30°C for 2 days.
3. Use a sterile loop to inoculate a colony and grow a 50-mL YAPD culture overnight at 30°C.
4. On the next day, harvest the yeast cells of the overnight culture in 50 mL tubes by centrifugation at 4,000 rpm for 1 min in a swing-out tabletop centrifuge at room temperature.
5. Discard the supernatant and resuspend the cells in 50 mL sterile distilled water by vigorous shaking, and repeat the centrifugation step.
6. Discard the supernatant fluid and resuspend the yeast cells in 1 mL 1× TE/1× LiAc prepared from the 10× stock solutions.

7. Transfer the yeast suspension to a 1.5 ml microcentrifuge tube, centrifuge for 30 s at maximum speed, and resuspend in 250 μL 1× TE/1× LiAc prepared from the 10× stock solutions.
8. For transformation, mix 100–500 ng of the isolated fragment from the pINT1 or pINT1-HIS3NB derived vector with 25 μg Yeastmaker carrier DNA (see Note 7). Next, add 50 μL of the yeast suspension and 300 μL of freshly prepared 40% PEG/1× TE/1× LiAc (see Note 13), and vortex vigorously to mix. Perform the procedure in standard 1.5 mL microcentrifuge tubes.
9. Incubate the yeast transformation mixes in the 1.5 mL microcentrifuge tubes at 30°C for 30 min with shaking. Next, incubate the tubes for 15 min in a 42°C waterbath and harvest the cells by 30 s centrifugation in a microcentrifuge.
10. Remove the supernatant and resuspend the cells in 1 mL YAPD and transfer the cells to a 15 or 50 mL tube. Incubate for 3–6 h shaking at 30°C (see Note 14).
11. After the recuperation step, the cells are transferred to 1.5 mL microcentrifuge tubes, and harvested by centrifugation in a microcentrifuge at full speed for 30 s. Resuspend the cells in 100 μL 1× TE prepared from the 10× stock solution, and plate on YAPD plates supplemented with 150 mg/mL G418. Incubate at 30°C for 2–3 days to see colonies.
12. Select and restreak four colonies (see Note 15) on YAPD plates supplemented with 150 mg/mL G418 and grow again at 30°C for 2–3 days to see colonies.
13. For long-term storage, prepare glycerol stocks (final concentration of 30%) by adding 500 μL 87% sterile glycerol to 1 mL YAPD cultures grown overnight at 30°C, and store the cultures at −80°C.

3.2. Assessment of the pINT1/pINT1-HIS3NB Bait Sequence Strains for Background Growth

1. Streak a G418-resistant colony of the pINT1/pINT1-HIS3NB bait sequence strain on SD His+ and SD His− plates, and SD His− plates with a range of 3-AT concentrations (5, 10, 25 to 50 mM). Grow the plates for a week at 30°C. Determine the 3-AT concentration required to reduce any possible background growth (see Note 16).

3.3. cDNA Library Transformation

1. Use a sterile loop, to inoculate a colony of the pINT1/pINT1-HIS3NB yeast reporter, and grow a 50-mL YAPD culture overnight at 30°C (see Note 17).
2. Dilute the overnight culture to an OD_{600} of approximately 0.25 in liquid YAPD pre-warmed to 30°C and grow the culture(s) for another 3 h with shaking at 30°C. Grow one culture for every ten planned library transformations (see Note 18).

3. Continue with steps 4–8 from Subheading 3.1 in this chapter for the LiAc-mediated transformation of the yeast reporter strain with a cDNA library (see Note 19).
4. After the 42°C heat-shock step of the transformation, remove the supernatant and resuspend the cells in 200 μL 1× TE prepared from the 10× stock solution.
5. Plate 100 μL of 100× and 1,000× dilutions on SD Leu- plates to determine the transformation efficiency of a representative number of transformation reactions (see Note 20).
6. Plate each transformation reaction on a large (15 cm) SD Leu- His- plate with the appropriate amount of 3-AT as determined using Subheading 3.2 described in this chapter.
7. Grow the library screening plates from step 6 up to 14 days at 30°C. During this period, putative positives will appear as faster growing colonies (see Note 21).
8. Carefully pick up colonies and re-streak them on fresh SD plates with the same medium as used in step 6, and grow again for 3–10 days at 30°C (see Note 22).

3.4. Direct PCR on Yeast

1. Perform PCR reactions using Phusion DNA Polymerase according to the manufacturers' instructions directly on the putative colonies. Use a pipette tip to pick some yeast cells from a colony and resuspend these in the PCR reaction mixture (see Note 23).
2. Sequence the PCR products with the primers by which the products were obtained, and run BLAST searches at NCBI (http://blast.ncbi.nlm.nih.gov/Blast.cgi) or other databases in order to identify the clones. See Subheading 1 of this chapter for tips how to recognize commonly found false positives.

3.5. Recovery of cDNA Clones from Putative Positive Yeast Colonies

1. Use a sterile loop, to inoculate a putative positive colony from a library screen and grow a 10-mL SD-Leu culture overnight at 30°C in a 50 mL disposable tube.
2. On the next day, harvest the yeast cells of the overnight culture in 50 mL tubes by centrifugation at 4,000 rpm for 1 min in a swing-out tabletop centrifuge at room temperature.
3. Discard the supernatant, resuspend the cells in 200 μL P1 buffer, and transfer to a 1.5 mL microcentrifuge tube
4. Add 100 μL lyticase buffer, mix and incubate at 37°C for 30 min in order to digest the cell wall and convert the yeast cells into spheroplasts.
5. Next, add 200 μL P2 buffer, mix and keep for 10 min at room temperature.

6. To the lysed cells, add 420 μL N3 buffer and mix. The lysates are centrifuged for 10 min in a microcentrifuge at full speed and supernatants are applied to the QIAprep spin columns.
7. Wash the QIAprep spin column by adding 500 μL buffer PB and centrifuge for 60 s in a microcentrifuge at full speed. The flow-through is discarded.
8. Repeat step 6 with 750 μL buffer PE and discard the flow-through. Centrifuge the QIAprep column 1 min to remove residual fluid.
9. Place the QIAprep column in a new 1.5 ml microcentrifuge tube. Elute the plasmid DNA by adding 50 μL buffer EB to the centre of each QIAprep spin column, let stand for 1 min, and centrifuge for 1 min. The DNA solution is now at the bottom of the microcentrifuge tube.
10. Transform the resulting plasmid DNA into chemically competent *E. coli* cells for further amplification and use in follow-up experiments.

4. Notes

1. YPD and SD media are best autoclaved at 110°C for 20 min; YPD tends to get very dark when autoclaved at 120°C and SD plates tend to be very soft (even when 2% agar is used) when the medium is autoclaved at 120°C.
2. Adenine hemisulphate (final concentration 20 mg/L) is normally added to YPD after autoclaving to prevent accumulation of a red substance in strains bearing the *ade2-101* mutation such as YPH500 or Y187.
3. The SD supplements can be sterilized by autoclaving at 120°C for 20 min. Storage is at room temperature. The tryptophan stock solution should be protected from light by wrapping the bottle with aluminium foil. It is normal that the tryptophan stock solution turns brown. When making SD plates, SD supplements are added after the medium is autoclaved and cooled down to about 60°C.
4. Do not autoclave the 3-AT stock, but sterilize the solution with filter bottles. Store at 4°C. Add the 3-AT to the SD media after cooling down to about 60°C.
5. The G418 solution should be filter sterilized, aliquoted, and kept at −20°C.
6. The 50% PEG solution is best sterilized by autoclaving at 120°C for 20 min and stored at room temperature. Bottles

should be closed air-tight to avoid evaporation of water and changes in concentration.

7. We have found Yeastmaker (BD Biosciences, San Jose, CA, USA) as carrier DNA to be excellent for yeast transformation and prefer this commercial preparation over making it ourselves, since this resulted in large variety in transformation frequencies between different preparations. Store at −20°C. Prior to use, aliquots of Yeastmaker should be denaturated by boiling for 10 min in a water bath and kept on ice for 2 min before use in yeast transformations.
8. This protocols assumes the use of a pACTII cDNA library that is maintained using LEU2 selection, thus leucine should be omitted from the plates. When using Y187 (see Table 2 for the genotype) as the genetic background to integrate the pINT1/pINT1-HIS3NB reporter, the SD plates to check the transformation efficiency should lack leucine, but be supplemented with histidine, tryptophan, adenine hemisulphate, and methionine. The library transformation plates should lack both leucine and histidine, and be supplemented with tryptophan, adenine hemisulphate, methionine, and the appropriate concentration of 3-AT as determined in Subheading 3.2.
9. Obviously, the primer sequences depend on the type of library vector. For pACT2 and pACTII, we use primers 4TH (CCCCACCAAACCCAAAAAAAG) and 3′AD (GTTGAAG TGAACTTGCG).
10. Lyticase powder is normally stored at −20°C and the lyticase solution is prepared freshly.
11. A linear fragment from pINT1 or pINT1-HIS3NB derivatives, without the pUC29 backbone is desirable to avoid single cross-over recombination in the yeast genome, which can be less stable.
12. Both Y187 and YM4271 give high transformation frequencies (10^4 to 10^5 transformants per μg library plasmid) and allow the use of several commonly used auxotrophic markers (LEU2, TRP1) for selection of the library vector. Both strains are routinely used in yeast one-hybrid screens. Y187 and YM4271 are *gal4gal80* mutant strains, which is desirable when using hybrid libraries containing the activation domain derived from the yeast transcription factor GAL4 (GAL4 AD). GAL4GAL80 wild type strains such as YPH500 can be used, but in that case galactose should be used as a carbon source instead of glucose in order to avoid inhibition of the GAL4 AD by GAL80. Galactose results in slower growth and is very expensive, therefore the use of *gal4gal80* mutant strains for this type of screens is favorable.

13. The 40% PEG/1× TE/1× LiAc solution is freshly prepared from the stock solutions in a 8:1:1 ratio.
14. This recuperation step is required to allow the yeast cells to express the dominant *APT1* marker gene carried on the pINT1 or pINT1-HIS3NB fragment to acquire resistance towards G418 and select for integration on the *PDC6* locus. The recuperation step in YAPD should not be done in microcentrifuge tubes because pressure is built up due to CO_2 formation, but preferably in 15 or 50 mL tubes.
15. Take care not to restreak very small colonies as these sometimes carry a spontaneous mitochondrial mutation (petites). These transformants are not useful in the screening because of their much slower growth. The obtained strains can be checked for correct integration of the reporter construct by PCR with *PDC6* (GACCTCAATCTTAGGTGATTGAG) and *HIS3* (TACTAGGGCTTTCTGCTCTG) primers or Southern blot analysis. From this stage onwards, it is no longer necessary to select the yeast strains for G418 resistance since the pINT1 insert is maintained stably.
16. The optimal 3-AT concentration required under screening conditions is also dependent on the plating density of yeast cells. Therefore, it can be useful to fine-tune the concentration of 3-AT in test transformations with an empty activation domain vector (such as pACTII) under the library screening conditions (selection for histidine and leucine) described in Subheading 3.3. It is not absolutely required to eliminate background growth completely, but it is important that the background colonies remain very small.
17. G418 selection is not necessary in the liquid culture since the reporter construct is stably integrated into the yeast genome.
18. For optimal transformation efficiency, it is important that the cells are harvested during the exponential phase (OD_{600} 0.4–0.8). The 3 h time frame is suitable for yeast strains Y187 and YM4271, but it may be necessary to adjust the incubation time when using other strains. Usually step 2 is not necessary for transformations of pINT1-HIS3NB or other plasmids, but we recommend including this step when preparing competent yeast cells for library transformations to increase transformation frequency.
19. Depending on the complexity of the cDNA expression library, set up a number of transformation reactions with 1 μg of cDNA library plasmid each. Increasing the amount of plasmid DNA in a reaction does not necessarily increase the number of yeast cells transformed. Each transformation usually results in up to 10^5 colonies transformed. One transformation

reaction can be plated on one large Petri dish (diameter 15 cm). If the library has a complexity of 10^6 independent clones, it will be sufficient to set up 20 transformation reactions for a saturating screening of that library.

20. After at least 3 days of incubation at 30°C, count the number of colonies on the transformation efficiency plates from step 5. When using Y187 or YM4271, transformation efficiency will range from 10^4 to 10^5 per µg plasmid library. Calculate the total number of transformants to determine if the screening was saturating.
21. Depending on the amount of background expression of the pINT1/pINT1-HIS3NB reporter strain and the concentration of 3-AT used in the SD plates, non-positive yeast cells can still grow at a slow rate resulting in very small colonies (less than 1 mm diameter) on the whole plate. However, true putative positive colonies should be clearly visible by their larger size, otherwise the 3-AT concentration should be raised. Usually, the background colonies stop growing and stay very small, whereas positives keep on growing.
22. Putative positive colonies should grow again. Occasionally, a small number of colonies will not grow again because they are false positives caused by a too high local density of yeast cells on the original screening plate.
23. Do not use wooden toothpicks since formaldehyde present in wood is known to interfere with PCR reactions.

Acknowledgements

This work was supported by grants CerealGene Tags (European Commission FP5 project QLG2-CT-2001-01453) and NWO 901-07-206 (Netherlands Organization of Scientific Research) to P.B.F.O. and TF-STRESS (European Commission FP5 project QLK3-CT-2000-00328) and ZF-Models (European Commission FP6 project LSHG-CT-2003-503496) to A.H.M.

References

1. Fields, S. and Song, O. (1989) A novel genetic system to detect protein-protein interactions. *Nature* **340**, 245–246.
2. Durfee, T., Becherer, K., Chen, P. L., Yeh, S. H., Yang, Y., Kilburn, A. E., Lee, W. H. and Elledge, S. J. (1993) The retinoblastoma protein associates with the protein phosphatase type 1 catalytic subunit. *Genes Dev.* **7**, 555–569.
3. Legrain, P., Dokhelar, M. C. and Transy, C. (1994) Detection of protein-protein interactions using different vectors in the two-hybrid system. *Nucleic Acids Res.* **22**, 3241–3242.
4. Meijer, A. H., Ouwerkerk, P. B. F. and Hoge, J. H. C. (1998) Vectors for transcription factor cloning and target site identification by means of genetic selection in yeast. *Yeast* **14**, 1407–1415.

5. Neuteboom, L. W., Lindhout, B. I., Saman, I. L., Hooykaas, P. J. J. and van der Zaal, B. J. (2006) Effects of different zinc finger transcription factors on genomic targets. *Biochem. Biophys. Res. Commun.* **339**, 263–270.
6. Menke, F. L., Champion, A., Kijne, J. W. and Memelink, J. (1999) A novel jasmonate- and elicitor-responsive element in the periwinkle secondary metabolite biosynthetic gene Str interacts with a jasmonate- and elicitor-inducible AP2-domain transcription factor, ORCA2. *EMBO J.* **18**, 4455–4463.
7. Meijer, A. H., de Kam, R. J., d'Erfurth, I., Shen, W. and Hoge, J. H. C. (2000) HD-Zip proteins of families I and II from rice: interactions and functional properties. *Mol. Gen. Genet.* **263**, 12–21.
8. Van der Fits, L., Zhang, H., Menke, F. L., Deneka, M. and Memelink, J. (2000) A *Catharanthus roseus* BPF-1 homologue interacts with an elicitor-responsive region of the secondary metabolite biosynthetic gene Str and is induced by elicitor via a JA-independent signal transduction pathway. *Plant Mol. Biol.* **44**, 675–685.
9. Deng, X., Phillips, J., Meijer, A. H., Salamini, F. and Bartels, D. (2002) Characterization of five novel dehydration-responsive homeodomain leucine zipper genes from the resurrection plant *Craterostigma plantagineum. Plant Mol. Biol.* **49**, 601–610.
10. Pauw, B., Hilliou, F. A., Martin, V. S., Chatel, G., de Wolf, C. J., Champion, A., Pre, M., van Duijn, B., Kijne, J. W., van der Fits, L. and Memelink, J. (2004) Zinc finger proteins act as transcriptional repressors of alkaloid biosynthesis genes in *Catharanthus roseus. J. Biol. Chem.* **279**, 52940–52948.
11. Diaz-Martin, J., Almoguera, C., Prieto-Dapena, P., Espinosa, J. M. and Jordano, J. (2005) Functional interaction between two transcription factors involved in the developmental regulation of a small heat stress protein gene promoter. *Plant Physiol.* **139**, 1483–1494.
12. Bundock, P. and Hooykaas, P. (2005) An Arabidopsis hAT-like transposase is essential for plant development. *Nature* **436**, 282–284.
13. Lopato, S., Bazanova, N., Morran, S., Milligan, A. S., Shirley, N. and Langridge, P. (2006) Isolation of plant transcription factors using a modified yeast one-hybrid system. *Plant Methods* **2**, 3.
14. Van der Fits, L. and Memelink, J. (2001) The jasmonate-inducible AP2/ERF-domain transcription factor ORCA3 activates gene expression via interaction with a jasmonate-responsive promoter element. *Plant J.* **25**, 43–53.
15. Banerjee, S., Fisher, O., Lohia, A. and Ankri, S. (2005) *Entamoeba histolytica* DNA methyltransferase (Ehmeth) is a nuclear matrix protein that binds EhMRS2, a DNA that includes a scaffold/matrix attachment region (S/MAR). *Mol. Biochem. Parasitol.* **139**, 91–97.
16. Viola, I. L. and Gonzalez, D. H. (2007) Interaction of the PHD-finger homeodomain protein HAT3.1 from *Arabidopsis thaliana* with DNA. Specific DNA binding by a homeodomain with histidine at position 51. *Biochemistry* **46**, 7416–7425.
17. Comelli, R. N., Viola, I. L. and Gonzalez, D. H. (2009) Characterization of promoter elements required for expression and induction by sucrose of the *Arabidopsis* COX5b-1 nuclear gene, encoding the zinc-binding subunit of cytochrome c oxidase. *Plant Mol. Biol.* **69**, 729–743.
18. Wanke, D. and Harter, K. (2009) Analysis of plant regulatory DNA sequences by the yeast-one-hybrid assay, in *Plant Signal Transduction: Methods and Protocols* (Pfannschmidt, T., ed.). Humana, Totowa, NJ, *Methods Mol. Biol.* **479**, 291–309.
19. Postma-Haarsma, A. D., Rueb, S., Scarpella, E., den Besten, W., Hoge, J. H. C. and Meijer, A. H. (2002) Developmental regulation and downstream effects of the knox class homeobox genes Oskn2 and Oskn3 from rice. *Plant Mol. Biol.* **48**, 423–441.
20. Memelink, J. (1997) Two yeast/*Escherichia coli* Lambda/plasmid vectors designed for yeast one- and two-hybrid screens that allow directional cDNA cloning. *Trends Genet.* **13**, 376.
21. Singh, M. V. and Weil, P. A. (2002) A method for plasmid purification directly from yeast. *Anal. Biochem.* **307**, 13–17.
22. Meijer, A. H., Schouten, J., Ouwerkerk, P. B. F. and Hoge, J. H. C. (2000) Yeast as a versatile tool in transcription factor research, in *Plant Molecular Biology Manual* (Galvin, S. B. and Schilperoort, R. A., ed.). Kluwer Academic Publishers, Dordrecht, Vol. E3, pp. 1–28.
23. Ouwerkerk, P. B. F. and Meijer, A. H. (2001) Yeast one-hybrid screening for DNA-protein interactions, in *Curr. Protoc. Mol. Biol.* (Ausubel, F.M., Brent, R., Kingston, R.E., Moore, D.D., Seidman, J.G., Smith, J.A. and Struhl, K., eds.), John Wiley & Sons, New York, **Chapter 12**, Unit 12 12.
24. Read, J. T., Cheng, H., Hendy, S. C., Nelson, C. C. and Rennie, P. S. (2009) Receptor-DNA interactions: EMSA and footprinting, in *Nuclear Receptor Superfamily: Methods and Protocols* (McEwan, I. J., ed.). Humana, Totowa, NJ, *Methods Mol. Biol.* **505**, 97–122.

25. Berendzen, K. W., Harter, K. and Wanke, D. (2009) Analysis of plant regulatory DNA sequences by transient protoplast assays and computer aided sequence evaluation, in *Plant Signal Transduction* (Pfannschmidt, T., ed.). Humana, Totowa, NJ, *Methods Mol. Biol.* **479**, 311–335.
26. Meijer, A. H., Scarpella, E., van Dijk, E. L., Qin, L., Taal, A. J., Rueb, S., Harrington, S. E., McCouch, S. R., Schilperoort, R. A. and Hoge, J. H. C. (1997) Transcriptional repression by Oshox1, a novel homeodomain leucine zipper protein from rice. *Plant J.* **11**, 263–276.
27. Harper, J. W., Adami, G. R., Wei, N., Keyomarsi, K. and Elledge, S. J. (1993) The p21 Cdk-interacting protein Cip1 is a potent inhibitor of G1 cyclin-dependent kinases. *Cell* **75**, 805–816.
28. Liu, J., Wilson, T. E., Milbrandt, J. and Johnston, M. (1993) Identifying DNA-binding sites and analyzing DNA-binding domains using a yeast selection system. *Methods* **5**, 125–137.
29. Sikorski, R. S. and Hieter, P. (1989) A system of shuttle vectors and yeast host strains designed for efficient manipulation of DNA in *Saccharomyces cerevisiae. Genetics* **122**, 19–27.
30. Wu, A. L. and Moye-Rowley, W. S. (1994) GSH1, which encodes gamma-glutamylcysteine synthetase, is a target gene for yAP-1 transcriptional regulation. *Mol. Cell. Biol.* **14**, 5832–5839.
31. Hadfield, C., Jordan, B. E., Mount, R. C., Pretorius, G. H. and Burak, E. (1990) G418-resistance as a dominant marker and reporter for gene expression in *Saccharomyces cerevisiae. Curr. Genet.* **18**, 303–313.

Chapter 17

Plant Metabolomics by GC-MS and Differential Analysis

Joel L. Shuman, Diego F. Cortes, Jenny M. Armenta, Revonda M. Pokrzywa, Pedro Mendes, and Vladimir Shulaev

Abstract

Metabolomics is a new genomics approach that aims at measuring all or a subset of metabolites in the cell. Several approaches to plant metabolomics are currently used in plant research. These include targeted analysis, metabolite profiling, and metabolic fingerprinting. Metabolic fingerprinting, unlike metabolite profiling, does not aim at separating or identifying all the metabolites present in the sample, but rather generates a fingerprint that characterizes a specific metabolic state of the plant system under investigation. This chapter describes the implementation of metabolic fingerprinting approach using gas chromatography coupled to mass spectrometry (GC-MS) and discriminant function analysis combined with genetic algorithm (GA-DFA). This approach enables the identification of specific metabolites that are biologically relevant, and which may go undetected if direct infusion-based fingerprinting approaches were used due to the sample complexity and matrix suppression effects.

Key words: *Arabidopsis*, Plant metabolomics, Polar metabolite, Extraction, Derivatization, GC-MS, Metabolic fingerprinting, Metabolite profiling, Targeted analysis, Bioinformatics, GA-DFA, Genetic algorithm, Discriminant function analysis

1. Introduction

The goal of the metabolomics study is to detect and measure all metabolites in the cell. Due to the complexity and diversity of cellular metabolites, the complete elucidation of the plant metabolome represents a significant challenge. It is estimated that the plant kingdom itself accounts for approximately 200,000 different metabolites, which vary significantly in their chemical nature. A few of the types of metabolites that can be found in plants include amino acids, flavonoids, anthocyanins, lipids, and organic acids. Differences in structure, molecular weight, and polarity limit the success of any single analytical platform in the interrogation of the entire set of metabolites present in the plant cell.

Andy Pereira (ed.), *Plant Reverse Genetics: Methods and Protocols*, Methods in Molecular Biology, vol. 678, DOI 10.1007/978-1-60761-682-5_17,

Progress in analytical instrumentation and bioinformatics has advanced our ability to measure numerous plant metabolites, evaluate metabolic changes in response to external stimulus, and elucidate metabolic pathways. Nevertheless, the analytical sensitivity and resolution required for the simultaneous separation and detection of the hundreds of thousands of metabolites that make up living organisms, have not yet been achieved. Currently, three strategies have been applied for the metabolomic analysis of plants, namely, targeted analysis, metabolite profiling, and metabolic fingerprinting (1). Targeted analysis focuses on the analysis of selected known metabolites. This approach has the advantage that it provides quantitative information, low limits of detection and high throughput. However, prior knowledge of the target compounds, as well as their highly purified standards are required. This limits the applicability of this method for the assessment of global metabolic changes. Metabolite profiling allows for the measurement of a large set of both known and unknown metabolites in a sample (2, 3). This technique offers medium throughput, is semiquantitative in nature, and does not provide identification of the majority of the peaks. Metabolic fingerprinting, on the other hand, exhibits the highest throughput, and does not aim at separating or identifying all the metabolites present in the sample, but rather generates a fingerprint that characterizes a specific metabolic state of the system under investigation.

Fingerprinting can be performed with nuclear magnetic resonance (NMR) (4), mass spectrometry (5), Fourier transform ion cyclotron resonance mass spectrometry, or Fourier transform infrared (FT-IR) spectroscopy (6). MS fingerprinting is usually done by direct infusion of the sample into mass spectrometer. This technique provides high analysis throughput but suffers several limitations, mainly due to a cosuppression effect, where the signal of many analytes can be lost at the mass spectrometer interface.

In our laboratory, we have developed a metabolic fingerprinting approach for plant analysis that minimizes the cosuppression effect (7). We actually carry out a chromatographic separation prior to the MS. This is similar to what is performed for metabolite profiling, except that we do not attempt to identify all the molecules responsible for the peaks in the separation, rather, we identify those that demonstrate to be discriminant between groups using multivariate statistics or machine learning algorithms. This chapter describes the implementation of our approach using gas chromatography coupled to mass spectrometry (GC-MS) and discriminant function analysis combined with genetic algorithm (GA-DFA). This approach enables us to center our attention on the specific metabolites that hold potential biological relevance, and which likely would go undetected if other approaches (i.e., direct infusion) were used as a result of sample complexity and matrix suppression effects. Further inspection of the mass spectrometric signals selected as discriminant against the mass spectral library renders compound identification.

2. Materials

2.1. Sample Preparation

1. Microcentrifuge tubes, 1.5 mL (FisherBrand, ThermoFisher Scientific, Inc., Catalog # 02-681-339) with screw caps and O-rings (FisherBrand, ThermoFisher Scientific, Inc., Catalog # 02-681-358).
2. Stainless steel beads, 2.3 mm, (Biospec Products, Inc. Bartlesville, OK, Catalog # 11079123ss).
3. Retsch Mixer Mill 300 (Retsch Inc. USA, Newtown, PA, Catalog # 85110) with a 2×24 mixer mill adapter set to hold twenty-four 1.5 or 2.0 mL microcentrifuge tubes (Qiagen Inc., Valencia, CA, Catalog # 69998).
4. Branson 3510 ultrasonic cleaner (Branson Ultrasonic, Danbury, CT, Catalog # B3510).
5. FreeZone 12 L console freeze dry system with stoppering tray dryer and purge valve (Labconco, Kansas City, MO, Catalog # 7759032).

2.2. Metabolite Extraction

1. Solvents: methanol (LC-MS grade); water (Milli-Q or LC-MS grade); chloroform (omnisolv or better. See Note 1).
2. Internal standard for semiquantitative untargeted analysis: the sugar alcohol ribitol (syn. adonitol, CAS 488-81-3) at 12 μg/mL in 3 methanol:1 water (v/v). Make a 1 mg/mL stock solution first, then dilute accordingly. Store solution at −20 to −80°C. Make enough for 250 μL×# samples. Other standards can be added for targeted and possibly untargeted analysis if needed.
3. A 1.1 mL glass vial is used for polar/nonpolar fractionation (SUN-Sri, Atlanta, GA, Catalog # 502-177) and sealed with 8 mm crimp caps with three-layer septa (Teflon/Silicon/Teflon) (National Scientific Inc., Rockwood, TN, Catalog # C4013-40A).
4. Centrivap benchtop centrifugal for drying polar extract.

2.3. Methoximation and Derivatization

1. Methoximation: make 20 mg/mL methoxylamine hydrochloride (MOX) (CAS 593-56-6) in dry pyridine (CAS 110-86-1) solution. Heat at 40°C to get MOX into solution and vortex briefly. Store at RT in an amber vial away from moisture (preferably ventilated chemical storage).

2. Derivatization (trimethylsilylation): *N*-Methyl-*N*-trimethylsilyl trifluoroacetamide + 1% trimethylchlorosilane (MSTFA + 1% TMCS).
3. A benchtop heated incubator/shaker such as the Barnstead MAX Q 4000-5 bench top shaker (Barnstead/Lab-Line, Catalog # SHKE4000-5).

2.4. GC-MS Analysis

1. Prepare a retention time mix of ten alkanes in hexane, 50 μg/mL each (see Table 1).
2. GC-MS system with autosampler. Our lab is equipped with a: Combi-PAL autosampler (Leap Technologies, Carrboro, NC) with a 10 μL SGE syringe (maximum syringe volume should not exceed 10× injection volume); a Trace GC (ThermoFisher Scientific, Milan, Italy) equipped with a hot split/splitless injector; and interfaced with a Trace DSQ dual-stage quadrupole mass selective detector (ThermoFisher Scientific, Austin, TX).
3. Metabolite separation by a nonpolar fused silica capillary Alltech AT-5 ms column of 30 m + 5 m guard × 0.25 mm ID × 0.25 μm film thickness. Stationary phase is 5% phenyl, 95% dimethylpolysiloxane. (Alltech Associates Inc., Deerfield, IL, Catalog # 15881). Substitute columns include J&W DB-5 ms, Restek Rtx-5 ms, and Zebron ZB-5 ms.

Table 1
Retention time mix for calculating retention indexes of metabolites

Name	Formula	MW	Retention index	CAS
Decane	$C_{10}H_{22}$	142.28	1,000	124-18-5
Dodecane	$C_{12}H_{26}$	170.33	1,200	112-40-3
Tetradecane	$C_{14}H_{30}$	198.39	1,400	629-59-4
Hexadecane	$C_{16}H_{34}$	226.44	1,600	544-76-3
Octadecane	$C_{18}H_{38}$	254.49	1,800	593-45-3
Eicosane	$C_{20}H_{42}$	282.56	2,000	112-95-8
Docosane	$C_{22}H_{46}$	310.60	2,200	629-97-0
Tetracosane	$C_{24}H_{50}$	338.66	2,400	646-31-1
Hexacosane	$C_{26}H_{54}$	366.71	2,600	630-01-3
Octacosane	$C_{28}H_{58}$	394.76	2,800	630-02-4

2.5. Data Analysis

1. Xcalibur 2.0 Software (ThermoFisher Scientific).
2. Ometer software (http://mendes.vbi.vt.edu/tiki-index.php?page=Ometer).
3. AMDIS software (http://www.amdis.net/).

3. Methods

The major steps involved in GC-MS based metabolomics analysis of *Arabidopsis* include: tissue collection and storage; sample preparation and extraction; sample fractionation into polar and nonpolar fractions; sample derivatization; separation and detection of metabolites by gas chromatography coupled with mass spectrometry; differential analysis to determine which ions are most discriminatory for the treatment/tissue type studied; and identification of the metabolites that correspond to the discriminant ions (see Fig. 1).

3.1. Sample Preparation

1. Sample storage: *Arabidopsis* tissue samples should be stored at –80°C. Metabolites should be extracted as soon as possible after harvest (within a few months). Metabolites can change, although quite slowly, even at –80°C. Freeze-dried tissue is stable for more than 2 months (see Note 2).
2. Weigh 10 mg of freeze-dried tissue (or 100 mg fresh tissue) into 1.5 mL microcentrifuge tubes (see Notes 3 and 4).
3. Add two 2.3 mm stainless steel beads to each microcentrifuge tube. If using fresh tissue, samples should be kept on dry ice throughout preparation and metabolite extraction.
4. Tightly close screw caps and flash-freeze vial w/tissue inside in liquid nitrogen (see Note 5).

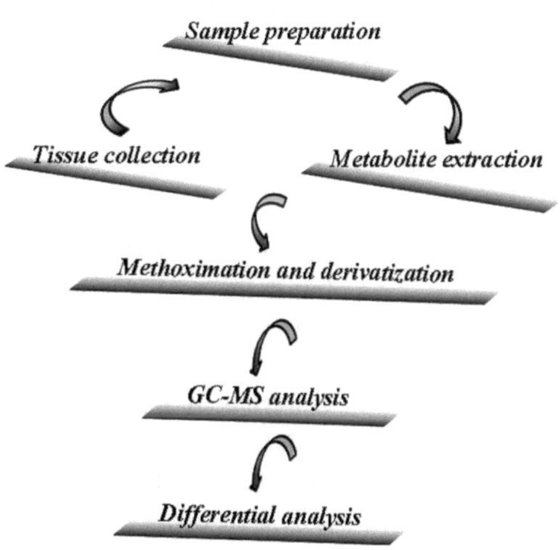

Fig. 1. General flowchart of entire process from sample tissue collection to differential analysis.

5. Place vials in 2 × 24 mixer mill adapter sets and grind tissue in mixer mill for 30 s at 30 Hz (=1,800 oscillations/min). A total of 48 vials can be processed at a time. Flash-freeze again. Rotate holder 180° from previous orientation and grind tissue again as before. Centrifuge at 4°C for 4 min at 13,000 × *g* to get powder to bottom of the tube. Return to dry ice (see Note 6).

3.2. Metabolite Extraction

1. Add 250 μL of ice-cold methanol-water (3:1, v:v) solution with ribitol at 12 μg/mL and any other internal standards, if needed, to each sample. Grind twice as in Subheading 3.1, item 5.
2. Add 250 μL ice-cold methanol-water (1:3, v:v) solution and grind two times.
3. Keep on dry ice for 15 min.
4. Centrifuge for 15 min at 13,000 × *g* at 4°C.
5. Transfer aqueous fraction to 1.1 mL glass chromacol vial. Do not remove particulate matter.
6. Vortex briefly. Add 250 μL chloroform to the aqueous fraction in the 1.1 mL glass chromacol vial using a syringe (see Note 7). Vortex briefly. Centrifuge at 4°C for 15 min at 13,000 × *g*.
7. Carefully remove upper, polar fraction and transfer it into a clean 1.1 mL glass chromacol vial. Do not disturb lower non-polar chloroform layer. The nonpolar fraction is sometimes analyzed but is not discussed here.
8. Dry polar fraction overnight at room temperature in the Centrivap centrifugal concentrator.

3.3. Sample Derivatization

1. Check MOX solution for suitability. To a clean, glass vial, add 80 μL MOX and 80 μL MSTFA + 1%TMCS. If a precipitate forms and solution rapidly heats up, the MOX is bad. Make new MOX (see Note 8).
2. Methoximate the metabolite in the sample by adding 80 μL MOX solution to the dried extract, crimp vial, and incubate at 30°C for 90 min in a benchtop incubator and gently shake at 100 rpm (see Note 9).
3. Derivatize (trimethylsilylate) samples by adding 80 μL MSTFA + 1%TMCS solution to methoximated sample. Incubate at 37°C for 30 min. If possible, use septum piercing syringe (see Note 10).

3.4. GC-MS Analysis

1. *Injection.* A 1 μL aliquot is injected into the GC-MS system with the Combi-PAL autosampler. The sample is vaporized in the 230°C inlet and most of the solvent is removed via the split vent (25:1).
2. *Gas chromatography.* Separation of metabolites is achieved using a 30 m capillary column in the Trace GC oven. Carrier

gas is helium supplied at a constant flow rate of 1 mL/min. Oven program is a thermal gradient of 5 min at 70°C, followed by a 5°C/min ramp to 310°C, 1 min at 310°C, and postrun equilibration of 5 min at 70°C prior to the next injection. Interface (transfer line) is kept at 250°C (see Note 11).

3. *Mass spectrometry*. As metabolites exit the column, they are ionized at 70 eV and detected by a dual-stage quadrupole Trace DSQ mass selective detector. Mass spectra are recorded at 2 scans/s from 50 to 650 m/z (see Note 12). GC-MS analysis results in a profile similar to the ones shown in Fig. 2 for the analysis of three different *Arabidopsis* genotypes.

4. *Metabolite identification*. Metabolites are identified based upon mass spectra and retention time or retention index. Retention index (RI) is calculated based upon where a given metabolite elutes relative to the alkane before and after it (8). RI can also be automatically calculated using the Automated Mass Spectral Deconvolution and Identification system (AMDIS), freely available from the National Institute of Standards and Technology (NIST) at this site (http://www.amdis.net/).

5. Semiquantitation of metabolites is performed by dividing the integrated peak area of the metabolite by the peak area of the internal standard (ribitol).

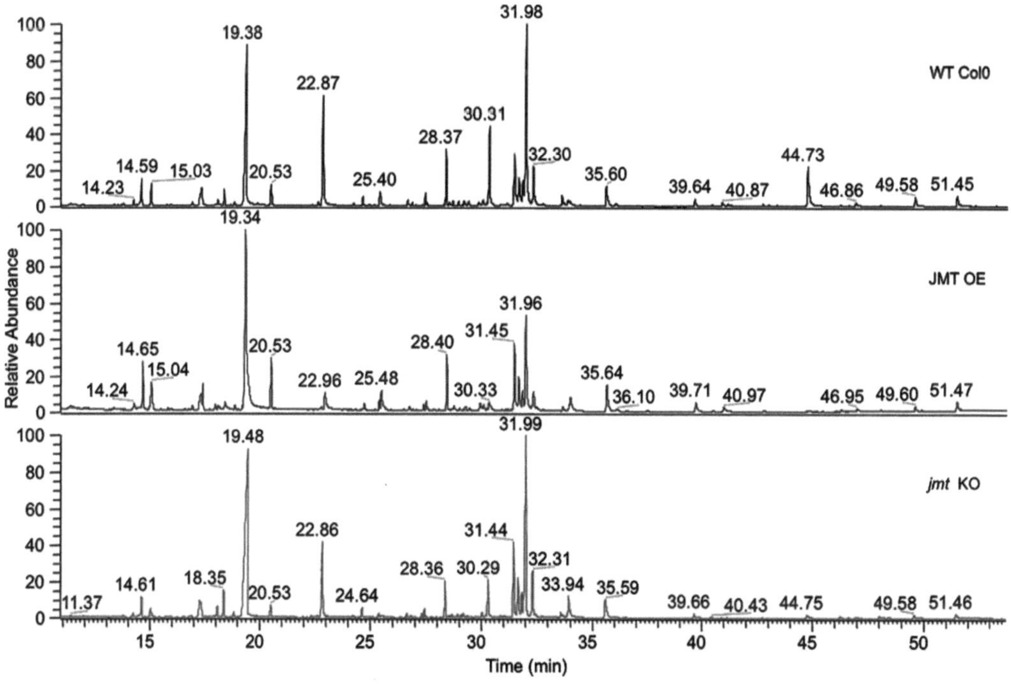

Fig. 2. GC-MS profiles (chromatograms) of polar extracts obtained from the leaves of *Arabidopsis* wild-type ecotype Columbia (*top*), jasmonic acid methyl transferase (JMT) overexpressing (*middle*), and JMT knockout (*bottom*) mutants.

3.5. Preprocessing of Mass Spectrometry Data (see Note 13)

1. These instructions assume the use of Xcalibur 2.0 Software (ThermoFisher Scientific) and the Qual Browser module for the visualization and preprocessing of the raw data files obtained from the GC-MS analysis.
2. Open the raw data file corresponding to a sample to be analyzed using Qual Browser module of the Xcalibur Software. By applying the default layout, the chromatogram is usually displayed in the upper portion of the window and the spectrum on the lower.
3. The GC-MS data array for each sample is reduced into a single MS vector by summing the ion counts of a given m/z ratio over the total scan time (see Fig. 3 and Note 14), then each MS vector, containing m/z ratios and their total intensity, was simplified to unit m/z ratio (ion counts of fractional m/z ratios were added to the nearest integer m/z). As a result, after this initial data reduction, an MS chromatogram with m/z range 50–650 will be reduced to a single vector having 601 m/z values with their correspondent intensity. Reduction of the data to a single MS vector is performed with Xcalibur software (v 2.0, ThermoFisher Scientific).
4. Every MS vector is normalized to the intensity of the ribitol internal standard so that different spectra can be compared quantitatively.
5. Repeat steps 2–4 for each sample to be included in the differential analysis. All the MS vectors in a dataset were exported and formatted before analysis into a single matrix of N objects and V variables, where N is the number of MS vectors or samples and V the m/z ratios in the mass range (see Note 15). This dataset serves as the input file for GA-DFA, and is specified last when using Ometer, stating the class to which each sample belongs according to its nature (for example: wild-type, overexpress or knockout).

3.6. Rapid Identification of Discriminating m/z Values Using GA-DFA

1. The overall flow chart for the GA-DFA analysis is shown in Fig. 4 (see Note 16). The steps for applying the GA-DFA algorithm to a given preprocessed dataset are described below.
2. Once the GC-MS dataset has been preprocessed to resemble a fingerprint, m/z values which discriminate between sample types can be quickly determined by employing the GA-DFA algorithm (6, 9). This algorithm couples the variable selection properties of genetic algorithms (GAs) to linear multiple discriminant function analysis (DFA). The software package Ometer (10) provides an easy to use implementation of this algorithm. It is available for Linux, Unix, Mac OS X, and MS Windows.

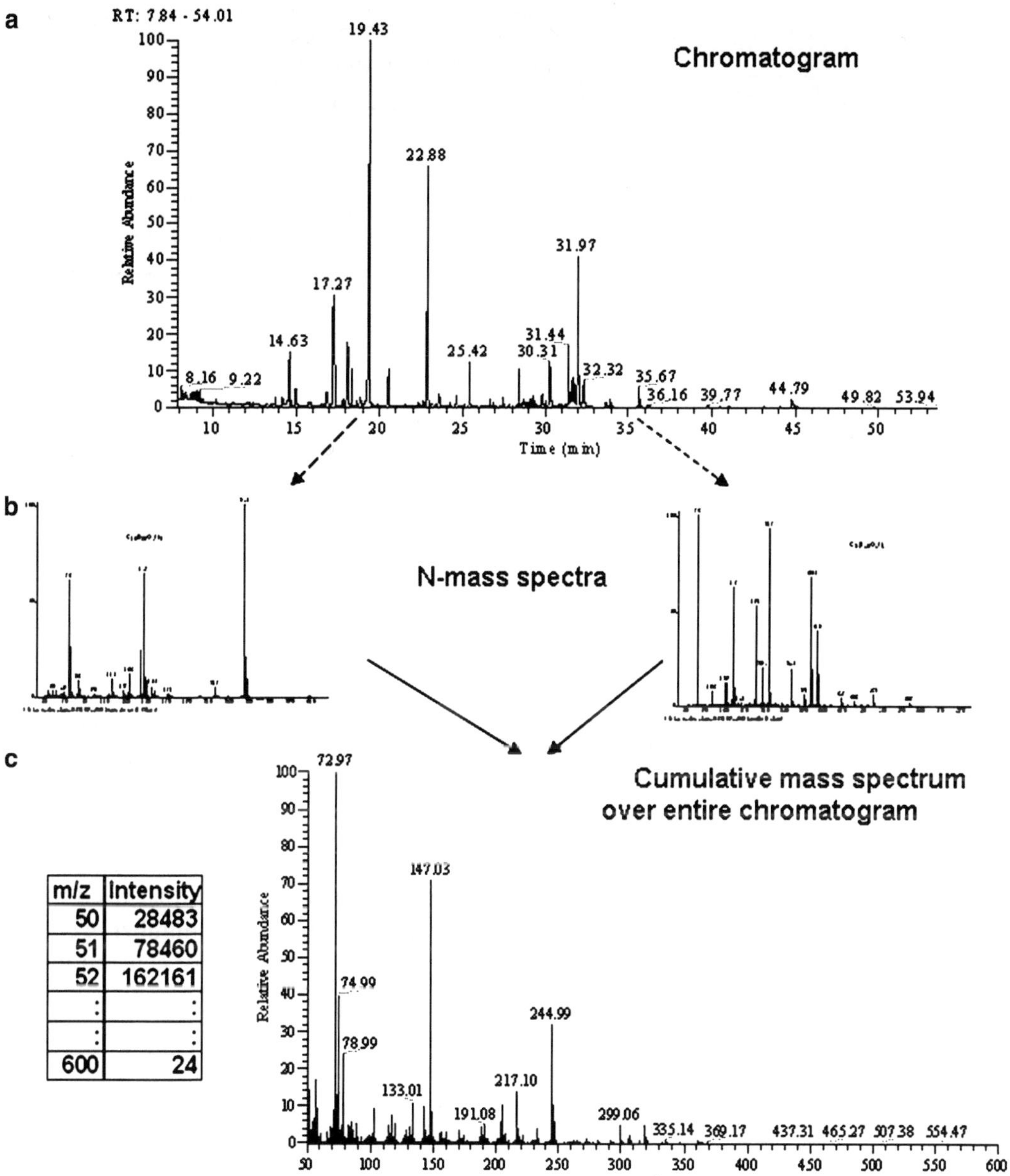

Fig. 3. Schematic illustration of dimension reduction of mass spectrometry data into a single MS vector. All peaks in the chromatogram (**a**) and their accompanying mass spectrum (**b**), e.g., fumaric acid (L) and inositol (R), are reduced to a single MS vector by summing the ion counts for each *m/z* ratio over the total scan time (**c**).

3. The program Ometer should be run from the command line. The command format that should be used for implementing GA-DFA is:

   ometer -m=ga-dfa --ga-genes=*gene number* --ga-generations=*number of generations* --ga-population=*population size* -o *name of result file input file*

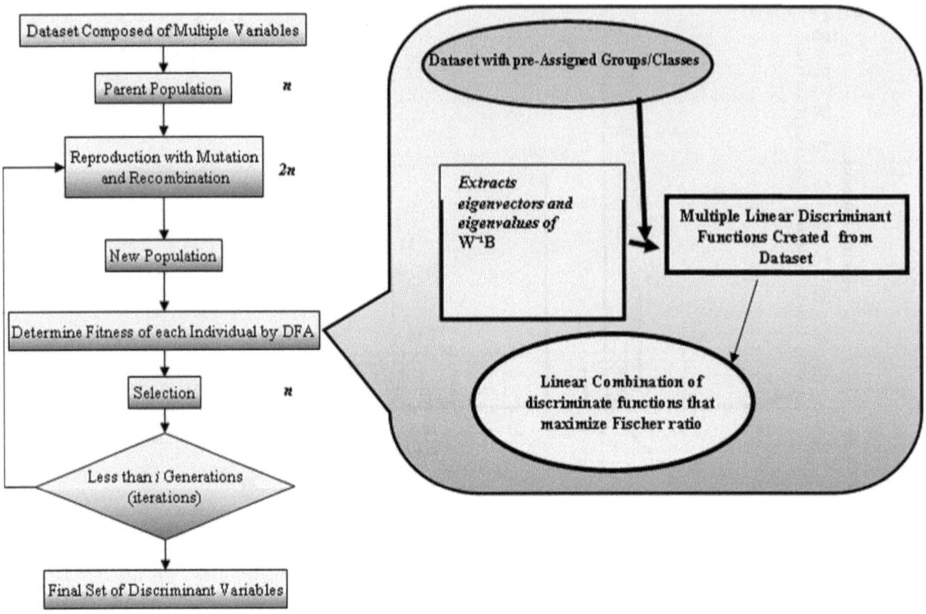

Fig. 4. Flowchart of GA-DFA method.

The values for these parameters are discussed in greater detail in the following steps. In addition, there are other parameters to those mentioned which may be used within Ometer to greater specify the types of results files generated by Ometer.

4. Determine the proper parameter values for applying GA-DFA to the preprocessed dataset. The parameters that need to be optimized for each individual dataset are the number of genes, the size of the population, and the number of generations.
   (a) The number of genes, specified as "--ga-genes=*gene number*" in Ometer, refers to the number of variables, in this case m/z values, that will be used to perform discrimination at each step. The minimum number of genes that must be selected is the number of classes minus one. The maximum number that may be selected is the entire set of variables in the dataset. Ideally, the number of genes should be the minimum number that may be selected while still achieving good discrimination. This value will differ for each dataset. Therefore, it is suggested that this number be determined empirically by trying several values and examining the classification results.
   (b) The population size, specified in Ometer as "--ga-population=population size" refers to the number of parallel searches to be performed at each time. The population will consist of the specified number of individuals each containing a set of variables, e.g. genes, which will be used to attempt discrimination. The population size should generally be larger than the number of genes, but small enough

to make the problem tractable. If the population size is too small, the GA will not be able to explore enough of the solution space to consistently find good solutions. Normally, the larger the population size, the better the solution will be. However, increasing the size of the population increases the length of time the program must run. Therefore, it is suggested that several values for population size be tried. A good initial starting value for population size is 30.

(c) The number of generations, specified in Ometer as "--ga-generations = number of generations", refers to the number of iterations that the GA-DFA should perform. Here, a larger number is preferred and can be optimized for a given preprocessed dataset. A good initial value is 500. Determine the optimal number of GA-DFA iterations that should be run. Because GAs may become trapped in a less optimal solution space, it is important to run them from multiple starting positions to increase the chance of a good solution. Running multiple iterations of GA-DFA can be done by employing a perl script, such as ga_iterate.pl (download from http://mendes.vbi.vt.edu/), which calls Ometer the specified number of times using the user defined parameter values. In order to run ga_iterate.pl, the parameters desired for each GA-DFA run, the total number of runs, and the name of the preprocessed MS dataset must be specified. The number of iterations that should be used should be large enough to create confidence in the discrimination ability of the m/z variables selected. The larger the number of GA-DFA iterations performed, the greater the confidence in the discrimination. A given m/z value is selected 95% of the time within 10,000 GA-DFA runs is to be trusted more than a m/z value selected 95% of the time in only 100 runs. Initially, it is a good idea to perform a smaller set of runs, such as 100 iterations in order to optimize all input parameters. Ideally, we wish to minimize the number of genes used in the discrimination and maximize the overall accuracy. The overall accuracy of each GA-DFA run is located in the specified result file, which is in html format. The result file also contains the variables that were used for discriminant analysis as well as the parameters utilized in the current GA-DFA run. Once optimum parameters have been identified, run GA-DFA for 10,000 iterations to ensure a good solution is found.

5. The ometer program will generate a number of files which may then be used to create a histogram of highly selected masses. For each individual GA-DFA run, a file containing the m/z values that were used in the discrimination will be created. These variable selection files, which may be identified

by their .varselect.dat extensions, can be combined to calculate the number of times that a given *m/z* value was selected. This process may be carried out via the use of the perl script *varselectstats.pl* (downloadable at http://mendes.vbi.vt.edu/). This script requires the input of all variable selection files for a given series of GA-DFA runs. The results of these calculations form a table containing the number of times each *m/z* value was selected in the specified GA-DFA runs. When a small number of GA-DFA runs are being used to identify optimum parameters, this table may be used to determine if the number of genes selected should be reduced. If the most highly selected *m/z* value is chosen less than 50% of the time, selection can often be improved by reducing the number of genes selected and increasing the population size and number of generations to maintain a high overall accuracy.

(a) Once a table containing selection counts for each *m/z* value has been created, a histogram for the distribution of frequencies can be generated. An example histogram for GA-DFA analysis of three *Arabidopsis* genotypes and graphical display of the data for the three most discriminant ions in discriminant space are presented in Figs. 5 and 6. In order to plot the histogram, use plotting software such as gnuplot or Excel. In both cases, read the table results into the program and have it plot the number of times each mass was selected as an impulse. By examining the resulting plots, highly selected masses which may be

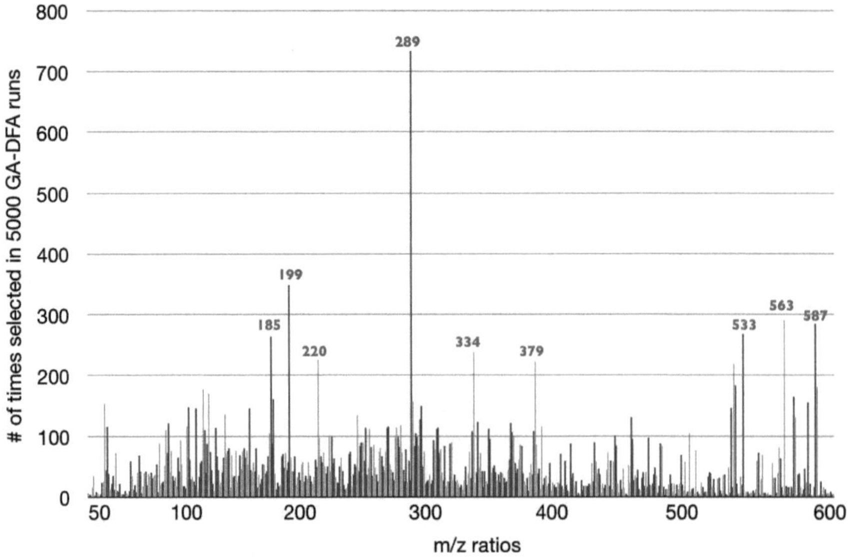

Fig. 5. Histogram for the distribution of frequencies of the *m/z* ratios chosen by GA-DFA on mass spectral data of leaf extracts for *Arabidopsis* JMT overexpressor, knockout and wild-type lines. Represented in the histogram are the frequency of selection for each *m/z* ratio to distinguish among the three classes in 5,000 GA-DFA runs. The top *m/z* ratios selected as the most discriminant were: 185, 199, 220, 289, 334, 379, 553, 563, and 587.

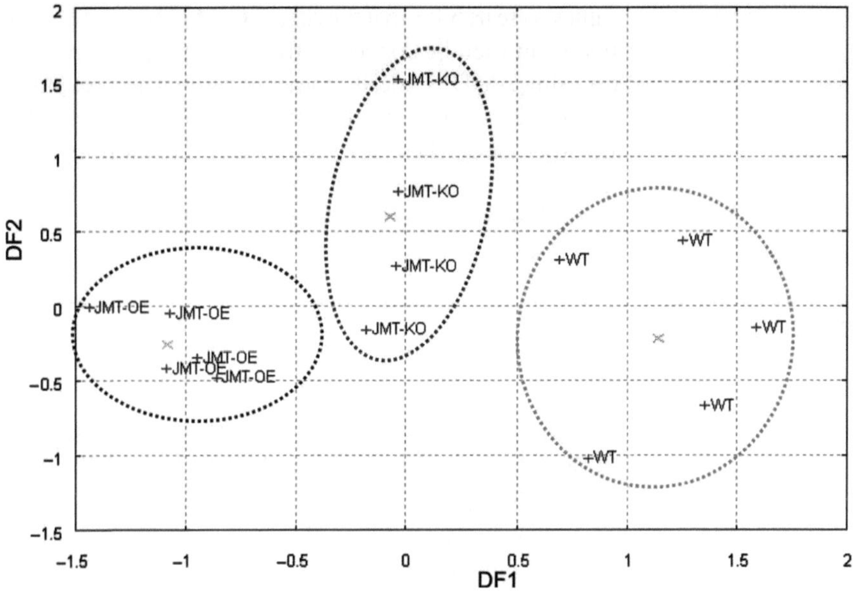

Fig. 6. Data projection from GA-DFA analysis using *m/z* 199, 289, 563 for the *Arabidopsis* wild-type, JMT overexpressor, and knockout mutants. Data labels: *JMT-KO* jasmonic acid methyl transferase knock-out mutant, *JMT-OE* jasmonic acid methyl transferase overexpressor, *WT* wild-type. Each label represents one biological replicate ($n = 5$).

useful in discriminating between the preexisting classes can be identified.

(b) The table containing the selection counts for each mass can also be used to determine which m/z values are most often selected together. When certain m/z values are coselected, it may mean that they represent the same chemical compound. To identify m/z values that are most often coselected, the perl script AssocMass.pl (downloadable at http://mendes.vbi.vt.edu/). The script AssocNotTop.pl, calculates the number of times that each m/z value is coselected. As inputs, it requires a file containing all selected m/z values and the number of times they were selected and the set of all variable selection files for a given GA-DFA series of runs. The resulting files containing the set of the most highly associated m/z values can then be used to identify m/z values that are potentially derived from the same molecule.

6. Once the most highly selected and coselected m/z values have been identified, this information can be used to quickly identify candidate metabolites in the original GC-MS spectra. In order to determine appropriate retention time search areas, the resultant m/z values can be plotted, for example, from each class within Xcalibur. Retention times where there is a significant difference between the m/z intensities of differing classes can then be identified and used in conjunction with the appropriate m/z

values to search for matches in a GC-MS library. These retention times can then be converted to their appropriate retention indices using AMDIS. Ideally, these m/z values and retention indices will match well with a known molecule in a GC-MS library. In other cases, these m/z values and retention indices will not match to any known molecule in the current GC-MS library, and will serve as excellent targets for further interrogation.

4. Notes

1. Avoid introducing excessive mass spectrometric background or chromatographic contaminants by using solvents that are HPLC grade or better.
2. Many researchers harvest their tissue by flash freezing in liquid nitrogen and then storing in aluminum foil at −80°C until use. If metabolomics analysis is to be performed on fresh tissue at a later date, it is more convenient to carefully weigh a 100 mg aliquot of fresh tissue into a 1.5 mL microcentrifuge tube prior to analysis. This avoids the issue of melting frozen tissue during weighing. If tissue is freeze-dried, vial weight, fresh weight before freeze-drying, and dry weight after freeze-drying are recorded for all samples. For *Arabidopsis*, approximately 90% of fresh weight is water. Rice is about 60–65%. If fresh tissue must be used, DO NOT ALLOW FRESH TISSUE TO MELT – metabolic changes will occur.
3. Freeze-dried tissue is often statically charged, especially in plastic vials. Weighing the charged tissue can be challenging. Reduce/remove charge with an antistatic gun. Flash-freezing may also help.
4. Several blanks should be included to account for contamination or false-positives which can be inadvertently added throughout the process from freeze-drying, solvents, extraction solutions, impure internal standards, dirty GC injection liners, and unconditioned septa or columns. The blanks should contain everything except plant material and undergo the entire process. If freeze-dried tissue is used, blanks should also undergo freeze-drying.
5. It is critical that screw caps are tightly closed and that they have O-rings. Otherwise, solvent and metabolites will be lost during the mixer mill process.
6. The mixer mill operates like a paint shaker. From our experience, it appears that tubes closest to the arm holding the 2×24 mixer mill adapter are less ground than tubes farthest from the arm. We speculate that this is due to a shorter

moment arm and thus less grinding energy than those tubes farthest away from the arm. We recommend rotating the adapter 180° for the second grinding cycle so that all tubes are exposed to similar grinding energy.

7. Chloroform should not be pipetted with plastic tips or centrifuged in plastic vials as chloroform denatures most plastic. Glass serological pipettes with volumes marked in paint should not be used either, especially from the original solvent: the paint will come off in the chloroform. This is less important for targeted analysis but introduces contaminants that are more problematic in untargeted analysis.
8. Pyridine and MSTFA are extremely moisture sensitive. Where possible, pyridine should be handled under an inert atmosphere such as nitrogen or argon to avoid the introduction of water. Use an adapter kit such as the EMD elbow adapter kit (Catalog # 692004-2). The bottle cap should not be removed but instead, pyridine should be removed through the resealable septum with an 18 or 20 gauge 6 in. long needle and syringe. Spinal tap needles work well. MOX is usually not suitable for use after prolonged storage (+2 months). When in doubt, fresh MOX should be made. MOX can be stored in vials with resealable septum and solution withdrawn using a dry syringe (flush with nitrogen). If MOX is exposed to the atmosphere, it should be flushed for 1 min under nitrogen or argon before recapping.
9. Methoximation prevents cyclization and stabilizes carbonyl groups in the β position of sugars (3, 11). Samples or analyses without sugars or metabolites not containing carbonyl groups may not need to be methoximated. Samples with high sugar content may benefit from incubation at elevated temperatures (e.g., 50°C). Dry heating blocks can also be used to heat vials but should be briefly vortexed every 15 min. Although risky, water baths can be used, provided vials are securely closed from moisture since MOX and MSTFA are extremely moisture sensitive. Oxime derivatives are relatively stable and can be prepared several days in advance of trimethylsilylation.
10. Derivatize (trimethylsilylate) only enough for 1–2 days worth of GC-MS analysis time – TMS derivatives are volatile and degrade, even after only a few days. Trimethylsilylation is the process where labile hydrogens on polar compounds are replaced with a –Si(CH3)3 group. TMCS aids derivatization of amides, secondary amines and hindered hydroxyls. MSTFA should be stored at 4°C but should be warmed up and dried in a desiccator before opening to avoid water.
11. If the instrument has been idle for several days, contaminants such as siloxanes from the septum can accumulate in the

injector and front of the column. Inject the ten alkane retention mix several times using the normal analysis method to remove any contaminants.

12. The mass spectrometer should be tuned for optimal sensitivity tris(perfluorobutyl)amine (CF43) as a reference gas.

13. The purpose of the data preprocessing is to extract from the chromatograms, the quantitative information of the metabolites, to generate variables that can be compared to each other between samples using data mining tools. Data from chromatographic-mass spectrometric detection analyses have four dimentions: m/z (mass to charge ratio), intensity, retention time, and the different observations (samples and replicates). One option, which is followed on this manuscript, is to reduce dimensionality to m/z and intensity by summation over the time domain. In this case, the output is similar to that obtained with direct infusion in metabolic fingerprinting procedures like NMR or fourier transform ion cyclotron resonance mass spectrometry (FTICR-MS), without the matrix cosuppression interference of this methods. The other option is to identify each one of the individual components (metabolites) through deconvolution of the total ion chromatogram. Then the metabolites (known or unknown) become the variables, which must be aligned across all samples, creating comparable variables.

14. In most cases, data collected in the first few minutes is not included in the resulting single MS vector to avoid methoximation and derivatization artifacts.

15. With data extraction and preprocessing a dataset containing the variables (m/z and intensities) in a format suitable for further analysis is generated. The generated data file is used as input for the discriminant analysis. Ensure that the preprocessed dataset is in the proper format; samples may be either in rows or columns and should be tab-delineated. If samples are specified by rows, it is important that the Ometer option "–f=2" is employed.

    (a) If samples are specified by columns, ensure that there is a row entitled "names" containing the sample names for each column. If samples are specified by rows, ensure that a names column exists.

    (b) If samples are specified by columns, ensure that there is a row entitled "classes" to specify the class of each sample. If samples are in rows, create a class column containing the same information. Sample classification is extremely important in allowing GA-DFA to function as expected and should be performed beforehand. A benefit of GA-DFA is that it allows the discrimination between manifold groups. Therefore, samples may be categorized into multiple classes based on properties such as treatment, genotype, time point, etc.

16. For an experimental design where samples are classified in two classes "A" and "B", it may be sufficient to compare and rank the list of metabolites according to *p*-values calculated by performing a *t*-test to determine if the means are equal or distinct. However, experimental designs are often more complex than two classes, e.g., time series, different developmental stages, different treatments or mutants. In experiments like these, a wide range of statistical tools, and machine-learning algorithms, generally called multivariate statistic methods, can be applied to metabolomics data. Multivariate analysis can be divided into supervised and nonsupervised classification methods. For the selection of metabolites discriminant among different mutants and the corresponding wild-type, we have been employing a novel approach to metabolic fingerprinting as recently described (12), with modifications. Rather than attempting to identify all the molecules obtained from metabolic profiling, we focus only on those metabolites that are determinant among sample groups. We applied the mutivariate statistical method called discriminant function analysis (DFA) coupled with genetic algorithm (GA) variable selection (GA-DFA), to process data generated through GC-MS. The classification software used in the analysis was OMETER v. 0.60 (http://mendes.vbi.vt.edu/). OMETER carries a number of multivariate statistical analysis and machine learning algorithms, applicable to functional genomics and systems biology data. Among the different tools provided by OMETER, there are two implementations of discriminant analysis with genetic algorithm variable selection. Both of them used as input a data file containing the information of the variables measured for each sample, and the classification of the samples into groups based on their biological characteristics or treatment. This information is used to minimize within-group variance and maximize between-group variance (12). In the first algorithm (GA-DFA), similar to the one described by Jarvis and Goodacre (9), GA selects a predetermined (fixed) number of variables from the full dataset, to formulate a robust model that gives an indication of those variables as the most important to discriminate among the different classes of samples being analyzed. In the second algorithm (GA-DFA2), the initial number of variables is arbitrary, but this number will increase or decrease during the analysis execution, in an attempt to find the optimal number of variables for the classification. With either algorithm a limited number of variables from the full dataset is selected and a model is built to distinguish between the different types of samples, expressed as a single linear function by DFA. Because of the stochastic nature of the

GA-DFA algorithms, it usually returns a different combination of variables as discriminant every time it is run. Therefore, the application of GA-DFA methods is best done by iterated run of the algorithm, followed by statistical analysis of the results to select the variables more frequently selected in the classification. The typical flow chart of the GA-DFA analysis is shown in Fig. 4. For the purpose of our research, we used GA-DFA2 to find discriminant variables when metabolite data for multiple classes, such as overexpressor, knockout and wild-type samples, were compared.

Acknowledgments

This work was funded by National Science Foundation grant 03122857 as part of the *Arabidopsis* 2010 Project.

References

1. Shulaev, V. (2006) Metabolomics technology and bioinformatics. *Brief. Bioinform.* 7, 128–139.
2. Rizhsky, L., Liang, H., Shuman, J., Shulaev, V., Davletova, S., et al. (2004) When defense pathways collide. The response of *Arabidopsis* to a combination of drought and heat stress. *Plant Physiol.* **134**, 1683–1696.
3. Roessner, U., Wagner, C., Kopka, J., Trethewey, R. N., Willmitzer, L. (2000) Simultaneous analysis of metabolites in potato tuber by gas chromatography-mass spectrometry. *Plant J.* **23**, 131–142.
4. Krishnan, P., Kruger, N. J., Ratcliffe, R. G. (2005) Metabolite fingerprinting and profiling in plants using NMR. *J. Exp. Bot.* **56**, 255–265.
5. Goodacre, R., York, E. V., Heald, J. K., Scott, I. M. (2003) Chemometric discrimination of unfractionated plant extracts analyzed by electrospray mass spectrometry. *Phytochemistry* **62**, 859–863.
6. Johnson, H. E., Broadhurst, D., Goodacre, R., Smith, A. R. (2003) Metabolic fingerprinting of salt-stressed tomatoes. *Phytochemistry* **62**, 919–928.
7. Shulaev, V., Cortes, D., Miller, G., Mittler, R. (2008) Metabolomics for plant stress response. *Physiol. Plant.* **132**, 199–208.
8. Kováts, E. (1958) Gas-chromatographische charakterisierung organischer verbindungen. Teil 1: retentionsindices aliphatischer halogenide, alkohole, aldehyde und ketone. *Helv. Chim. Acta* **41**, 1915–1932.
9. Jarvis, R.M. and Goodacre, R. (2005) Genetic algorithm optimization for pre-processing and variable selection of spectroscopic data. *Bioinformatics* **21**, 860–868.
10. Mendes, P. (2006) Available from: http://mendes.vbi.vt.edu/tiki-index.php?page=ometer.
11. Schweer, H. (1982) Gas chromatography–mass spectrometry of aldoses as O-methoxime, O-2-methyl-2-propoxime and O-n-butoxime pertrifluoroacetyl derivatives on OV-225 with methylpropane as ionization agent: II. *Hexoses. J. Chrom. A* **236**, 361–367.
12. Henriques, I. D., Aga, D. S., Mendes, P., O'Connor, S. K., Love, N. G. (2007) Metabolic footprinting: a new approach to identify physiological changes in complex microbial communities upon exposure to toxic chemicals. *Environ. Sci. Technol.* **41**, 3945–3951.

Chapter 18

Gramene Database: A Hub for Comparative Plant Genomics

Pankaj Jaiswal

Abstract

The rich collection of known genetic information and the recent completion of rice genome sequencing project provided the cereal plant researchers a useful tool to investigate the roles of genes and genomic organization that contribute to numerous agronomic traits. Gramene (http://www.gramene.org) is a unique database where users are allowed to query and explore the power of genomic colinearity and comparative genomics for genetic and genomic studies on plant genomes. Gramene presents a wholesome perspective by assimilating data from a broad range of publicly available data sources for cereals like rice, sorghum, maize, wild rice, wheat, oats, barley, and other agronomically important crop plants such as poplar and grape, and the model plant *Arabidopsis*. As part of the process, it preserves the original data, but also reanalyzes for integration into several knowledge domains of maps, markers, genes, proteins, pathways, phenotypes, including Quantitative Trait Loci (QTL) and genetic diversity/natural variation. This allows researchers to use this information resource to decipher the known and predicted interactions between the components of biological systems, and how these interactions regulate plant development. Using examples from rice, this article describes how the database can be helpful to researchers representing an array of knowledge domains ranging from plant biology, plant breeding, molecular biology, genomics, biochemistry, genetics, bioinformatics, and phylogenomics.

Key words: Comparative plant genomics, Genome, Genetic map, Synteny, Ontology, Comparative map, Genetic markers, Plant pathway, Rice, Maize, Sorghum, Grape, Poplar, *Arabidopsis*, QTL

1. Introduction

Rice is arguably humankind's most important staple food source, particularly in poor and developing countries. In 2005, the world consumed 519,749,020 tonnes of rice (83% of all rice production), compared to 410,080,280 tonnes of wheat and 134,768,000 tonnes of maize (FAOSTAT, 11 May 2007). Although agricultural production has increased significantly despite dwindling arable acreage, the production per capita has not improved due to population growth rates. It has become evident that new and

Andy Pereira (ed.), *Plant Reverse Genetics: Methods and Protocols*, Methods in Molecular Biology, vol. 678, DOI 10.1007/978-1-60761-682-5_18,

innovative use of genetic diversity information is essential to our ability to improve our crop production, nutritional quality, pest resistance, and environmental hardiness in order to meet the increasing nutritional and caloric needs of the world's expanding population. With the help of new cheap sequencing technology, it has been possible to sequence the complete genomes of almost any species. The genome of rice (1, 2) was the first to be done among crop plants because it has the smallest genome of the cereals (about 430 Mbp). It is followed by the recent report of *Sorghum* genome getting sequenced (3–5) and the upcoming completion of maize (6, 7) and *Brachypodium* (8, 9). In addition to these, plant researchers have also sequenced the genomes of dicots, namely, the *Arabidopsis* species *A. thaliana* also a model plant species, and *A. lyrata* and others like poplar (10), grape (11), and the upcoming ones for soybean (12), *Medicago* (13–15), and several specialty crops from Rosaceae (e.g., peach, strawberry, peach, etc.) (16).

This deluge of new large-scale datasets and essentially of new types as well, have propelled researchers to undertake a holistic perspective on how to utilize interdisciplinary tools in order to integrate and analyze complex data from multiple experiments. In addition, comparisons are made to similar datasets from other species, to find the common landmarks and the developmental, adaptive natures of a plant species by proxy to another species with much more mature datasets. This data includes the information on genetic diversity, genetic stocks, or germplasms and the availability of mutants, genome sequences, and genes identified in the germplasms, as well as the functional characteristics of genes, their interactions, and contributions toward the quantitative and qualitative phenotypes associated with the plant's ability to develop, adapt and inherit important agronomic traits, and last but not least, the mRNA and protein expression profiles. Gramene database (www.gramene.org) is a one-stop database (Fig. 1) where users are allowed to query and explore the power of genomic colinearity and comparative genomics for genetic and genomic studies on plants in general. Though the database still emphasizes on cereal crop plants, more and more noncereal species datasets are being integrated. It collects sequenced genome annotations, genes, gene products (proteins) from the sequencing project initiatives, develops its own de novo and/or curated resources on metabolic pathways, genetic markers and maps, QTL, genetic diversity, and literature (Table 1). The annotation and curation process involves many of the softwares and database components that are developed by other projects and/or developed in-house (Table 2), then seamlessly maps and links these pieces of information with the sequenced genomic coordinates for querying, hypothesis building and testing in potential research projects. Gramene also includes similar information from nonrice

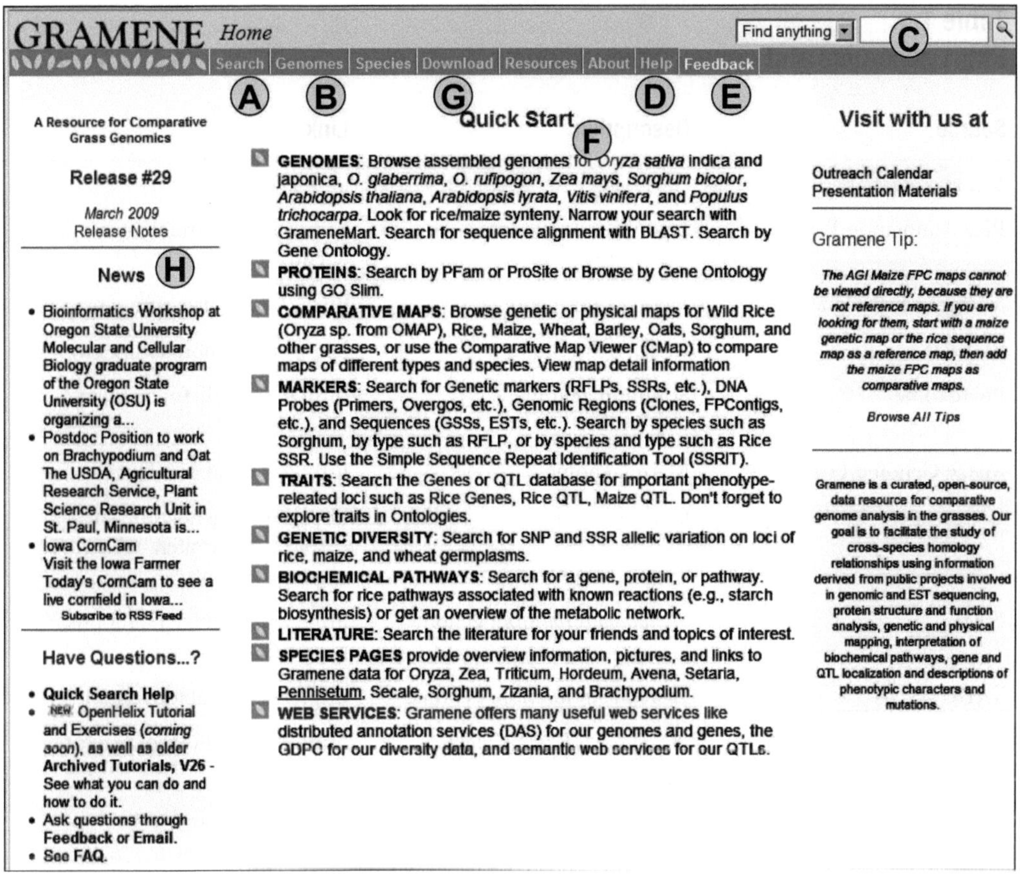

Fig. 1. A screen shot of the Gramene database Web site available at http://www.gramene.org. Users can hover their mouse or click by following icons from the figure to look for (**a**) links to individual search pages and details about each of the data modules, such as genomes, QTL, markers, maps, pathways, etc. (**b**) List of genomes and links to their genome browser pages. (**c**) A simple search. Users can select the data types. (**d**) Click here for help documents. (**e**) If you are unable to find something or want to report error, click on the "Feedback" button on any given page you are visiting and send us the feedback by filling a small form (F0). A brief outline of what is presented in each of the data module. (**g**) Click on download button to fetch a dataset of choice of the whole database for local installation. (**h**) Browse the Gramene News portal for latest announcements and various news items.

monocots, including maize (*Zea mays*), sorghum, oats (*Avena sp.*), barley (*Hordeum sp.*), wheat (*Triticum sp.*), rye (*Secale sp.*), and the millets (*Setaria sp.* and *Pennisetum sp.*) because, together with rice, they have an established colinearity in their genome structure and gene orders as defined by the macro- and microlevel synteny (17–19). Since a large amount of genomics initiatives are centered on the model species *Arabidopsis*, the database also presents connections between the genomic coordinates of *Arabidopsis*, grape, poplar, and those from cereals by way of gene-to-gene homology and whole genome alignments. Gramene is a curated, free for use; Web-accessible MySQL-based database resource for comparative genome analysis in the plants (20, 21).

Table 1
Plant genomics and genetics data sources

Source	Description	Link
Sequence databases		
Rice Annotation Project	Japonica rice genome	http://rice.plantbiology.msu.edu/
Michigan State University	Japonica rice genome	http://rapdb.dna.affrc.go.jp/
BGI RIS	Indica rice genome	http://rise.genomics.org.cn/
Phytozome	Sorghum genome	http://www.phytozome.net/sorghum
Maize Genome Project	Maize genome	http://www.maizesequence.org
Genoscope	Grape genome	http://www.genoscope.cns.fr/spip/Vitis-vinifera-whole-genome.html
NASC	*Arabidopsis thaliana* genome	http://atensembl.arabidopsis.info/index.html
TAIR	*Arabidopsis thaliana* genome	http://www.arabidopsis.org
Phytozome	*Arabidopsis lyrata genome*	http://www.phytozome.net/alyrata
Phytozome	Poplar genome	http://www.phytozome.net/poplar
Phytozome	Soybean genome	http://www.phytozome.net/soybean
NCBI	Nucleoide sequences	http://www.ncbi.nlm.nih.gov/
Uniprot	Protein sequences	http://www.uniprot.org/
dbEST	EST sequences	http://www.ncbi.nlm.nih.gov/dbEST/
dbSNP	SNP sequences	http://www.ncbi.nlm.nih.gov/projects/SNP/
Unigene home	EST clusters	http://www.ncbi.nlm.nih.gov/unigene
Gene index	EST clusters	http://compbio.dfci.harvard.edu/tgi/
PlantGDB	EST clusters	http://www.plantgdb.org
Plant genetic diversity databases/datasets		
Panzea	Maize diversity project	http://www.panzea.org
Haplotype polymorphism in polyploid wheats and their diploid ancestors	Wheat SNP genotyping data	http://wheat.pw.usda.gov/SNP/new/index.shtml

(continued)

Table 1 (continued)

Source	Description	Link
Nordborg lab	Genomic polymorphism data in *Arabidopsis thaliana*	http://walnut.usc.edu/2010/2010-project
Rice diversity	Rice diversity project	http://www.ricediversity.org/
Gramene: genetic diversity	A resource for curated genotype and phenotype association study dataset from rice	http://www.gramene.org/db/diversity/diversity_view
Model plant databases		
MaizeGDB	Maize genetics and genomics dataset provider	http://www.maizegdb.org
GrainGenes	Triticeae genetics and genomics dataset provider	http://www.graingenes.org
SOL genomics network	Solanaceae genetics and genomics dataset provider	http://sgn.cornell.edu
Soybase	Soybean genetics and genomics dataset provider	http://www.soybase.org
Barchybase	*Brachypodium* genetics and genomics dataset provider	http://www.brachybase.org
Medicago	*Medicago* genetics and genomics dataset provider	http://www.noble.org/medicago/
GDR	Genomics and genetics database for Rosaceae	http://www.bioinfo.wsu.edu/gdr/

2. Materials

The various data sources, tools, analysis softwares, and database schemas that were used in putting together the Gramene database are listed in Table 1. Even though the Gramene has data on several different species, this article focuses on the examples from genus *Oryza* (rice). All the datasets and the database infrastructure presented in this article refers to the Gramene database version 29, released in March 2009.

3. Methods

3.1. Annotation of Genomics and Genetic Datasets

Annotations include the identification and mapping of novel and known genes from the sequenced genomes and the biological role they play in exhibiting biochemical functions and associated

Table 2
Bioinformatics tools and software

Source	Description	Link
Ensembl	Genome browser, Compara: ortholog finder, BioMart, Ensembl genome analysis pipeline	http://www.ensembl.org
CMap	Comparative map viewer	http://www.gmod.org
MySQL	Open source database	http://www.mysql.com
Gene ontology (GO)	Controlled vocabulary for annotating gene products' molecular function(s), role in biological process(es) and location in cellular component(s)	http://www.geneontology.org
Plant ontology (PO)	Controlled vocabulary for annotating plant gene product and phenotype's localization of spatial and temporal expression pattern	http://www.plantontology.org
Trait ontology (TO)	Controlled vocabulary for annotating plant gene product and phenotype's associated traits	http://www.gramene.org/plant_ontology/ontology_browse.html#to
Interpro	A database of protein families and domains	http://www.ebi.ac.uk/interpro/
BioCyc	Pathway tools	http://bioinformatics.ai.sri.com/ptools/
Genomic diversity and phenotype data model (GDPDM)	Genotype and phenotype association database schema	http://www.maizegenetics.net/gdpdm/
BLAT	The BLAST-like alignment tool	http://www.kentinformatics.com/index.asp
Perl	Programming language	http://www.perl.org/
SSRIT	Simple Sequence Repeat Identification Tool	http://www.gramene.org/db/markers/ssrtool
Genomic diversity and phenotype connection (GDPC)	The genomic diversity and phenotype connection	http://www.maizegenetics.net/gdpc/

phenotypes; mapping genetic and sequence-based markers, such as restriction fragment length polymorphism (RFLPs), simple sequence repeats (SSRs), insertions and deletions (indels), and single nucleotide polymorphism (SNPs), for detecting polymorphic sites based on in-silico analysis and/or in-lab experiments; finding alternative spliced forms; mapping the expressed sequence tags (ESTs), EST clusters or theoretical gene contigs (e.g., Unigenes);

functional assignments based on domain detection and/or confirmation by the literature citation; and building associations to the metabolic pathways, reactions, and genetic diversity. In order to provide our users a one-stop shopping portal to research on the above aspects, as well as associations to the experimental data reported in the peer-reviewed publications for any evidence-based hypothesis building, we work with several collaborators and have established the protocols on analyses, annotation, and integration. For the Gramene Markers database, the various sequenced entities listed as markers, such as bacterial artificial clones (BAC), BAC ends, protein coding sequence (CDS), messenger RNAs (mRNAs), ESTs, etc., are imported from various databases such as NCBI's core nucleotide databases (22). EST clusters from Unigenes (23), PlantGDB (24), The Gene Index project (25), SNPs collected from dbSNP (26), and microarray and serial analysis of gene expression (SAGE) tags are obtained from numerous other sources. Proteins are downloaded from Uniprot (27) (Table 1). The information on various genetic maps, curated gene sets, proteins, and quantitative trait locus (QTL) is gathered from the published literature and updated based on personal communication with the authors of the peer reviewed published articles (28). We then run computational analyses on sequenced entities using the Blast-like algorithm (BLAT) (29) and Basic Local Alignment Search Tool (BLAST) (30) together with the Ensembl analysis pipeline (31) and/or in-house protocols available from the alignment docs. These documents are updated often with every build to accommodate new data sets and analysis parameters. The various tools we use are listed in Table 2 and some of the analysis types and their documents are listed in Table 1. Manual analysis is done in random cases for quality assurance. After this, the links are established and a second round of quality assurance is run among the manually curated information extracted from the published literature. Following this, the datasets are assigned to the various ontologies (see Subheading 3.2.6). Finally, the data in various sections as described below is linked to each other for interoperability and referencing, after which the data is made accessible to the users via the website. However, the process does not stop here; the users are encouraged to send their feedback, which will provide the basis for improvements in subsequent database releases.

3.2. Accessing Information from Gramene Database

Data at Gramene is stored in several smaller sections (Fig. 1, Table 3) for the purpose of clarity of presentation, namely, the Markers, CMap, Genomes, BLAST, Genes, Proteins, QTL, Ontologies, Diversity, Pathways, and Literature. However, each section (or module) connects to other sections when there is a relevant mapping link, e.g., markers are linked to the maps via map location on the genetic maps and sequenced assembly of the rice genomes. They are also associated to genes and QTL either

Table 3
List of useful links from Gramene database web site and respective modules described in Subheading 4

Source	Description	Link
Gramene	Web site and database	http://www.gramene.org
	Genomes (rice, wild rice, maize, sorghum, poplar, grape, *Arabidopsis thaliana*, *Arabidopsis lyrata*)	http://www.gramene.org/genome_browser/index.html
	Genome synteny (genetic colinearity)	http://www.gramene.org/Oryza_sativa_japonica/Location/Synteny?otherspecies=Zea_mays&r=6
	Genes	http://www.gramene.org/rice_mutant/index.html
	Proteins	http://www.gramene.org/qtl/index.html
	Pathways (RiceCyc, SorghumCyc, AraCyc, SolCyc, MedicCyc)	http://www.gramene.org/pathway/
	OMICs viewer	http://pathway.gramene.org/expression.html
	Ontologies (GO, PO, TO, EO, GAZ, etc.)	http://www.gramene.org/plant_ontology/
	Markers	http://www.gramene.org/markers/index.html
	Rice SSR marker resource	http://www.gramene.org/microsat/index.html
	Comparative maps	http://www.gramene.org/cmap/
	Genetic diversity	http://www.gramene.org/db/diversity/diversity_view
	QTL	http://www.gramene.org/qtl/index.html
	Literature	http://www.gramene.org/literature/index.html
	BLAST	http://www.gramene.org/Multi/blastview
	GrameneMart	http://www.gramene.org/Multi/martview
	Sequence alignment documents	http://www.gramene.org/documentation/Alignment_docs/
	Download from FTP site	ftp://ftp.gramene.org/pub/gramene/
	About the project	http://www.gramene.org/about/
	Species pages	http://www.gramene.org/species/
	Tutorials	http://www.gramene.org/db/help?state=display_topic_in_context&topic_name=Tutorials+and+Exercises&sticky=0
	Frequently asked questions (FAQs)	http://ww.gramene.org/db/help?state=display_topic_in_context&topic_name=FAQ
	Contact	gramene@gramene.org
	Sitemap	http://www.gramene.org/sitemap.html
	Mail archive/discussion list	http://www.gramene.org/mailarch.html
	Collaborators	http://www.gramene.org/collaborators/index.html

by their overlapping location or by direct evidence based on the inferences derived by the association of genotype and phenotype experiments. Similarly, genes are connected to the reactions and pathways. This is expected to help the users in making queries such as "find QTL on a given genetic map, its inferred mapping to the sequenced genome and a list of genes underlying that stretch of genome for a desired gene function and/or phenotype and/or pathway." The various subsections are described below and listed in Table 3. By using rice examples, a few important features are also highlighted. However, some useful tips and a few complex queries that users may like to perform using Gramene are listed in the Subheading 4. These are just the examples, and there are many ways and means one can use Gramene database.

3.2.1. Markers

Gramene's Marker module is the primary source of information on markers and mappings. Although the information displayed may vary depending upon the marker type, all markers show the name, synonyms (if any), marker type, source species, and list of map positions with links to the actual map displays. When available it will also display the germplasm data, associations to other markers (for example, a QTL may be associated to a gene/gene model, or individual ESTs are part of the EST cluster or Unigenes), sequence, location on the genome, source information, and literature citations. Finally, if available, images showing gels with polymorphisms are displayed. Gramene's Marker module contains about 47.1 M markers (Table 4), of which about 15% (7.2 M) are from rice.

Marker names and IDs can be searched either one at a time or in multiples, each separated by commas or spaces. To further refine a query, select a marker type (e.g., "RFLP") and/or a species (e.g., "*Oryza sativa*"). As an example, search the Markers module for "RG103" and marker type "RFLP". From the marker detail pages, a user can get an overview of the mappings, suggesting that the marker is mapped on 33 different maps (Fig. 2a) and has associations to 96 other markers in the database either as colocalized, neighbor or with a related sequence (Fig. 2b).

The SSR resources page provides access to the Simple Sequence Repeat Identification Tool (SSRIT) (32), the panel of 50 standard SSR markers used by the Generation Challenge Program for rice diversity analysis, and a Table of SSR Primers from (33), in addition to links for several other resources.

3.2.2. Maps and CMap (The Comparative Mapping Viewer)

A map is a linear array of interconnected genomic features, and represents a single linkage group (as in the case of a genetic map), a single contig (as in physical maps) or a chromosome from the assembled genome. All the individual maps from across a study or genome assembly project are grouped into map sets, such as a set of linkage groups produced by a genetic mapping study. The maps are displayed using the CMap tool developed by Gramene. The software is available through the Generic Model Organism

Table 4
Number and type of markers for all plant species present in the Gramene database in the March 2009 release #29

Type	Count
Clone	12,967,503
EST	15,016,126
EST cluster	6,103,358
Gene prediction	326,882
Gene primer	19
Genomic DNA	1,617,855
GSS	7,935,371
mRNA	293,697
Oligo	2,396,466
OVERGO	24,464
Primer	69,997
STS	2,678
AFLP	8,150
Breakpoint interval	303
Centromere	23
Chromosomal segment	1
Deletion	333
FISH probe	37
FPC	16,446
Gene	13,085
Insertion	310
Maize bin	114
Microarray probe	260,623
Point	126
QTL	11,624
RAPD	174
RFLP	18,853
SSR	19,673
Undefined	1,285
Total all species	47,105,576
Total rice species	7,266,251

a

⊟ Map Positions (33)

Oryza sativa Japonica Group (1)

Map Type	Map Set	Name	Map	Position	Ext. Links
Sequence	Gramene Annotated Nipponbare Sequence 2006	RG103	Chr. 11	20,301,909-20,302,418 (-) bp	View in Genome Browser View Comparative Map
		⊞ Details ⊞ CMap Preview			

Oryza sativa (28)

Map Type	Map Set	Name	Map	Position	Ext. Links
Sequence	Gramene IRGSP Assm 2005	RG103	Chr. 11	22,645,738-22,646,247 (-) bp	View Comparative Map
		⊞ Details ⊞ CMap Preview			
Genetic	Cornell SSR 2001	RG103	11	93.6-93.6 (-) cM	View Comparative Map
		⊞ Details ⊞ CMap Preview			
	IGCN 1998	RG103	11	79.8-79.8 (-) cM	View Comparative Map
		⊞ Details ⊞ CMap Preview			
	TTU IRIR 2000	RG103	11	136.6-136.6 (-) cM	View Comparative Map
		⊞ Details			

b

⊟ Associations (96)

Direction	Name	Type	Species	Analysis	Assoc. Type
From	CQAE18	QTL	Oryza sativa	colocalized_qtl_marker	qtl_marker
From	CQAE12	QTL	Oryza sativa	colocalized_qtl_marker	qtl_marker
From	CQAE22	QTL	Oryza sativa	colocalized_qtl_marker	qtl_marker
From	CQAE25	QTL	Oryza sativa	colocalized_qtl_marker	qtl_marker
From	AQGC038	QTL	Oryza sativa	colocalized_qtl_marker	qtl_marker
From	AQDR063	QTL	Oryza sativa	colocalized_qtl_marker	qtl_marker
From	AQHQ012	QTL	Oryza sativa	colocalized_qtl_marker	qtl_marker
From	AQHQ021	QTL	Oryza sativa	colocalized_qtl_marker	qtl_marker
To	RG103	Clone	Oryza sativa Indica Group	gss_poaceae	shared_synonym
To	G73750	Genomic DNA	Oryza sativa Indica Group	RFLP_SEQUENCE	related_sequence
To	AQ074219	GSS	Oryza sativa Indica Group	RFLP_SEQUENCE	related_sequence
To	G73750	Genomic DNA	Oryza sativa Indica Group	fix-genbank-accs.pl	contained_in_sequence

Fig. 2. Map position and association sections of the detail information page for RFLP marker RG103. (**a**) A screen shot of the map positions for the marker RG103. The Map position view provides information on the species, map type, mapset name, the chromosome/linkage group number and the start–stop positions. To view the position of the marker on a given map, click on the "view Comparative Map" or click on the map set name itself to get complete information about the given map study. (**b**) Associations view provides list-related markers, their type, species and analysis type that were used to infer the relationship. The marker name links to the detail view of the particular marker.

Database project (GMOD). Map sets, maps, and marker positions are generated from the entries stored in the Markers Module, e.g., RFLP marker RG103 (Fig. 2). These correspondences are assigned either automatically based on marker identity or manually by the curator based on published information. Currently, the maps module hosts a total of 145 map sets from *Oryza* genus belonging to the class, namely, sequence, physical, genetic, and QTL. QTL maps are genetic maps that were classified as QTL maps in order to identify them separately from the rest of the genetic maps that are often used as reference maps. In the Maps module, researchers can view correlations and genetic colinearity between and within species by adding a comparative map to a reference map.

There are various ways to view a map. Start by selecting the map to view by going to the map module. Another way is to enter from the marker entry of your choice by selecting the mapped position and view comparative map options (Fig. 2a). CMap allows the addition of any number of maps to the right and left of the selected map for comparative view, and the lines between the maps identify the correlations between the two maps (Fig. 3) (see Note 1). This allows the users to find not only the position and display on the maps, but also a way to find additional markers present on comparative maps that they may like to add to their investigations, for example, SSR markers, or to find QTL of desired traits mapped in the region that are similar to the one under investigation.

3.2.3. Genomes

Gramene's Genome section hosts two completely sequenced rice (*O. sativa*) genomes from genotype Nipponbare (2, 34) and 93–11 (35) representing the two subspecies japonica and the indica, respectively. The japonica genome hosted by Gramene is a mirror of the TIGR/MSU assembly (34) of the completely sequenced genome of *Oryza sativa subsp. Japonica* cv. Nipponbare (2). The other rice species genomes include, physical map for *Oryza rufipogon* which was generated using the OR_CBa BAC library and the short arm of the Chromosome 3 of *Oryza glaberrima*. Both non-*O. sativa* datasets were provided by the OMAP project (36).

Search and browse this module to get information on the genomic sequence, genes, gene structures, transcripts, peptides, repeats, and mapped ESTs, markers, insertion elements, FSTs, etc. One has an option to view the summary of features present or mapped on individual chromosomes, such as *Xa21* gene (TIGR Loci LOC_Os11g35500) (Fig. 4a) or browse the syntenic region, e.g., between rice chr-11 and maize genome or individual gene pages to find information on transcript, peptide, orthologs (Fig. 4b) and the variation, such as SNPs in genetic diversity (Fig. 4c) (see Note 2). Gramene also hosts the genomes of monocots *Sorghum bicolor*, and *Zea mays* via the maize genome sequencing project (37) and dicots, namely, the *Arabidopsis thaliana*, *Arabidopsis lyrata*, *Populus trichocarpa*, and *Vitis vivifera* (grape) and includes organelle (mitochondria and chloroplast) genomes from rice, maize, and *A. thaliana*. Advanced level users can also

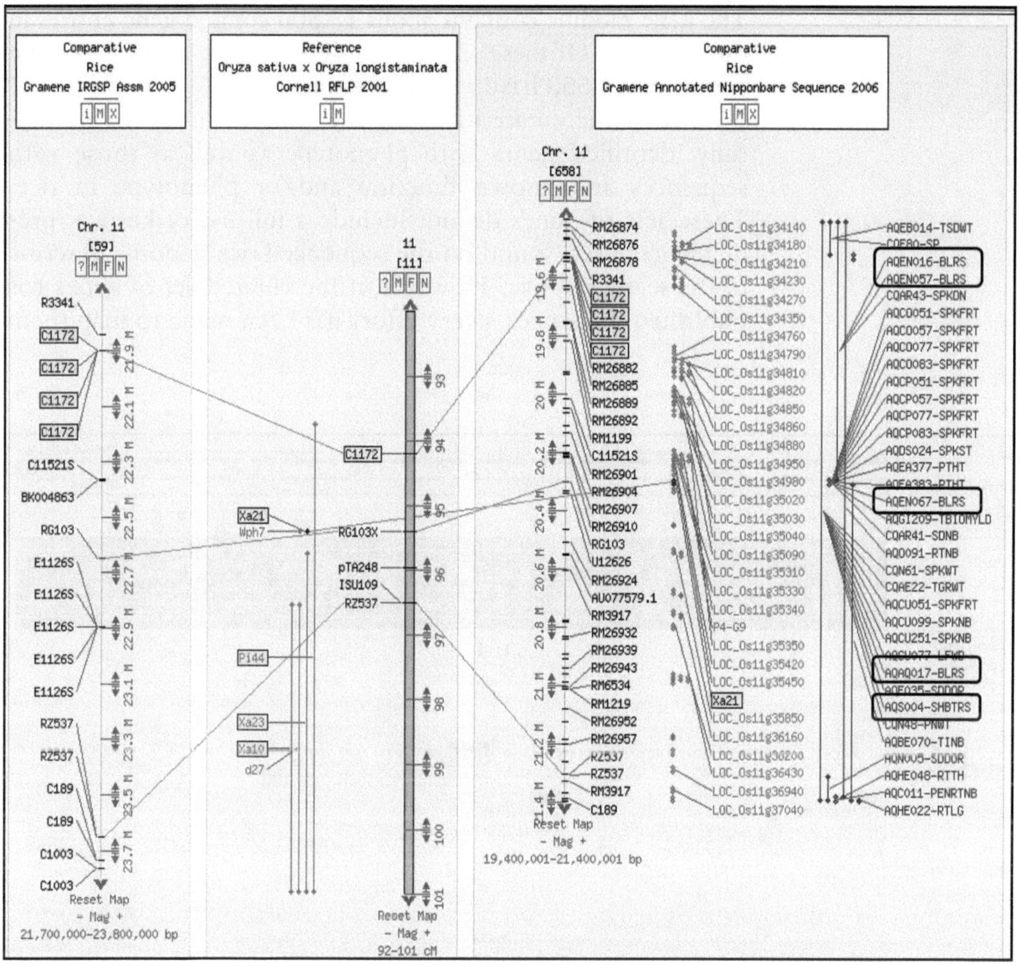

Fig. 3. View of the Comparative Map viewer (CMap). From this example view researchers can find that, on the Cornell RFLP 2001 genetic map, the RFLP marker C1172 overlaps the gene *Pi44* identified genetically. Also, marker RG103X/RG103 cosegregates with *Xa21* and *bph7* genes. In this example, we added the Gramene's annotated version of the TIGR's rice genome assembly build-5 on the *right hand side* and IRGSP's rice genome assembly build-4 for comparative views. In the original view on the Gramene Web site, users can see the *gray lines* connecting the maps that suggest related markers, also highlighted by *red* colored marker name. In the figure *boxes* around QTLs on the Gramene's annotated version of TIGR's rice genome assembly suggest the disease resistant trait QTL overlap the same region of the genome, where several disease resistance genes, such as *bph7*, *Pi44*, *Xa21*, *Xa23*, and *Xa10*, were mapped in the Cornell RFLP 2001 genetic map.

use GrameneMart described later to find all the genes mapped in the given chromosomal region for example of rice (see Note 3). Find functional annotations (Gene Ontology and Interpro domain assignments) of the genes and does the genome have other genes with similar functional annotations (see Note 4)? Find rice gene Orthologs for the species of my interest (e.g., *Arabidopsis*, Poplar)/find all the homologous (ortholog and paralog) gene pairs of the gene of interest (see Note 5).

3.2.4. Gene Search

The gene module contains about 13,500 curated gene entries in the database. Of these, 2,713 rice gene entries, 659 are protein coding and 1,553 have known phenotypes.

Search the curated list of genes that includes the genetically identified genes with phenotype as well as those with sequences and known function and/or phenotype in rice. These sets of genes do not include a full list of known/predicted gene sets found on the sequenced rice genome describe in the section above. However, if the curated set of genes has published sequences, every effort has been made to map them

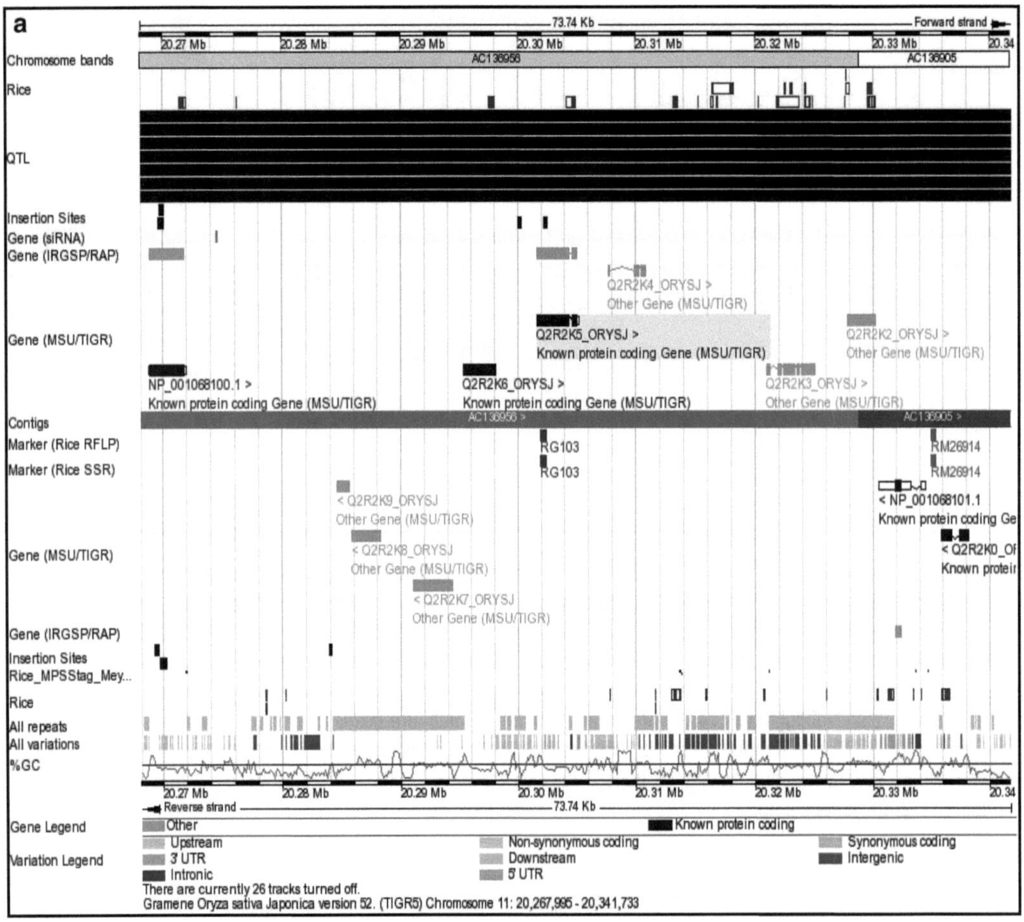

Fig. 4. The Genome browser detail view for rice japonica subspecies. (**a**) The genomic location and gene structure of the *Xa21* gene. The browser also displays various other markers, such as SSR, SNPs, Repeats, RFLP, ESTs, BAC clones and BAC ends, Microarray probesets and QTL mapped in the genomic location of rice *Xa21* gene and its neighborhood. (**b**) Alignment of the SNPs mapped to the rice *Xa21* gene. The SNP variation view also presents information on the SNP source, alleles, ambiguity, synonymous and nonsynonymous classification and their alignments to various functional domains present on the gene. (**c**) Phylogenetic tree of the gene–gene homologies (includes orthologs and paralogs) from numerous plant and model eukaryote species. Individual nodes can be collapsed/expanded to see the members of the node. The homologies were computed by the Compara method provided by the Ensembl (www.ensembl.org).

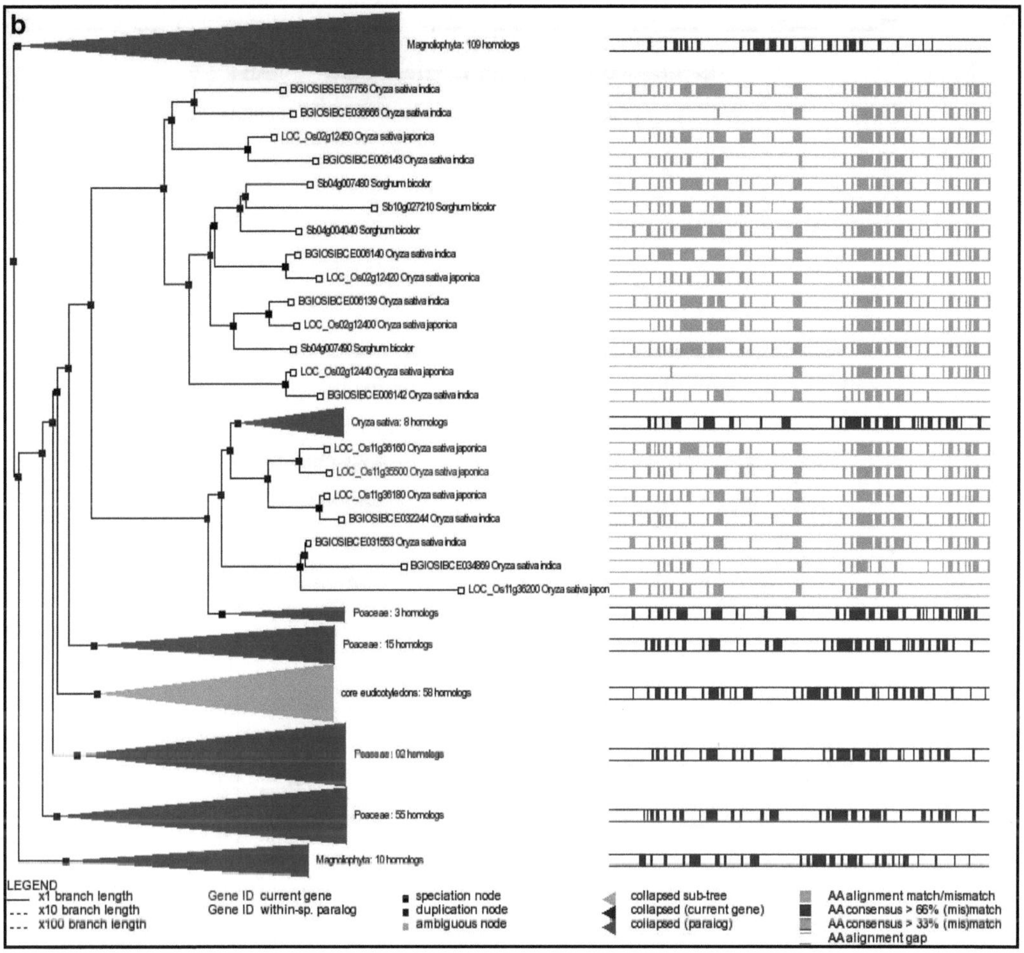

Fig. 4. (continued)

to the gene loci found on the sequenced genome. The gene detail pages include name, symbol, associated sequence(s), description, alleles, germplasm, and ontology annotations to describe the phenotype, expression and treatments, cross references to published citations, and links to the Oryzabase database (38). For example, the *Xa21* gene information provides in-depth information about its function, phenotype, and map locations (Fig. 5).

3.2.5. QTL

A QTL is a polymorphic locus on a genetic map, which contains alleles that differentially affect the expression of a quantitative phenotype trait, such as bacterial blight disease resistance, and days to flowering type. Gramene hosts 11,624 curated set of QTLs from rice, maize, oats, barley, wheat, pearl millet, foxtail millet, wild rice (*Zizania*). Of these 8,646 belong to rice (*O. sativa*) that were annotated to 237 traits representing nine major trait categories,

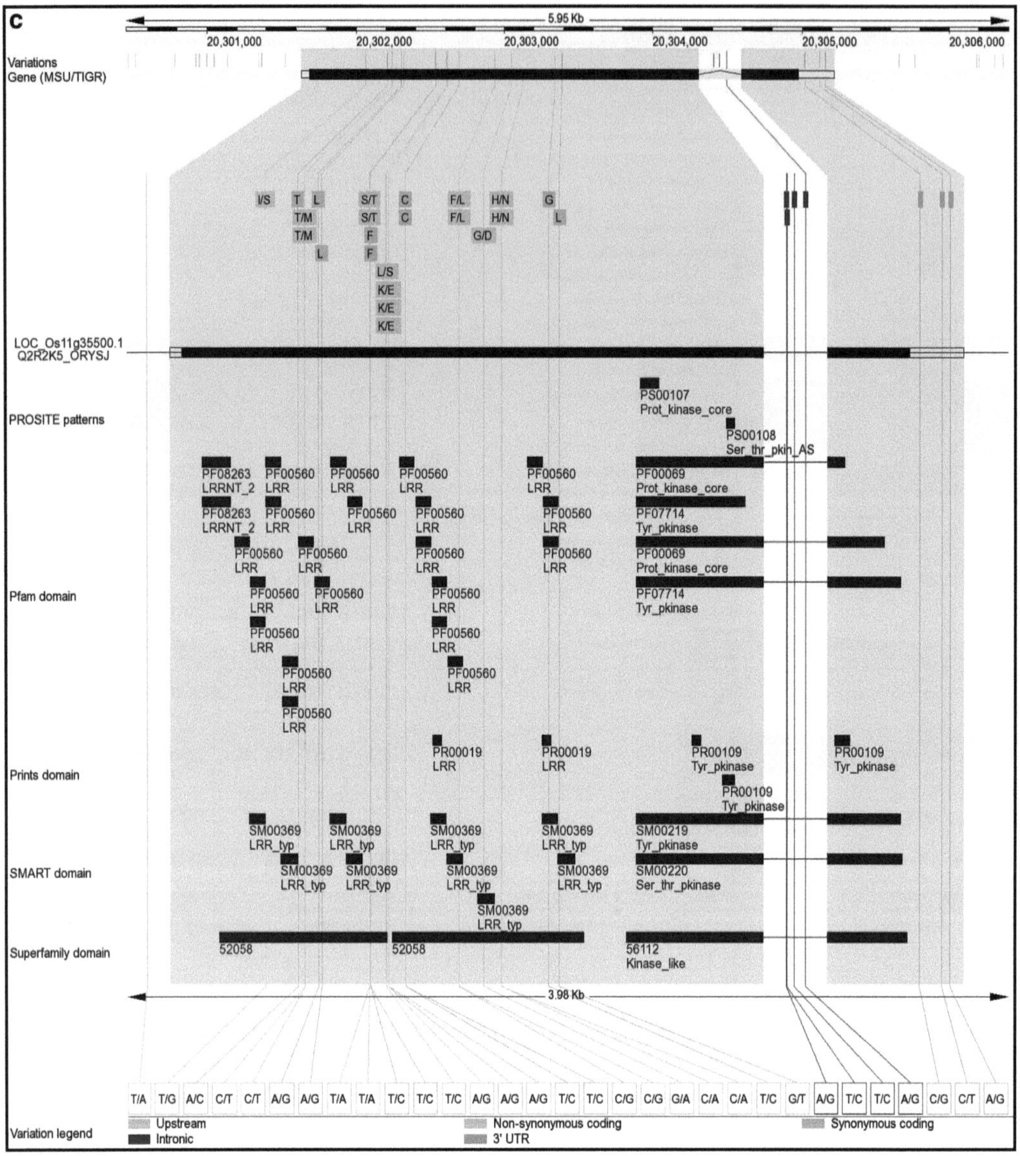

Fig. 4. (continued)

namely, abiotic stress, biotic stress, anatomy, biochemical, development, quality, sterility or fertility, vigor, and yield (28).

Search the QTL information available from the QTL module that includes associated trait name and symbol, mapped position on the original genetic map that was described in the publication and an inferred position on the sequenced rice genome assembly, published QTL symbol, cited reference, and curator comments. A standardized method is followed throughout Gramene when assigning the QTL trait symbols, names, and mapping to the trait ontology (TO) for finding other QTL

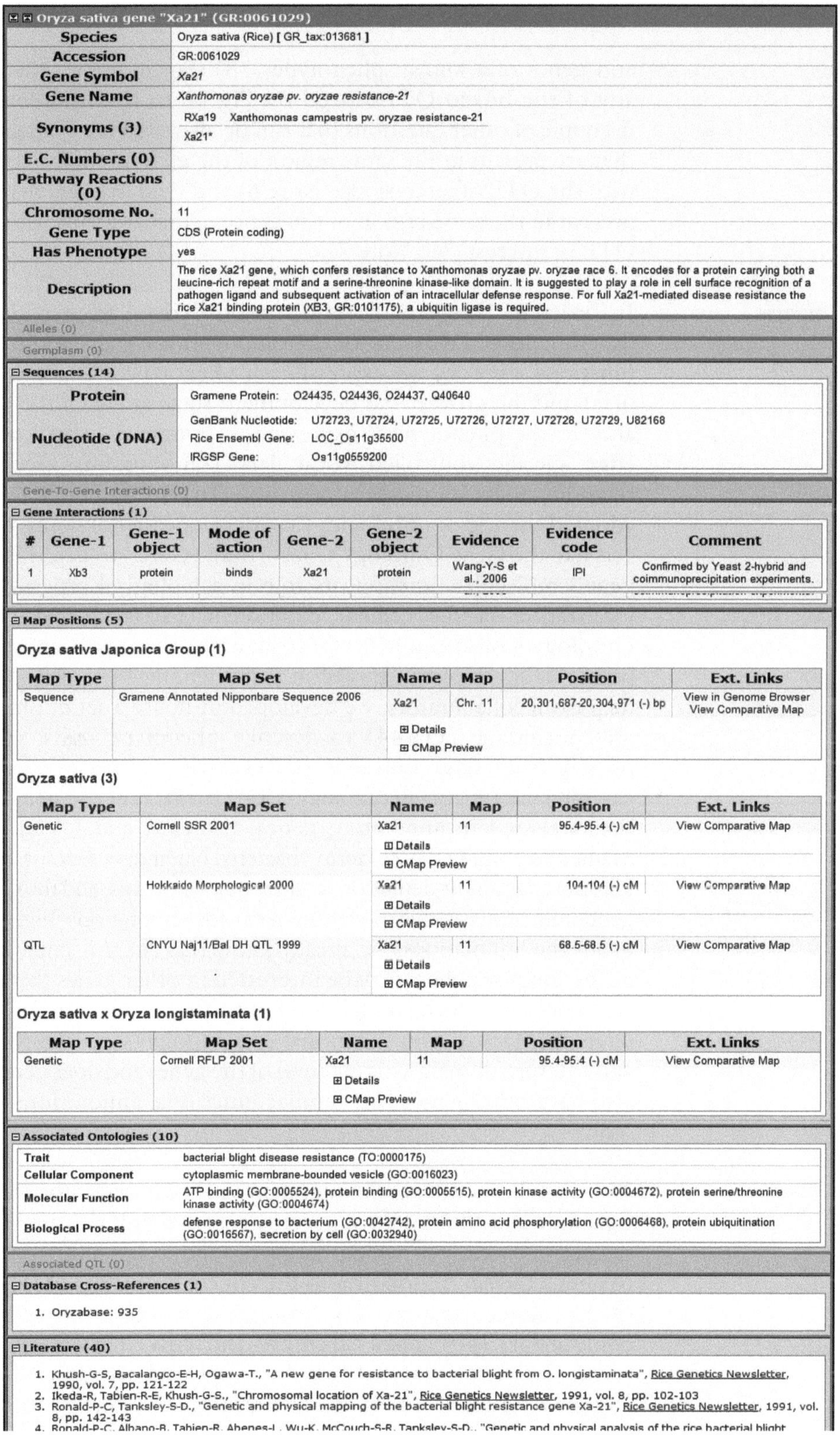

Oryza sativa gene "Xa21" (GR:0061029)

Species	Oryza sativa (Rice) [GR_tax:013681]
Accession	GR:0061029
Gene Symbol	*Xa21*
Gene Name	*Xanthomonas oryzae pv. oryzae resistance-21*
Synonyms (3)	RXa19 Xanthomonas campestris pv. oryzae resistance-21 Xa21*
E.C. Numbers (0)	
Pathway Reactions (0)	
Chromosome No.	11
Gene Type	CDS (Protein coding)
Has Phenotype	yes
Description	The rice Xa21 gene, which confers resistance to Xanthomonas oryzae pv. oryzae race 6. It encodes for a protein carrying both a leucine-rich repeat motif and a serine-threonine kinase-like domain. It is suggested to play a role in cell surface recognition of a pathogen ligand and subsequent activation of an intracellular defense response. For full Xa21-mediated disease resistance the rice Xa21 binding protein (XB3, GR:0101175), a ubiquitin ligase is required.

Alleles (0)

Germplasm (0)

Sequences (14)

Protein	Gramene Protein: O24435, O24436, O24437, Q40640
Nucleotide (DNA)	GenBank Nucleotide: U72723, U72724, U72725, U72726, U72727, U72728, U72729, U82168 Rice Ensembl Gene: LOC_Os11g35500 IRGSP Gene: Os11g0559200

Gene-To-Gene Interactions (0)

Gene Interactions (1)

#	Gene-1	Gene-1 object	Mode of action	Gene-2	Gene-2 object	Evidence	Evidence code	Comment
1	Xb3	protein	binds	Xa21	protein	Wang-Y-S et al., 2006	IPI	Confirmed by Yeast 2-hybrid and coimmunoprecipitation experiments.

Map Positions (5)

Oryza sativa Japonica Group (1)

Map Type	Map Set	Name	Map	Position	Ext. Links
Sequence	Gramene Annotated Nipponbare Sequence 2006	Xa21 ⊞ Details ⊞ CMap Preview	Chr. 11	20,301,687-20,304,971 (-) bp	View in Genome Browser View Comparative Map

Oryza sativa (3)

Map Type	Map Set	Name	Map	Position	Ext. Links
Genetic	Cornell SSR 2001	Xa21 ⊞ Details ⊞ CMap Preview	11	95.4-95.4 (-) cM	View Comparative Map
	Hokkaido Morphological 2000	Xa21 ⊞ Details ⊞ CMap Preview	11	104-104 (-) cM	View Comparative Map
QTL	CNYU Naj11/Bal DH QTL 1999	Xa21 ⊞ Details ⊞ CMap Preview	11	68.5-68.5 (-) cM	View Comparative Map

Oryza sativa x Oryza longistaminata (1)

Map Type	Map Set	Name	Map	Position	Ext. Links
Genetic	Cornell RFLP 2001	Xa21 ⊞ Details ⊞ CMap Preview	11	95.4-95.4 (-) cM	View Comparative Map

Associated Ontologies (10)

Trait	bacterial blight disease resistance (TO:0000175)
Cellular Component	cytoplasmic membrane-bounded vesicle (GO:0016023)
Molecular Function	ATP binding (GO:0005524), protein binding (GO:0005515), protein kinase activity (GO:0004672), protein serine/threonine kinase activity (GO:0004674)
Biological Process	defense response to bacterium (GO:0042742), protein amino acid phosphorylation (GO:0006468), protein ubiquitination (GO:0016567), secretion by cell (GO:0032940)

Associated QTL (0)

Database Cross-References (1)

1. Oryzabase: 935

Literature (40)

1. Khush-G-S, Bacalangco-E-H, Ogawa-T., "A new gene for resistance to bacterial blight from O. longistaminata", Rice Genetics Newsletter, 1990, vol. 7, pp. 121-122
2. Ikeda-R, Tabien-R-E, Khush-G-S., "Chromosomal location of Xa-21", Rice Genetics Newsletter, 1991, vol. 8, pp. 102-103
3. Ronald-P-C, Tanksley-S-D., "Genetic and physical mapping of the bacterial blight resistance gene Xa-21", Rice Genetics Newsletter, 1991, vol. 8, pp. 142-143
4. Ronald-P-C, Albano-B, Tabien-R, Abenes-L, Wu-K, McCouch-S-R, Tanksley-S-D., "Genetic and physical analysis of the rice bacterial blight

Fig. 5. Curated gene entry display citing the *Xa21* example. It provides complete information on the phenotype, associated sequences, gene type, ontology associations, map positions, full list of citations, and if available the list of germplasms and alleles.

and genes of a known phenotype (28). For example, click on any of the boxed QTLs listed on Fig 2b to find the details. A couple of other questions that can be answered are Find genes that are present in the same region of the genome that overlaps with the QTL of interest (see Note 6) and Find markers that are present in the same region of the genome that overlaps with the QTL of interest (see Note 7).

3.2.6. Ontologies

In order to compare the information derived from multiple experiments and datasets for rice on phenotype traits, molecular functions, sites of gene expression or phenotype evaluation in a plant and the growth and development stage at which the gene and/or the phenotype is expressed, it is essential that knowledge specific controlled vocabulary (ontology) is used for recording these key pieces information. To achieve this, Gramene adopted the use of ontologies (39) developed by ongoing projects like the Gene Ontology Consortium (GOC) to describe the gene's molecular function, its role in a biological process and localization in the cellular component (40) and the Plant Ontology Consortium (POC) to describe the spatial and temporal relationships to expression and phenotype (41–43). With help from collaborators, we developed in-house a set of ontologies, including TO (44) to describe phenotype traits, cereal plant growth stage ontology (GRO) that are species specific, environment/treatment ontology (EO) (Jaiswal et al. unpublished) that are used in annotating genes, proteins, and QTLs. For example, searching the TO term "bacterial blight disease resistance" tells a user about its definition, comments, synonyms, and the QTL, gene and proteins that used this term for annotation (Fig. 6a). From the annotations, we already know that *Xa21* is one of the genes, but researchers may be interested in other genes that also are associated with bacterial blight disease resistance. Use GrameneMart to find functional annotations (Gene Ontology and Interpro domain assignments) of the genes and does genome also have other genes with similar functional annotations (see Note 4).

3.2.7. Proteins

The protein module at Gramene provides curated information on about 197,000 Uniprot protein entries for all the species of grasses (Poaceae) of which majority (about 144,000) belong to genus *Oryza*. Annotation to the concepts of Gene Ontology (GO) (molecular function, biological process, and cellular component) (40), plant structure and GRO (PO) (41–43), cereal plant GRO and TO (44) are supported with an evidence code or reference citation. Protein information can be retrieved either by following the protein sequence links in the Genes page, or by searching Gramene's Protein module. Protein information includes hits to TIGR rice gene models; links to

NCBI's BLink tool (45) to find homologs from nonrice species and protein features (Fig. 6b). Another question one can ask is; how can I get a mapping between the rice gene/locus ID provided in the rice genome to that of Uniprot entries? (see Note 8).

3.2.8. Diversity

The Gramene Diversity database implements the Genomic Diversity and Phenotype Data Model (GDPDM) database schema. It provides experimentally scored genotypes, phenotypes, passport descriptors data on 954 germplasm accessions or stocks from the three rice species, *O. sativa*, *O. glaberrima*, and *O. rufipogon* reported in eight published reports. The current database release holds rice polymorphism data on three rice species accounted for 21 experiments, 1,617 genetic markers, such as SNP, , and RFLPs, used for genotyping a total of 836 germplasms and 28 phenotype traits associations. In addition, users can find information on

a

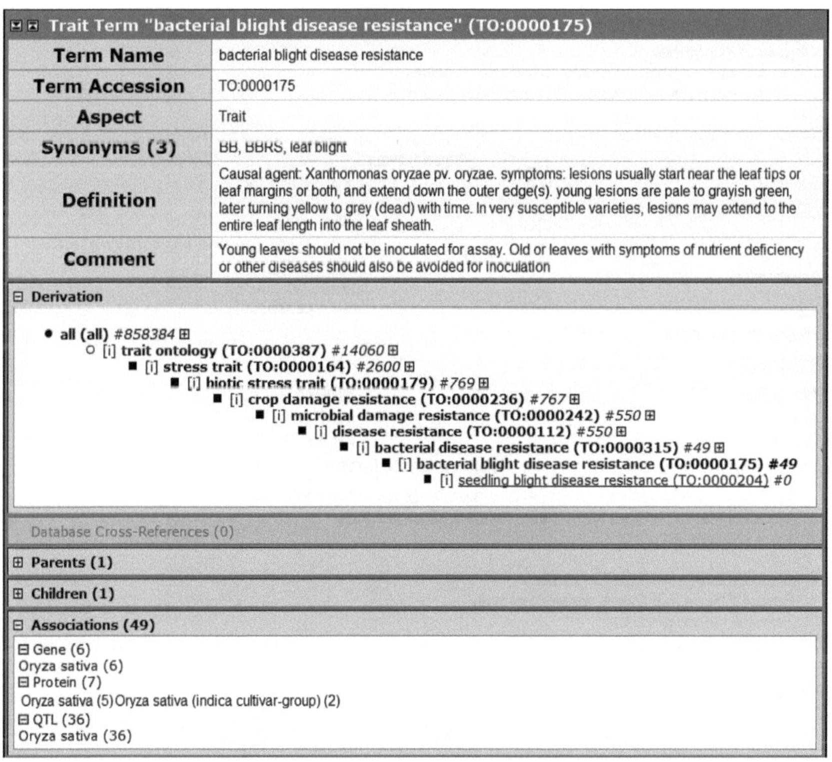

Fig. 6. Trait Ontology and Protein entry display. (**a**) The Trait Ontology (TO) view provides information on the controlled vocabulary term name, definition, comments, a hierarchy relationship between the child and the parent term, and the number of annotations. Click on the number of annotations to view the full list. (**b**) The modified protein entry page for XA21. It provides the details of the experiment types (evidence) and references used in inferring various ontology term assignments (associations section), links to NCBI's BLink tool to find orthologs, a list of Pfam, Prosite and Transmembrane features present on the protein and the list of literature citations. It also provides a link back to the gene entry in the genome browser via "Best hit to the TIGR gene model" section.

b

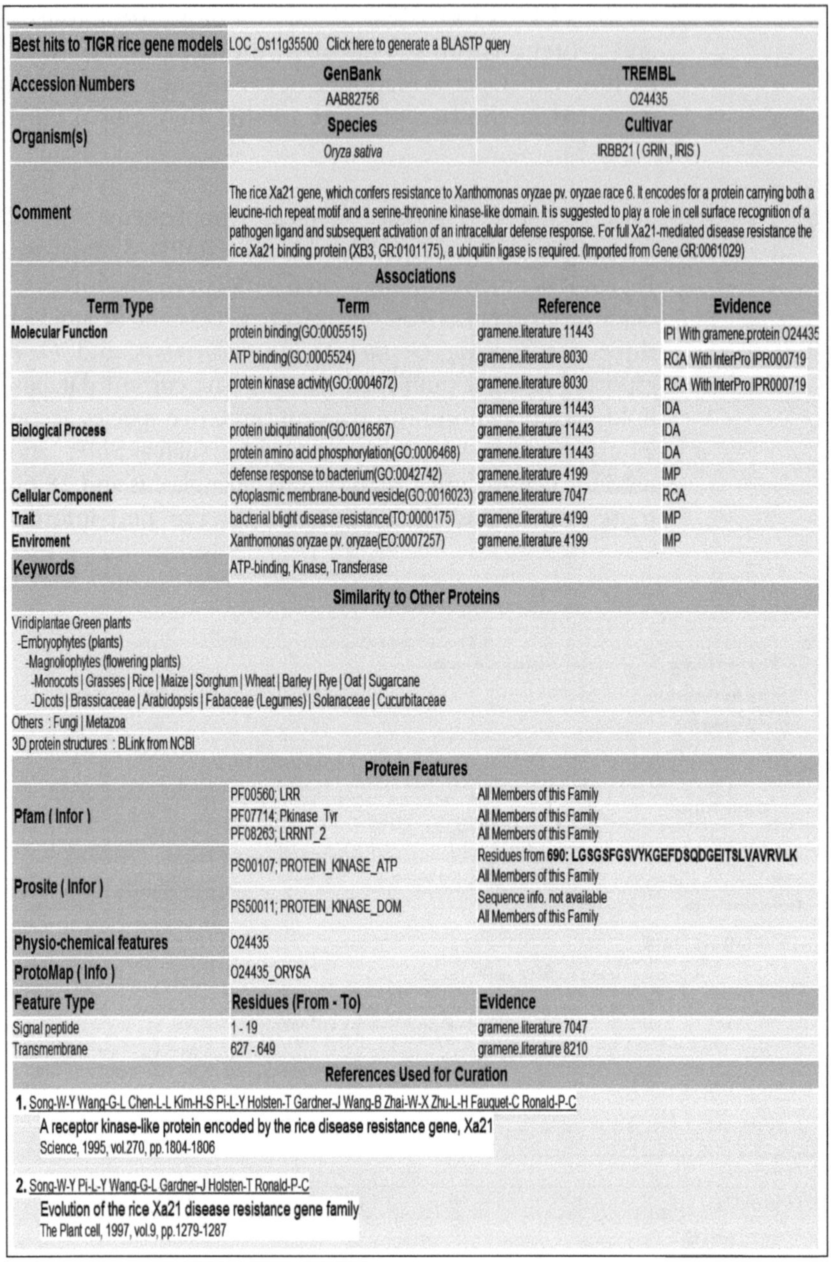

Best hits to TIGR rice gene models	LOC_Os11g35500 Click here to generate a BLASTP query		
Accession Numbers	GenBank	TREMBL	
	AAB82756	O24435	
Organism(s)	Species	Cultivar	
	Oryza sativa	IRBB21 (GRIN , IRIS)	
Comment	The rice Xa21 gene, which confers resistance to Xanthomonas oryzae pv. oryzae race 6. It encodes for a protein carrying both a leucine-rich repeat motif and a serine-threonine kinase-like domain. It is suggested to play a role in cell surface recognition of a pathogen ligand and subsequent activation of an intracellular defense response. For full Xa21-mediated disease resistance the rice Xa21 binding protein (XB3, GR:0101175), a ubiquitin ligase is required. (Imported from Gene GR:0061029)		
Associations			
Term Type	**Term**	**Reference**	**Evidence**
Molecular Function	protein binding(GO:0005515)	gramene.literature 11443	IPI With gramene.protein O24435
	ATP binding(GO:0005524)	gramene.literature 8030	RCA With InterPro IPR000719
	protein kinase activity(GO:0004672)	gramene.literature 8030	RCA With InterPro IPR000719
		gramene.literature 11443	IDA
Biological Process	protein ubiquitination(GO:0016567)	gramene.literature 11443	IDA
	protein amino acid phosphorylation(GO:0006468)	gramene.literature 11443	IDA
	defense response to bacterium(GO:0042742)	gramene.literature 4199	IMP
Cellular Component	cytoplasmic membrane-bound vesicle(GO:0016023)	gramene.literature 7047	RCA
Trait	bacterial blight disease resistance(TO:0000175)	gramene.literature 4199	IMP
Enviroment	Xanthomonas oryzae pv. oryzae(EO:0007257)	gramene.literature 4199	IMP
Keywords	ATP-binding, Kinase, Transferase		

Similarity to Other Proteins

Viridiplantae Green plants
-Embryophytes (plants)
-Magnoliophytes (flowering plants)
-Monocots | Grasses | Rice | Maize | Sorghum | Wheat | Barley | Rye | Oat | Sugarcane
-Dicots | Brassicaceae | Arabidopsis | Fabaceae (Legumes) | Solanaceae | Cucurbitaceae
Others : Fungi | Metazoa
3D protein structures : BLink from NCBI

Protein Features

Pfam (Infor)	PF00560; LRR	All Members of this Family
	PF07714; Pkinase_Tyr	All Members of this Family
	PF08263; LRRNT_2	All Members of this Family
Prosite (Infor)	PS00107; PROTEIN_KINASE_ATP	Residues from **690: LGSGSFGSVYKGEFDSQDGEITSLVAVRVLK** All Members of this Family
	PS50011; PROTEIN_KINASE_DOM	Sequence info. not available All Members of this Family
Physio-chemical features	O24435	
ProtoMap (Info)	O24435_ORYSA	
Feature Type	**Residues (From - To)**	**Evidence**
Signal peptide	1 - 19	gramene.literature 7047
Transmembrane	627 - 649	gramene.literature 8210

References Used for Curation

1. Song-W-Y Wang-G-L Chen-L-L Kim-H-S Pi-L-Y Holsten-T Gardner-J Wang-B Zhai-W-X Zhu-L-H Fauquet-C Ronald-P-C
   A receptor kinase-like protein encoded by the rice disease resistance gene, Xa21
   Science, 1995, vol.270, pp.1804-1806
2. Song-W-Y Pi-L-Y Wang-G-L Gardner-J Holsten-T Ronald-P-C
   Evolution of the rice Xa21 disease resistance gene family
   The Plant cell, 1997, vol.9, pp.1279-1287

Fig. 6. (continued)

experiment design and protocol with which the data was generated and a summary of the data. Users can compare genotypes on multiple loci for a germplasm or genotypes of multiple germplasms for a given locus/marker (see Note 9). Besides rice, the section also provides datasets collected from the maize (46), wheat (47), and sorghum genetic diversity projects (48). Advanced users read more on search options using GDPC (see Note 12).

3.2.9. Pathways

Gramene is the home of RiceCyc, a Web-based tool for searching various known or predicted biochemical pathways present in the rice plant. It provides a mapping of the genes identified in the sequenced rice genome to the biochemical reactions from 350 pathways in rice. For example, the momilactone biosynthesis pathway in rice (Fig. 7a). Another interesting feature of the RiceCyc module is the Omics viewer. This tool enables data from expression, proteomics, and metabolomics experiments

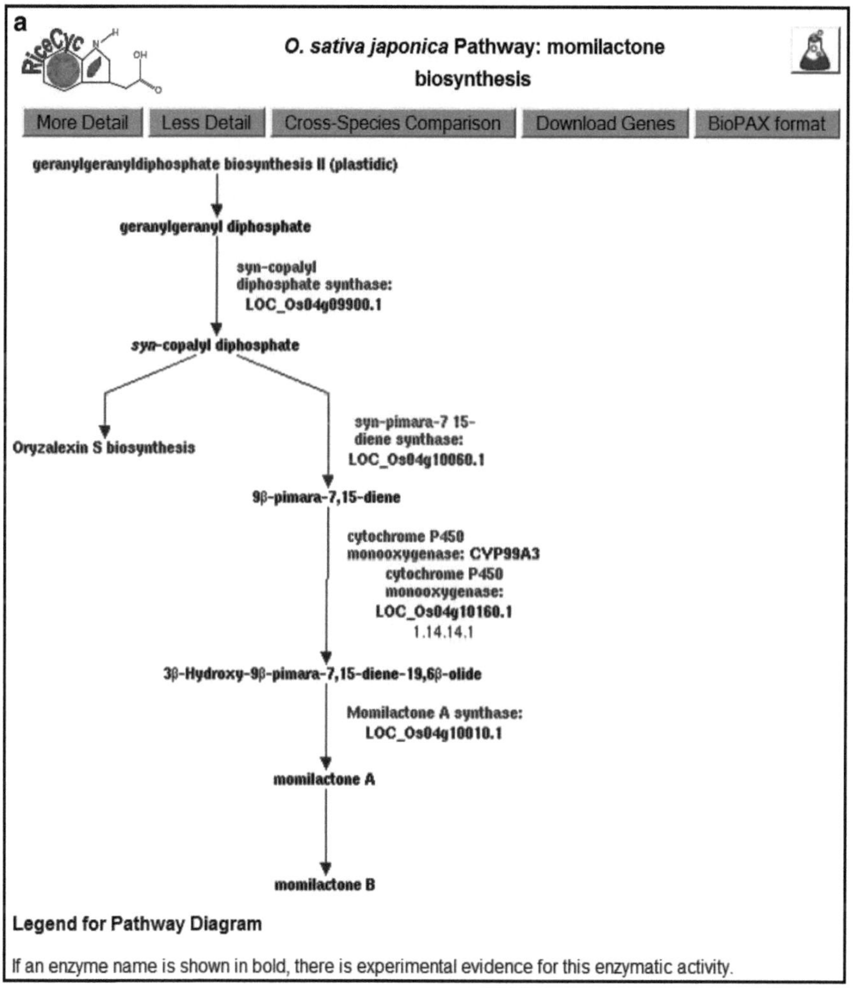

Fig. 7. RiceCyc pathway database screen shot. (**a**) A view of the Momilactone biosynthesis pathway from rice. The individual gene IDs, e.g., LOC_Os04g09900.1 links to the genome section described earlier. And the compound names links to details about the compounds. Users can use the buttons more/less detail to zoom-in/out. Click on the cross species button to select this pathway and compare the same from any other species pathway hosted in Gramene for comparison. For Bioinformatics enthusiasts, the BioPAX format button provides the pathway data in the standard BioPAX/XML format. (**b**) A view of the cellular overview of the pathways present in the database that were highlighted with expression (microarray) datasets. Users are encouraged to use the OMICS viewer tool provided from pathway database to upload their expression data and generate views like this.

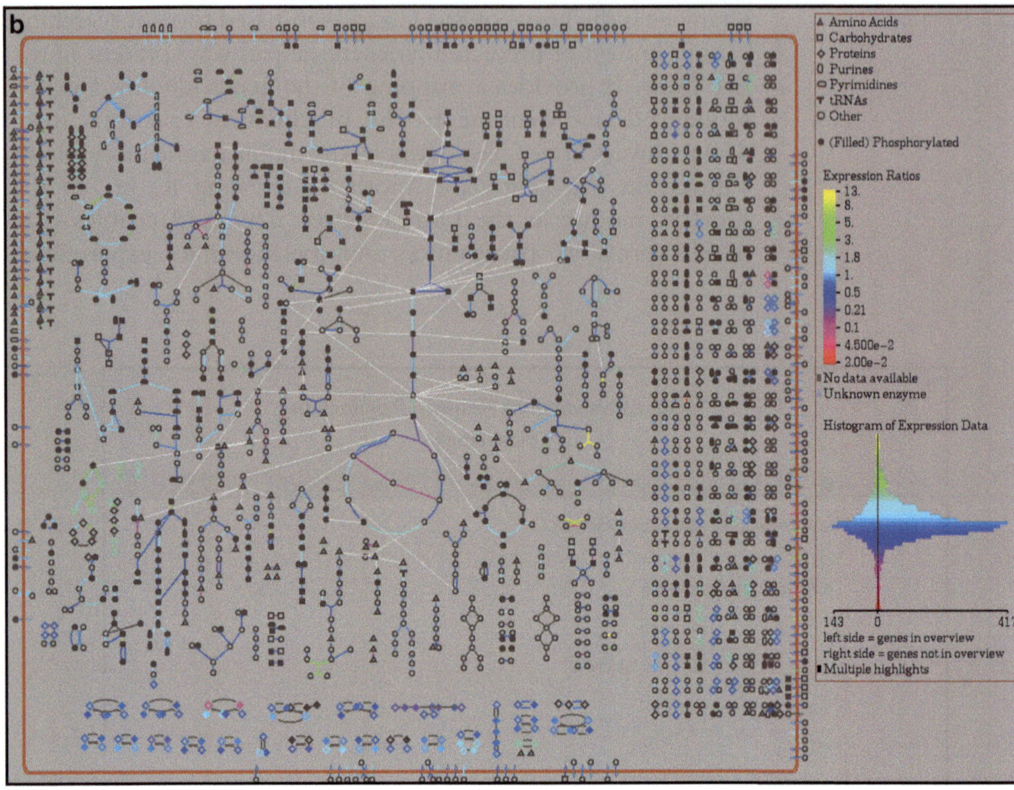

Fig. 7. (continued)

(see Note 10) to be overlaid onto pathway overview diagrams, thus capacitating the visualization of changes in pathway regulation (Fig. 7b). Besides rice, Gramene hosts eight other organisms in our pathways module. We add new pathways, make improvements in usability, and update RiceCyc. Last year, we added a beta release of SorghumCyc, and in our next release this will be christened our first official release based on the release of the sorghum genome and annotations from JGI. Our mirror databases allow the users to compare between the pathways known/predicted and listed in the organism-specific pathway database. This approach allows, e.g., the researchers to draw comparisons of rice (RiceCyc) and sorghum (SorghumCyc) databases developed and maintained by us to compare against each other as well as the dicots, namely, the *Arabidopsis*, tomato, potato, coffee, pepper, and *Medicago*.

3.2.10. Literature

The Literature module at Gramene contains citation information for the published references used for the curated datasets at Gramene. This is a good location to begin a search for recent literature on a topic of interest.

3.2.11. Sequence BLAST and GrameneMart Searches

In addition to the basic and advanced search interfaces available in the above described subsections of Gramene, we also provide one of the most popular types of searches, i.e., the BLAST search and the advanced step-by-step data mining tool, GrameneMart. The BLAST search allows for nucleotide or peptide searches with multiple query sequences against sequence datasets from multiple target species. GrameneMart, a Gramene version of the BioMart, is a generic data mining tool that allows the researchers to find items such as rice genes with nonsynonymous SNP mutations mapped them in a given region of the genome (see Note 2), and have a desired molecular function or phenotype (see Note 4). It also allows selecting output as sequence or tabulated list data in either HTML, tab delimited text or Microsoft Excel formats.

3.2.12. Species Pages

General information about the rice crop and the plant can be accessed through the "Species" link in the navigation bar at the top of the page, or by clicking on the rice grain image on the species bar along the bottom of the page. Rice information (Table 3 species pages) includes the history and uses of rice, anatomy diagrams, taxonomic nomenclature information, and useful summary on its planted acreage, production, and consumption around the world. Besides rice, there are some interesting information provided for many other species as well. These are available on respective species pages provided through species section.

4. Notes

Integration of the dataset and modules described above, along with inter- and intraspecific gene to gene homology, whole genome alignments, and common genetic landmarks, such as RFLP markers, used in genetic studies allow a researcher to address following questions and many more. The answer(s) then found by using Gramene can be applied for various purposes such as:

1. To discover new markers that can be used to enrich the existing genetic maps. Start with selecting your first map by going to http://www.gramene.org/db/cmap/map_search, for example, select species Rice and genetic map JRGP RFLP 2000 and click on the "submit" button. A list of linkage maps will be returned in the browser window. Select the linkage group/chromosome number (e.g., 1) by clicking it and it will display a map image. Scroll down the page and under the "map options" add a map to the left, e.g., Genetic: sorghum – Paterson 2003. The next view will show you which linkage

groups from the new map to be added have correspondences between this and the previous map, e.g., sorghum A with 21 associations. Select the linkage group 21 and click on the button "Add maps". The map image on the top of the page is refreshed and shows the lines connecting the two maps. Users can zoom into a region by clicking on the arrow (<-|||->) icons on the image. The views will help in enriching the mare sets that can be used from one map to another. Similar views can be generated between the sequenced and QTL maps, for example, to find all the genes and markers underlying a region of the QTL.

2. Find SNPs from a given genomic region or those underlying a gene/chromosomal and their positions and potential phenotypic/functional significance for nonsynonymous polymorphism. Start the same way as in #C and in the attribute sections, select for SNPs. Follow the steps to retrieve your results as described above.

3. Find all the genes mapped in the given chromosomal region for example of rice.

   (a) Go to http://www.gramene.org/biomart/martview/

   (b) "CHOOSE DATABASE" = "Gramene XX Ensembl Genes" (where XX = release #)

   (c) "CHOOSE DATASET" = "*Oryza sativa japonica* genes"

   (d) Click on the "Filters" and open the "region" section by clicking on the [+] icon

   (e) Select the chromosome and the region start

   (f) Stop positions-Select the "Attributes" on the left, select the options on features-open the "gene" section by clicking on the [+] icon a check the box for gene ID

   (g) After selecting all these, hit the "result" button in the navigation bar in black color

   (h) It will give you a results table

   (i) Select your export format to get the complete results

4. Find functional annotations (Gene Ontology and Interpro domain assignments) of the genes and does genome also have other genes with similar functional annotations. Start the same way as in #C and in the attribute sections, select for Features and open "external" section and check mark GO ID and GO description. Follow the steps to retrieve your results as described above.

5. How do I find rice gene Orthologs for the species of my interest (e.g., *Arabidopsis*, Poplar)?/Find all the homologous (ortholog and paralog) gene pairs of the gene of interest.

(a) Go to http://www.gramene.org/biomart/martview/
(b) "CHOOSE DATABASE" = "Gramene XX Ensembl Genes" (where XX = release #)
(c) "CHOOSE DATASET" = "*Oryza sativa japonica* genes"
(d) Then click on "attributes" link in the gray-colored column and select homologs
(e) Select species and mark geneID to get orthologs from desired species
(f) Click on the results section to view options to download

6. Find genes that are present in the same region of the genome that overlaps with the QTL of interest. Go to the QTL of interest, e.g., as mentioned in Fig. 2b or by searching at http://www.gramene.org/db/qtl/qtl_display, for example, type "CQAE18", select the QTL accession ID and species Rice. The detail QTL page shows the information about the QTL, including the map position and associated markers. From the map position section, note down the map/chromosome number and the start–stop positions. Now, go to the GrameneMart and follow the steps described from #5(c) onwards.
7. Find markers that are present in the same region of the genome that overlaps with the QTL of interest. Go to the QTL of interest, e.g., as mentioned in Fig. 2b or by searching at http://www.gramene.org/db/qtl/qtl_display, for example, type "CQAE18", select the QTL accession ID and species Rice. The detail QTL page shows the information about the QTL, including the map position and associated markers.
8. How can I get a mapping between the rice gene/locus ID provided in the rice genome to that of Uniprot entries?
   (a) Go to http://www.gramene.org/biomart/martview/
   (b) "CHOOSE DATABASE" = "Gramene XX Ensembl Genes" (where XX = release #)
   (c) "CHOOSE DATASET" = "*Oryza sativa japonica* genes"
   (d) Select the "Filters" on the left, then open the "GENE" filter shown on the right
   (e) For the "ID list limit," select the "Gramene Transcript ID(s)" and put your "LOC*" identifiers into the box
   (f) Select the "Attributes" on the left, then open the "EXTERNAL" attributes box on the right
   (g) Select the UniProt IDs you want under "External References"; note the max of three allowed

9. Find the genotypes of known germplasms based on the genetic diversity analysis that happen to be stored in the database. Go to the diversity database (http://www.gramene.org/db/diversity/diversity_view), type the marker name, e.g., RM101 in the search box, select option marker and species, and click on the search glass icon. The results will give you the list of experiments and table with linked information. From the table, click on the link "show allele data" and you will get a complete list of germplasms, where the same marker was used in genotyping experiments, along with its genotype profile.
10. For uploading the expression data, for example, from a microarray experiment, you must use a tab-delimited text file similar to the one provided at http://pathway.gramene.org/expr-examples/sample.dat. Make sure that the first column has gene IDs. And second column onward you can add expression values. It can be from one to many columns. If it is a time series or multiple treatments, each column represents single time point or treatment. Use the OMICs viewer validator tool (http://www.gramene.org/db/omics/validator) first, by uploading your file and validating it. Use the validated file to upload it to the OMICs viewer (http://pathway.gramene.org/expression.html). Please be patient to get the results back. It may take minutes depending on the number of genes listed in your file.
11. Database users are encouraged to browse the tutorials and the frequently asked questions (FAQs) (http://www.gramene.org/db/help). They have an option to use the search and/or browse function of Gramene to find the answers. However, for large and complicated queries, such as those described in Note 1, we encourage our users to use the GrameneMart that is built using the BioMart source software.
12. For querying the genetic diversity datasets from Gramene and many more projects outside of Gramene, users have the option to undertake more complicated quires, and we encourage the users to use Java-based GDPC by visiting the GDPC Web site http://gramene.org/diversity/gramene_gdpc.html and following the instructions. It allows advanced searches and download tool that works as a stand-alone tool from your personal computer.
13. Since the datasets and data organization and Web site views continue to evolve with updates and additions, the Gramene database makes every effort to inform its users by organizing workshops at several large meetings and symposiums as well as provides online tutorials and helps documents through http://www.gramene.org/db/help. For an explanation of how to search a module and what information is available there, please refer to the tutorial links provided on this page

or from links in every module's navigation bar on top. For an example of moving between modules, refer to the hands-on exercises with real examples.

14. We make every effort to provide the best quality and most updated information for our users. However, if the users encounter problems, in datasets and/or accessing the database users are encouraged to send the e-mail to Gramene (gramene@gramene.org) or to use the "Feedback" link at the top right hand side of every page for questions, suggestions, and reporting errors.
15. In order to get the Gramene news and announcements, users are encouraged to sign up for the mailing lists available at http://www.gramene.org/mailarch.html.
16. From time to time, we provide updates and new features on the Gramene Web site and the database. These announcements are made through the group mailing list mentioned above as well as the Gramene News Blog available through http://news.gramene.org.
17. If you have the datasets that you think are important for the comparative work and can be shared with the researchers via its integration in Gramene, please write to us and we will be happy to work with you.

Acknowledgments

This work was initially supported (2001–2004) by the USDA Initiative for Future Agriculture and Food Systems (IFAFS) (grant no. 00-52100-9622) and USDA-Agricultural Research Service specific cooperative agreement (grant no. 58-1907-0-041). For the years 2004–current, this work is supported by National Science Foundation (NSF) awards #0321685 and 0703908. We are thankful to our project team members and numerous collaborators and data contributors for help in curation, sharing the datasets and tools. For a full list of our collaborators, please visit http://www.gramene.org/collaborators.

References

1. Yu, J., S. Hu, J. Wang, G.K. Wong, S. Li, B. Liu, *et al.*, (2002), A draft sequence of the rice genome (*Oryza sativa* L. ssp. indica). *Science*, 296(5565): p. 79–92.
2. IRGSP. (2005), The map-based sequence of the rice genome. *Nature*, 436: p. 793–800.
3. DoE-JGI. (2007), Sorghum genome project. *Available from*: http:/s/www.phytozome.net/sorghum.
4. Bedell, J.A., M.A. Budiman, A. Nunberg, R.W. Citek, D. Robbins, J. Jones, *et al.*, (2005), Sorghum genome sequencing by

methylation filtration. *PLoS Biol*, 3(1): p. e13. Epub 2005 Jan 4.

5. Paterson, A.H., J.E. Bowers, R. Bruggmann, I. Dubchak, J. Grimwood, H. Gundlach, *et al.*, (2009), The Sorghum bicolor genome and the diversification of grasses. *Nature*, 457(7229): p. 551–6.
6. Project, M.G.S. (2007), The Maize Genome Sequencing Project. *Available from*: http://www.maizesequence.org/overview.html.
7. Maziesequence.org. (2008), *Available from*: http://www.maizesequence.org.
8. DoE-JGI. (2007), Why Sequence Brachypodium? *Available from*: http://www.jgi.doe.gov/sequencing/why/CSP2007/brachypodium.html.
9. Brachypodium_Sequencing_Project. (2009), Brachypodium genome, *BrachyBase*, www.brachybase.org.
10. Tuskan, G.A., S. Difazio, S. Jansson, J. Bohlmann, I. Grigoriev, U. Hellsten, *et al.*, (2006), The genome of black cottonwood, *Populus trichocarpa* (Torr. & Gray). *Science*, 313(5793): p. 1596–604.
11. Travis, J., (2008), Uncorking the grape genome. *Science*, 320(5875): p. 475–7.
12. DoE-JGI. (2009), Soybean genome. *Available from*: http://www.phytozome.net/soybean.
13. Town, C.D., (2006), Annotating the genome of *Medicago truncatula*. *Curr Opin Plant Biol*, 9(2): p. 122–7.
14. Cannon, S.B., J.A. Crow, M.L. Heuer, X. Wang, E.K. Cannon, C. Dwan, *et al.*, (2005), Databases and information integration for the *Medicago truncatula* genome and transcriptome. *Plant Physiol*, 138(1): p. 38–46.
15. Bell, C.J., R.A. Dixon, A.D. Farmer, R. Flores, J. Inman, R.A. Gonzales, *et al.*, (2001), The Medicago Genome Initiative: a model legume database. *Nucleic Acids Res*, 29(1): p. 114–7.
16. Jung, S., M. Staton, T. Lee, A. Blenda, R. Svancara, A. Abbott, *et al.*, (2008), GDR (Genome Database for Rosaceae): integrated web-database for Rosaceae genomics and genetics data. *Nucleic Acids Res*, 36(Database issue): p. D1034–40.
17. Tarchini, R., P. Biddle, R. Wineland, S. Tingey, and A. Rafalski, (2000), The complete sequence of 340 kb of DNA around the rice Adh1-adh2 region reveals interrupted colinearity with maize chromosome 4. *Plant Cell*, 12(3): p. 381–91.
18. La Rota, M., R.V. Kantety, J.K. Yu, and M.E. Sorrells, (2005), Nonrandom distribution and frequencies of genomic and EST-derived microsatellite markers in rice, wheat, and barley. *BMC Genomics*, 6(1): p. 23.
19. Keller, B. and C. Feuillet, (2000), Colinearity and gene density in grass genomes. *Trends Plant Sci*, 5(6): p. 246–51.
20. Jaiswal, P., J. Ni, I. Yap, D. Ware, W. Spooner, K. Youens-Clark, *et al.*, (2006), Gramene: a bird's eye view of cereal genomes. *Nucleic Acids Res*, 34(Database issue): p. D717–23.
21. Liang, C., P. Jaiswal, C. Hebbard, S. Avraham, E.S. Buckler, T. Casstevens, *et al.*, (2008), Gramene: a growing plant comparative genomics resource. *Nucleic Acids Res*, 36(Database issue): p. D947–53. Epub 2007 Nov 4.
22. Wheeler, D.L., B. Smith-White, V. Chetvernin, S. Resenchuk, S.M. Dombrowski, S.W. Pechous, *et al.*, (2005), Plant genome resources at the national center for biotechnology information. *Plant Physiol*, 138(3): p. 1280–8.
23. Wheeler, D.L., T. Barrett, D.A. Benson, S.H. Bryant, K. Canese, D.M. Church, *et al.*, (2005), Database resources of the National Center for Biotechnology Information. *Nucleic Acids Res*, 33(Database issue): p. D39–45.
24. Dong, Q., S.D. Schlueter, and V. Brendel, (2004), PlantGDB, plant genome database and analysis tools. *Nucleic Acids Res*, 32(Database issue): p. D354–9.
25. Lee, Y., J. Tsai, S. Sunkara, S. Karamycheva, G. Pertea, R. Sultana, *et al.*, (2005), The TIGR Gene Indices: clustering and assembling EST and known genes and integration with eukaryotic genomes. *Nucleic Acids Res*, 33(Database issue): p. D71–4.
26. Sherry, S.T., M.H. Ward, M. Kholodov, J. Baker, L. Phan, E.M. Smigielski, *et al.*, (2001), dbSNP: the NCBI database of genetic variation. *Nucleic Acids Res*, 29(1): p. 308–11.
27. Apweiler, R., A. Bairoch, C.H. Wu, W.C. Barker, B. Boeckmann, S. Ferro, *et al.*, (2004), UniProt: the Universal Protein knowledgebase. *Nucleic Acids Res*, 32(Database issue): p. D115–9.
28. Ni, J., A. Pujar, K. Youens-Clark, I. Yap, P. Jaiswal, I. Tecle, *et al.*, (2009), Gramene QTL database: development, content and applications. *Database* Vol. 2009:bap005; doi:10.1093/database/bap005.
29. Kent, W.J., (2002), BLAT – the BLAST-like alignment tool. *Genome Res*, 12(4): p. 656–64.
30. Altschul, S.F., W. Gish, W. Miller, E.W. Myers, and D.J. Lipman, (1990), Basic local alignment search tool. *J Mol Biol*, 215(3): p. 403–10.

31. Potter, S.C., L. Clarke, V. Curwen, S. Keenan, E. Mongin, S.M. Searle, *et al.*, (2004), The Ensembl analysis pipeline. *Genome Res*, 14(5): p. 934–41.
32. Temnykh, S., G. DeClerck, A. Lukashova, L. Lipovich, S. Cartinhour, and S. McCouch, (2001), Computational and experimental analysis of microsatellites in rice (*Oryza sativa* L.): frequency, length variation, transposon associations, and genetic marker potential. *Genome Res*, 11(8): p. 1441–52.
33. McCouch, S.R., L. Teytelman, Y. Xu, K.B. Lobos, K. Clare, M. Walton, *et al.*, (2002), Development and mapping of 2240 new SSR markers for rice (*Oryza sativa* L.). *DNA Res*, 9(6): p. 199–207.
34. Yuan, Q., S. Ouyang, A. Wang, W. Zhu, R. Maiti, H. Lin, *et al.*, (2005), The institute for genomic research Osa1 rice genome annotation database. *Plant Physiol*, 138(1): p. 18–26.
35. Zhao, W., J. Wang, X. He, X. Huang, Y. Jiao, M. Dai, *et al.*, (2004), BGI-RIS: an integrated information resource and comparative analysis workbench for rice genomics. *Nucleic Acids Res*, 32(Database issue): p. D377–82.
36. Wing, R.A., J.S. Ammiraju, M. Luo, H. Kim, Y. Yu, D. Kudrna, *et al.*, (2005), The *Oryza* map alignment project: the golden path to unlocking the genetic potential of wild rice species. *Plant Mol Biol*, 59(1): p. 53–62.
37. maizesequence.org. (2009), Release 4a.53: Summary of intentions. *Available from*: http://maizesequence.org/summary_of_intentions.html.
38. Kurata, N. and Y. Yamazaki, (2006), Oryzabase. An integrated biological and genome information database for rice. *Plant Physiol*, 140(1): p. 12–7.
39. Yamazaki, Y. and P. Jaiswal, (2005), Biological ontologies in rice databases. An introduction to the activities in Gramene and Oryzabase. *Plant Cell Physiol*, 46(1): p. 63–8. Epub 2005 Jan 19.
40. Harris, M.A., J. Clark, A. Ireland, J. Lomax, M. Ashburner, R. Foulger, *et al.*, (2004), The Gene Ontology (GO) database and informatics resource. *Nucleic Acids Res*, 32(Database issue): p. D258–61.
41. Pujar, A., P. Jaiswal, E.A. Kellogg, K. Ilic, L. Vincent, S. Avraham, *et al.*, (2006), Whole plant growth stage ontology for angiosperms and its application in plant biology. *Plant Physiol*, 142: p. 414–428.
42. Jaiswal, P., S. Avraham, K. Ilic, E.A. Kellogg, A. Pujar, L. Reiser, *et al.*, (2005), Plant sOntology (PO): a controlled vocabulary of plant structures and growth stages. *Comp Funct Genomics*, 6(7–8): p. 388–97.
43. Ilic, K., E.A. Kellogg, P. Jaiswal, F. Zapata, P.F. Stevens, L.P. Vincent, *et al.*, (2007), The plant structure ontology, a unified vocabulary of anatomy and morphology of a flowering plant. *Plant Physiol*, 143(2): p. 587–99.
44. Jaiswal, P., D. Ware, J. Ni, K. Chang, W. Zhao, S. Schmidt, *et al.*, (2002), Gramene: development and integration of trait and gene ontologies for rice. *Comp Funct Genomics*, 3(2): p. 132–136.
45. Wheeler, D.L., T. Barrett, D.A. Benson, S.H. Bryant, K. Canese, V. Chetvernin, *et al.*, (2007), Database resources of the National Center for Biotechnology Information. *Nucleic Acids Res*, 35(Database issue): p. D5–12.
46. Canaran, P., L. Stein, and D. Ware, (2006), Look-Align: an interactive web-based multiple sequence alignment viewer with polymorphism analysis support. *Bioinformatics*, 22(7): p. 885–6. Epub 2006 Feb 10.
47. Dvorak, J., J.A. Anderson, O.D. Anderson, M.T. Clegg, J. Dubcovsky, B. Gill, *et al.*, (2009), Haplotype polymorphism in polyploid wheats and their diploid ancestors, *Available from*: http://wheat.pw.usda.gov/SNP/project.html.
48. The Sorghum Diversity Database. (2009), The Sorghum Diversity Database. *Available from*: http://sorghumdiversity.org/sorghum/database.html.

Index

Andy Pereira (ed.), *Plant Reverse Genetics: Methods and Protocols*, Methods in Molecular Biology, vol. 678,
DOI 10.1007/978-1-60761-682-5,

MIX
Papier aus verantwortungsvollen Quellen
Paper from responsible sources
FSC® C105338

If you have any concerns about our products,
you can contact us on
ProductSafety@springernature.com

In case Publisher is established outside the EU,
the EU authorized representative is:
Springer Nature Customer Service Center GmbH
Europaplatz 3, 69115 Heidelberg, Germany

Printed by Libri Plureos GmbH
in Hamburg, Germany